Disinfection and Reuse Symposium 2018

Linking Water Quality and Public Health

Portland, Oregon, USA
29 – 31 July 2018

ISBN: 978-1-5108-7159-5

Printed from e-media with permission by:

Curran Associates, Inc.
57 Morehouse Lane
Red Hook, NY 12571

Some format issues inherent in the e-media version may also appear in this print version.

Printed by Curran Associates, Inc. (2018)

Water Environment Federation (WEF)
601 Wythe Street
Alexandria, Virginia 22314
USA

Phone: 1-800-666-0206
Fax: 1-703-684-1545

csc@wef.org

Additional copies of this publication are available from:

Curran Associates, Inc.
57 Morehouse Lane
Red Hook, NY 12571 USA
Phone: 845-758-0400
Fax: 845-758-2633
Email: curran@proceedings.com
Web: www.proceedings.com

WATER ENVIRONMENT FEDERATION

Disinfection & Reuse Symposium

July 29–31, 2018 | Portland, OR

CONFERENCE PROGRAM

Water Environment Federation®
the water quality people™

Disinfection and Reuse Symposium 2018

Linking Water Quality and Public Health

July 29-31, 2018
Portland Crowne Plaza
Portland, Oregon

PNCWA

THE Water Research FOUNDATION

The symposium is held by the Water Environment Federation in cooperation with the Pacific Northwest Clean Water Association and The Water Research Foundation.

July 29, 2018

Dear Colleagues,

Welcome to Portland, Oregon. The Water Environment Federation in cooperation with Pacific Northwest Clean Water Association and the Water Research Foundation are honored to have you join us for the Disinfection and Reuse Symposium 2018. On their behalf, we invite you to fully participate in this exceptional opportunity for education, collaboration and networking opportunity with industry experts!

This symposium will bring together environmental professionals from around North America for a showcase on reuse and disinfection. This exciting symposium is designed to educate practitioners, facility owners, operators, researchers, and public administrators about current reuse and disinfection issues, including regulatory requirements and methods for analyzing problems and finding innovative solutions. With rapid development of Potable Reuse we believe that this Symposium is a timely event to share experience, exchange findings, information and lessons learned in this area.

The focus for this event will be on research, practical implementation and collaboration to promote the rapid application of new innovative solutions for water reuse and disinfection. This Symposium will provide valuable information for researchers, students, regulators, designers, technology developers, operators, municipal agencies, industrial dischargers, and others seeking to understand the full picture of the latest developments and practical experiences on these important topics, water reuse and disinfection.

The speakers and attendees for this event come from a wide variety of backgrounds including regulatory, research, design, implementation, and utility operations. The conference program is composed of 30 minute presentations in the following topics:

- Disinfection Fundamentals
- Wet Weather Disinfection – Current and Emerging Practices
- Innovative Disinfection Methods and Technology Trends
- Potable Reuse (DPR and IPR) - Safety, Water Quality Control, Demonstration
- Expanding Reuse Strategies and Markets
- Disinfection Optimization and Control of Disinfection Byproducts (DBPs)
- New Water Reuse Approaches to Manage Tomorrow's Challenges
- Supporting Public Health with Water, Sanitation and Hygiene (WaSH) Projects
- Residuals, Sanitation and Hygiene: Concerns, Design and Application for Low Technology Processing
- Optimizing a Future for DPR – Design, Implementation and Operation of Robust Solutions
- Disinfection of WRRF Effluents for Surface Water Discharges

While you network with other attendees and presenters, make a point to share your experiences. We look forward to interacting with each of you and to the many interesting exchanges that will take place in the next few days – and beyond.

We hope you enjoy this informative event and find many fruitful interactions.

Sincerely,

Robert English
Palintest
Co-Chair

Val Frenkel
Greeley and Hansen
Co-Chair

TABLE OF CONTENTS

Conference Steering and Program Committee
Registration
Presenter and Moderator Information
Reception and Meal Functions
Continuing Education
Online Proceedings
Technical Publication
Sponsors
WEF Policies & Positions
WEF Social Media Policy
Symposium Safety and Security
Pre-Conference Workshop
Opening General Session
Technical Program
Technical Program at-a-Glance
Presentations at-a-Glanc
Exhibition Information
Exhibitor Directory
Floor Plan
Presenter and Moderator Directory
Conference Schedule at-a-Glance
Upcoming Education and Training Event

SYMPOSIUM COMMITTEE

Robert English
Palintest
Symposium Co-Chair

Val Frenkel
Greeley and Hanses
Symposium Co-Chair

DISINFECTION AND REUSE SYMPOSIUM STEERING COMMITTEE

Akin Babatola
City of Santa Cruz

Cyndy Bratz
Tetra Tech

Daniela Castaneda
Stantec

James Crook
Self-employed

Bertrand Dussert
Xylem, Inc.

Robert English
Palintest

Val Frenkel
Greeley and Hansen

Brian Hilts
CDM Smith

Matt Noesen
CH2M

Zuhal Ozturk
AECOM

Patrick Regan
Evoqua WAter Technologies

Jenny Reina
CH2M

Andrew Salveson
Carollo

Samendra Sherchan
Tulane University

Jay Swift
Gray and Osborne

Venkat Venkatasubbiah
Marathon Petroleum Company

Michael Watts
Garver

SYMPOSIUM COMMITTEE

DISINFECTION AND REUSE SYMPOSIUM PROGRAM COMMITTEE

Akin Babatola
City of Santa Cruz

Charles Bott
HRSD

Kari Brisolara
LSUHSC

Daniela Castaneda
Stantec

Jessica Cochran
Taylor Engineering

Jeffrey Cunningham
University of South Florida

Steven Farabaugh
Isle Utilities

Val Frenkel
Greeley and Hansen

Ajit Ghorpade
Veolia

Liping Han
Exxonmobil

Stuart Humphries
Orthos Liquid Systems

Melanie Mann
Hazen and Sawyer

Erica Marti
University of Nevada, Las Vegas

Justin Mattingly
The Water Research Foundation

Brenley McKenna
The Water Research Foundation

Michaela McKenzie
DuPont

Steve Ravel
Mott MacDonald

Patrick Regan
Evoqua Water Technologies

Robert Reimers
Tulane University

Jenny Reina
CH2M

Scott Schaefer
AE2S

Donald Seagle
Water Environmental Consultants

Bahman Sheikh
Water Reuse Consulting

Samendra Sherchan
Tulane University

Vijay Sundaram
Stantec

Ryujiro Tsuchihashi
AECOM

Art Umble
Stantec

Venkat Venkatasubbiah
Marathon Petroleum Company

Justin Waples
Central Contra Costa Sanitary District

Rick Warner
Washoe County

Michael Watts
Garver

Paul Westerhoff
Arizona State University

Christine Yi
Hazen and Sawyer

REGISTRATION

All events are held in the Portland Crowne Plaza Hotel.

The Registration Desk is located in the Windsor Foyer and will be open during the following times:

Hours:

Sunday, July 29	12:00 PM – 4:30 PM
Monday, July 30	7:30 AM – 4:45 PM
Tuesday, July 31	8:00 AM – 5:00 PM

PRESENTER AND MODERATOR INFORMATION

All presenters, alternates, and moderators should pick up their badge from the Registration Desk and attend their assigned briefing.

Presenters, alternates, and session moderators, participating Monday and Tuesday should attend their assigned briefing. Presenters should attend only once unless they are speaking on different days.

Volunteer Room Monitors are also encouraged to attend.

The Speaker Briefing and room schedule is as follows:

Sessions 1 through 5

Monday, July 30
7:45 AM – 8:15 AM
Room: Windsor C

Sessions 6 through 11

Tuesday, July 31
8:00 AM – 8:15 AM
Room: Windsor C

RECEPTIONS AND MEAL FUNCTIONS

Networking Reception **Bellmont Foyer**

Get the chance to try beer and wine made from recycled water! Join us in the exhibit hall to mingle with your fellow attendees while you view the exhibits, renew acquaintances, and make new contacts. Light hors d'oeuvres and refreshments will be served. Beer will be available first come, first served with a cash bar to follow. Don't miss out!

Monday, July 30
4:45 PM – 6:15 PM

Sponsored by: Xylem

Luncheons **Bellmont C**

Lunch will be provided for all full symposium, daily registrants, exhibitors, and students. Please advise staff as soon as possible if you have any special dietary requirements.

Monday, July 30
11:45 AM – 1:30 PM

Tuesday, July 31
11:45 AM – 1:30 PM

Networking Breaks

The Networking Breaks will be held:

Monday, July 30
Bellmont Foyer
10:00 AM – 10:45 AM and 3:00 PM – 3:45 PM

Tuesday, July 31
Bellmont Foyer
10:00 AM – 10:45 AM and 3:00 PM – 3:45 PM

CONTINUING EDUCATION

How Do I Receive Credit For this Symposium?
To receive credit for all technical sessions and pre-Symposium workshop, please fill out a Continuing Education Request Form and be sure to have a room monitor initial for verification.

Attendees will have to submit their CE Request Form to the Registration Desk at the end of the Symposium. Please request a continuing education form when you check-in and ask WEF staff at the Registration Desk if you have further questions.

Pre-Symposium Workshop:
WEF offers Continuing Education Units (CEUs) for participation in workshop. **One CEU is the equivalent to 10 hours** of training or formal instruction. These are distributed for structured, relevant professional training above and beyond that of initial certification or employment in a particular field.

Technical Sessions:
WEF offers Professional Development Hours (PDHs) for participation in technical sessions. **A PDH is defined as one hour** spent engaged in an activity that contributes to the advancement or enhancement of professional skills or scientific knowledge of a professional engineer or operator.

When Will I Receive Credits For this Symposium?
Certificates and transcripts for this event will be mailed within 8 weeks of the Symposium.
Please keep in mind that although WEF does provide these files, most states will require the individual licensee to report continuing education credits.

Note: *Educational Credits will not be recorded and documentation will not be distributed unless the attendee is a confirmed registrant of this event and the proper steps are is completed as indicated in the directions provided.*

Are WEF Continuing Education Credits Approved in My State?
WEF applies for approval in many states and will be happy to work with individuals and Member Associations for additional state or agency approvals upon request. In addition, WEF has been approved as a Training Provider through the following:
The Florida Board of Professional Engineers, the New York State Department of Education, and the Ohio EPA. Many other states accept WEF PDH credits and WEF CEU credits as long as subjects and content meets with state requirements. For example: California (CWEA), Nevada, and New Jersey.

CONTINUING EDUCATION

What Else Do I Need to Know?

WEF follows the International Association of Continuing Education and Training (IACET) guidelines along with state-specific regulations to achieve strict policies and procedures regarding its Continuing Education Program. WEF calculates education credits following a standardized method that is the most widely accepted by certification and licensing agencies. However, many states differ in the type and/or number of credits they will approve for educational events. Because of this, participants are responsible for exploring their state requirements and for ensuring that WEF Symposium credits are accepted.

Service and Support...

In keeping with IACET guidelines, WEF maintains a database of all continuing education files for a minimum of 7 years. You may contact WEF's Customer Service Team between the hours of 8:30am and 5:00pm EST, Monday through Friday to request these files. Please call 1-800-666-0206 or submit an email request to csc@wef.org.

State Credit Calculations:

*Some state licensing boards will accept CEUs for session under 3 hours in length. Some use different acronyms for training credits. In most instances the credits issued by WEF can be converted to meet state specific requirements that vary from the system used by WEF. This is usually managed at the state level using the following conversion method:

1.0 CEU = 10 Hours of session time
1.0 PDH = 1 Hour of session time
1.0 Contact Hour = 1 Hour of session time
For example: **1.2 CEU** Credits = **12.0 PDH** Credits or **17.0 PDH** Credits could equate to **1.7 CEU Credits** depending on individual state regulations.

**CEU & PDH credits are available for Workshops to Professional Engineers licensed in the state of New York (NYSED).*

For more information regarding WEF's Continuing Education Program, please visit http://www.wef.org/DisinfectionReuse.

ONLINE PROCEEDINGS

Symposium proceedings, consisting of manuscripts for each presentation, have been made available through an online portal. Advance registrants within Full Symposium, Daily, and Student categories will receive access to the Disinfection and Reuse 2018 online proceedings on the day prior to the Symposium. Onsite registrants will receive an access link following the event's conclusion.

Copies of proceedings may be ordered after the Symposium at the member rate of $100 USD/nonmember rate of $150 USD. All orders will be processed after conclusion of the Symposium. You may call 1-800-666-0206 or visit www.wef.org/ShopWEF and ask for Stock Number – CPDF1806.

TECHNICAL PUBLICATION

Disinfection and Reuse
The Water Reuse Roadmap
Member: $67.20
Non-member: $84.00

The Water Reuse Roadmap is an overview of all the opportunities and issues faced by water reuse projects. Developed to help water managers facilitate successful water reuse programs, from concept development through implementation and maintenance, the guidance in this book enables water managers to determine the feasibility of water reuse options in their specific situation, initiate a water reuse program and expand an existing reuse program based on new approaches and opportunities for innovation.

***Symposium attendees save 20%

SPONSORS

We would like to thank the following sponsoring companies for their contributions to the Symposium and program.

HDR

www.hdrinc.com
Bronze Elite

---

xylem
Let's Solve Water

www.xylem.com
Networking Reception

WEF POLICIES & POSITIONS

WEF's Vision
A community of empowered professionals creating a healthy global water environment.

Core Values
Leadership, Passion, Scholarship, and Collaboration, and Service

WEF Policies
WEF respects and takes the broadest view of human diversity and inclusion and is committed to providing a professional, safe, and welcoming environment at its events for all water professionals and their guests. WEF expects all sponsors, speakers, attendees, media, exhibitors and other participants to uphold our commitment to diversity and inclusion by helping us provide a positive Symposium environment for everyone.

For more information, please see WEF's Diversity and Inclusivity Policy, as well as WEF's Non-Discrimination and Harassment Policy at www.wef.org/about/about-wef.

Reporting Concerns
If you have any concerns during this Symposium, please stop by the Symposium registration desk in the Grand Ballroom Foyer of the Hilton North Raleigh Midtown Hotel or you may email the WEF Executive Director Dr. Eileen O'Neill at eoneill@wef.org.

WEF SOCIAL MEDIA POLICY

WEF strongly encourages the use of social media to share your experiences at our event. This includes sharing interesting quotes or information, taking pictures with colleagues, and using the event hashtag. However, to protect intellectual property, videotaping, filming, or live-streaming of any workshop or technical session presentation, or exhibit booth is prohibited. Any participant violating this policy must relinquish the media and may be removed from the Symposium. Also, promotional or commercial use of photographs taken at WEFTEC and other WEF Symposiums is strictly prohibited. If you are interested in content, materials, or products, please consider talking to the speaker or exhibitor, who may provide the information or grant permission.

SYMPOSIUM SAFETY AND SECURITY

WEF works hard to provide a comfortable and safe environment during our meetings and events, including contracting with private security firms and off-duty police officers when required. Your help also is needed to provide a safe and secure environment.

WEF recommends that you follow these safety guidelines:

1. **Share your plans**. Advise a family member or friend of your travel plans, including the dates you will be gone, how you will travel to the meeting or event, and where you are staying;

2. **Protect your identity**. Wear your name badge only inside the meeting or event venue; remove and secure your name badge when going to and from the venue; when disposing of your name badge, scratch out or destroy your name and/or the QR code to protect your personal information;

3. **Plan for an emergency**. Be aware of your surroundings. Know where the nearest exists are located. Use the buddy system. Exchange contact information with another event attendee. In case of a venue evacuation, agree where you will meet up. Call for help if your buddy doesn't meet up at the agreed location within a reasonable amount of time.

4. **See something, say something.** If you see something that raises a safety or security concern, please follow the instructions printed on the back of your name badge to alert security personnel.

Portland Crowne Plaza Emergency Plan:

Portland Crowne Plaza
1441 NE 2nd Avenue
Portland, OR 97232
Phone: (503) 233-2401

To report a medical emergency, fire, safety or security concern, call the hotel main phone number at 503-233-2401 or pick up any house phone in the meeting room area which will automatically call the hotel operations department for assistance.

PRE-SYMPOSIUM WORKSHOP

Workshop A: Emerging Issues with Pathogens and CECs in Recycled Water and Discharges to Surface Water
Sunday, July 29th 1:00 PM – 4:30 PM
Room: Windsor A
(Additional Fees Apply)

Moderators: Matt Noesen, Jenny Reina, Jacobs; Jay Swift, Gray and Osborne

Description: Emerging chemical and biological constituents in water and wastewater are a concern for the public and regulators alike. It is fundamental for our wastewater and water reuse industry to increase the scientific understanding of the risks of constituents of concern (CECs) and prioritize their assessment, as well as to become familiar with treatment alternatives that can mitigate a range of these emerging compounds. This workshop will examine hormones, pharmaceuticals, personal care products, per- and poly-fluorinated alkyl compounds (PFAS) (including fire-fighting foams), pathogens, and antibiotic resistant bacteria in wastewater, non-potable reclaimed water, and purified reclaimed water for potable water reuse. The workshop team will share the latest research from pilot and full-scale treatment and monitoring approaches and provide attendees with the necessary tools to protect public health and water quality.

1:00 PM	**Welcome and Introductions**
1:05 PM	**USEPA Regulatory Considerations and Research on Contaminants of Emerging Concern (CECs) during De Facto Water Reuse** Dr. Susan Glassmeyer, U.S. EPA
1:30 PM	**Removal of Pathogen and CECs with Advanced Membrane Technologies** Jim Lozier, Jacobs
1:55 PM	**PFAS Analytical Challenges: Remedial Insights** Dr. Jennifer Field, Oregon State University
2:20 PM	**Coffee Break in Room**
2:35 PM	**Antibiotic Resistance in Municipal Wastewater** Dr. Daniel Gerrity, University of Nevada
3:00 PM	**Emerging Pollutants Monitoring for Direct Potable Reuse** Andrew Salveson, Carollo
3:25 PM	**PURE WATER BREW: Telling the story of ONE WATER a pint at a time** Ting Lu and Mark Jockers, Clean Water Services
3:50 PM	**Q&A and Interactive Discussion of Future Research Needs**
4:30 PM	**Workshop Adjourns**

TECHNICAL PROGRAM

Opening General Session
Monday, July 30
8:30 AM – 10:00 AM
Room: Bellmont A

8:30 AM	**Welcome to the 2018 Symposium!** Bob English, *Palintest, 2018 Symposium Co-Chair* Val Frenkel, *Greeley and Hansen, 2018 Symposium Co-Chair*
8:40 AM	**WEF Welcome** Mark Poling, *Clean Water Services, WEF Board of Trustees*
8:45 AM	**Welcome from Pacific Northwest Clean Water Association** Adam McClymont, *PNCWA Vice President*
8:50 AM	**Dealing with Clean Water and Sanitation in Underserved Communities around the Globe** Catherine Leslie, *Engineers Without Borders - USA*
9:15 AM	**The Emergence of On Site Reuse** Jay Garland, Environmental Protection Agency
9:40 AM	**The Evolution of the World's largest Advanced Water Purification** **Project for Potable Reuse** Denis Bilodeau, Orange County Water District
10:00 AM	**Networking Break in Exhibit Floor**

TECHNICAL PROGRAM

Session 01: Disinfection Fundamentals
Monday, July 30
10:45 AM – 11:45 AM
Room: Bellmont B

Moderators: Jay Swift, Gray and Osborne
Bob English, Palintest

10:45 AM	**Intro to Disinfection, including recent WEF Disinfection Practices Survey** Scott Schaefer, AE2S	N/A
11:00 AM	**Chlorination** Stan Shmia, DeNora Water Technologies	N/A
11:15 AM	**Ultra Violet** Bill Sotirakos, Carollo	N/A
11:30 AM	**Peracetic acid** Josh Goldman, CDM Smith	N/A
11:45 AM	**Session Adjourns for Luncheon**	

TECHNICAL PROGRAM

Session 02: Wet Weather Disinfection – Current and Emerging Practices
Monday, July 30
10:45 AM – 11:45 AM
Room: Bellmont A

Moderators: Matt Noesen, CH2M
Ajit Ghorpade, Veolia

10:45 AM	**Evaluation of Disinfection Alternatives to Meet Main Plant and CSO Permit Criteria** Robert Sharp, Manhattan College; Maria Grieco, Keith Mahoney, New York City Department of Environmental Protection (NYCDEP); Xiao Lin, Sarah Galst, Hazen and Sawyer	1
11:15 AM	**Bayonne, NJ Combined Sewer Overflow Treatment Demonstration Project: Disinfection Results** Jurek Patoczka, Mott MacDonald; John Dening	10
11:45 AM	**Session Adjourns for Luncheon**	

TECHNICAL PROGRAM

Session 03: Innovative Disinfection Methods and Technology Trends
Monday, July 30
1:30 PM – 4:45 PM
Room: Bellmont B

Moderators: Bertrand Dussert, Xylem, Inc.
Erica Marti, University of Nevada

1:30 PM **Can MBR Reduce or Eliminate the Need for Disinfection in Water Reuse Projects** 31
Ufuk Erdal, Zeynep Erdal, Beverley Stinson, AECOM

2:00 PM **PAA, UV, and PAA/UV: Finding the Right "Bowl of Porridge" For Texas Type I Reuse Disinfection** 40
Kaylee Dusek, Michael Watts, Garver; Matthew Jalbert, Trinity River Authority

2:30 PM **Disinfection of Wastewater with PAA And UV Combined Treatment: A Pilot Study** 41
Achal Garg, Metropolitan Sewer District, City of Cincinnati; Vasudevan Namboodiri, EPA; Bruce Smith, Abdulaziz Al-Anazi, Brindha Murugesan, Tyler Bowman, Metropolitan Sewer District, City of Cincinnati

3:00 PM **Networking Break**

3:45 PM **Advanced Dose Control Strategy for Disinfection of Wastewater By Peracetic Acid: Pilot-Scale Validation And Performance** 55
Kyriakos Manoli, Siva Sarathy, Western University - Trojan Technologies; Paris Neofotistos, USP Technologies; Domenico Santoro, Western University - Trojan Technologies

4:15 PM **UV-C LED – Advances in Technology Foster Implementation** 62
Oliver Lawal, AquiSense Technologies; Gary Hunter, Black & Veatch; Raymond Ehrhard, Wahington University

Session 03 Continues on following page

Session 03 Continues from the previous page

Alternate 1	**Re-Birth of Ozone Based IPR Treatment Trains and the Resulting Integrated Controls for an Efficient Process** Steve Green, Kevin Flis, Xylem; Charles Bott, Chris Wilson, Germano Salazar Benites, Lauren Zuravnsky, Matt Poe, HRSD	80
Alternate 2	**Peracetic Acid Disinfection** Christopher Jepson, Van Cleef Engineering Associates	86
4:45 PM	**Session Adjourns for Networking Reception in Bellmont Foyer**	

TECHNICAL PROGRAM

Session 04: Potable Reuse (DPR and IPR) - Safety, Water Quality Control, Demonstration
Monday, July 30
1:30 PM – 3:00 PM
Room: Bellmont A

Moderators: Val Frenkel, Greeley and Hansen
Stuart F. Humphries, Orthos Liquid Systems, Inc.

1:30 PM	**Are TOC and COD Limits Appropriate for Regulating Potable Reuse? Analyzing Organics through the Domestic Water Cycle: Results from WE&RF Project-15-04** Larry Schimmoller, James Lozier, James Rosenblum, CH2MHill	88
2:00 PM	**Using Large Data Sets and Statistical Tools to Define "Failure" and Demonstrate the Safety of DPR** Austa Parker, Carollo Engineers; Yan Qu, David Hokanson, Trussell Technologies; Jeff Soller, Soller Environmental; Eric Dickenson, Southern Nevada Water Authority; Robert Angelotti, Upper Occoquan Service Authority; Julian Inoue, Carollo Engineers; Shane Trussell, Trussell Technologies; Andrew Salveson, Carollo Engineers	95
2:30 PM	**A Framework for Regulating Raw Water Augmentation - Results from a Pathogen Benchmarking Study** Dave MacNevin, Matt Azarian, Andrea Netcher, Tetra Tech	97
3:00 PM	**Session Adjourns for Networking Break**	

TECHNICAL PROGRAM

Session 05: Expanding Reuse Strategies and Markets
Monday, July 30
3:45 PM – 4:45 PM
Room: Bellmont A

Moderators: Patrick Regan, Evoqua Water Technologies
Matt Noesen, CH2M

3:45 PM	**Stormwater Reuse: Aquifer Storage and Recovery Evaluation Using Municipal Residential Stormwater in the Pacific Northwest** Jason Melady, GSI Water Solutions, Inc.; Jadene Stensland, Clean Water Services, OR; Andrew Davidson, GSI Water Solutions, Inc.; David Winship, City of Beaverton, OR	116
4:15 PM	**Telling the Story of ONE WATER A Pint At A Time: What Four Years Of Drinking Beer Made From Wastewater Has Taught Us About The Language Of Water** Mark Jockers, Mark Poling, Clean Water Services, OR	137
4:45 PM	**Session Adjourns for Networking Reception**	

TECHNICAL PROGRAM

Session 06: Disinfection Optimization and Control of Disinfection Byproducts (DBPs)
Tuesday, July 31
8:30 AM – 11:45 AM
Room: Bellmont A

Moderators: Scott Schaefer, AE2S
Val Frenkel, Greeley and Hansen

8:30 AM	**Disinfection By-Product (DBP) Formation and Control in Potable Reuse Applications** Caroline Russell, Carollo Engineers	141
9:00 AM	**Evaluation of Over 100 Disinfection Reactor Tracer Studies in Washington State** Keith Stewart, Russ Porter, Gray & Osborne	147
9:30 AM	**Ranitidine – A Potential Significant NDMA Precursor for Potable Reuse** Erica Marti, University of Nevada, Las Vegas; Caitlin Glover; Eric Dickenson, Southern Nevada Water Authority	166
10:00 AM	**Networking Break**	
10:45 AM	**Getting a Jump on Disinfection Byproducts: A Pro-Active Approach for Optimizing Reclaimed Water Disinfection** Andrea Netcher, Tetra Tech; Gary Revoir, Reiss Environmental Consulting Engineers	174
11:15 AM	**Experiences and Successful Strategies for Reuse Water Chorine Residual Control** Bing Lin, Alfredo Suarez, EPCOR Water Services	192
11:45 AM	**Session Adjourns for Luncheon**	

TECHNICAL PROGRAM

Session 07: New Water Reuse Approaches to Manage Tomorrow's Challenges
Tuesday, July 31
8:30 AM – 10:00 AM
Room: Bellmont B

Moderators: Jenny Reina, CH2M
Justin Mattingly, Water Research Foundation

8:30 AM	**PureWaterSF: Direct Potable Reuse Demonstration in San Francisco** Manisha Kothari, Paula Kehoe, San Francisco Public Utilities Commission; Andrea Corral, Carollo Engineers; Nicola Fontaine; Mark Millan; Andrew Salveson, Carollo Engineers	199
9:00 AM	**Quantification and Prediction of Degradation and Deactivation Kinetics of Antibiotic Resistance Genes During Disinfection/Oxidation Processes: From Bench-scale to Full-scale** Huan He, Peiran Zhou, Kyle Shimabuku, Xuzhi Fang, University of Washington; Annika Anderson, Washington University in St. Louis; Michael Dodd, University of Washington	206
9:30 AM	**Assessing Chemical Disinfectant Alternatives Using I-CT Dose-Response Kinetics for Multiple Microorganisms** Roberta Maffettone, Siva Sarathy, Western University - Trojan Technologies; Ronald Gehr, McGill University Dept of Civil Engineering; Paris Neofotistos, USP Technologies; Domenico Santoro, Trojan Technologies	213
10:00 AM	**Session Adjourns for Networking Break**	

TECHNICAL PROGRAM

Session 08: Supporting Public Health with Water, Sanitation and Hygiene (WaSH) Projects
Tuesday, July 31
10:45 AM – 11:45 AM
Room: Bellmont B

Moderators: Michael Watts, Garver

10:45 AM	**Managing Chlorinated Water for Ebola Disinfection** Lee Gary, Robert Reimers, Tulane University; Kari Brisolara, LSUHSC	221
11:15 AM	**Increasing In-Home Water Availability in Remote Alaska's Homes with Water Reuse** Aaron Dotson, University of Alaska Anchorage	222
11:45 AM	**Session Adjourns for Luncheon**	

TECHNICAL PROGRAM

Session 09: Residuals, Sanitation and Hygiene: Concerns, Design and Application for Low Technology Processing
Tuesday, July 31
1:30 PM – 4:45 PM
Room: Bellmont B

Moderator: Leonard Casson, University of Pittsburgh

1:30 PM	**Sampling and Disinfection of Wastewater relative to Parasites and the Role of associated Surrogates, Indicators and Non-Vital Models** Dwight Bowman, Cornell University	228
2:00 PM	**Effective Treatment of Biosolids and Manures – Multiple Additives/Factors for Disinfection, Stabilization and Product Quality** Robert Reimers, Tulane University	229
2:30 PM	**Low-Cost, Low-Tech Biosolids Treatment via combined Long-Tern Storage (Lagoon) and Air Drying: A Comparison of Two Pilot-Scale Studies** Jennifer Becker, Michigan Tech University	236
3:00 PM	**Networking Break**	
3:45 PM	**Effectiveness of Tropical Lagoon Treatment for Disinfection of Domestic Septage** Joel Nkiama Konde, Kinshasa School of Public Health, University of Kinshasa	239
4:15 PM	**Usage of Low-Cost Bio Electro Technology to convert Class B Biosolids Processes (aerobic and anaerobic digestion) to Class A Disinfected and Stabilized Biosolids** Adam Faschan, Ardurra Group	243
4:45 PM	**Symposium Adjourns**	

TECHNICAL PROGRAM

Session 10: Optimizing a Future for DPR – Design, Implementation and Operation of Robust Solutions
Tuesday, July 31
1:30 PM – 3:00 PM
Room: Bellmont A

Moderators: Andrew Salveson, Carollo
Melanie Mann, Hazen and Sawyer

1:30 PM	**Accounting for Future Potable Reuse When Implementing Disinfection Solutions Today** June Leng, Jacobs	245
2:00 PM	**From Collection Systems to Tap: Resilience of Treatment Processes for Potable Reuse** Justin Mattingly, Water Research Foundation; Sharon Waller, Sustainable Systems, LLC – Consulting; Patti Craddock, Short Elliott Hendrickson, Inc.; Stuart Khan, University of New South Wales	257
2:30 PM	**Design of a Managed Aquifer Recharge and Advanced Treatment Facility** Andrew Newbold, Hazen and Sawyer; Lauren Zuravnsky, Charles Bott, Germano Salazar-Benites, Chris Wilson, HRSD; Dwayne Amos, David Briley, Hazen and Sawyer	266
Alternate	**Planning for a Potable Reuse Future While Maximizing Recycled Water Today** Jasmine Diaz, Water Systems Consulting, Inc.; Michael Falk, HDR Inc	272
3:00 PM	**Session Adjourns for Networking Break**	

Session 11: Disinfection of WRRF Effluents for Surface Water Discharges
Tuesday, July 31
3:45 PM – 4:45 PM
Room: Bellmont A

Moderator: Cyndy Bratz, Tetra Tech

3:45 PM **Aberdeen, WA Water Resource Recovery Facility Disinfection System Evaluation** 286
Jay Swift, Gray and Osborne

4:15 PM **Using Dynamic Dilution Methods to Derive More Accurate Effluent Total Residual Chlorine Limits for WWRFs** 316
Robert Sharp, Manhattan College; Xiao Lin, Sarah Galst Hazen and Sawyer; Kathy Ammari, Manhattan College; Laura Grieco, New York City Department of Environmental Protection (NYCDEP)

4:45 PM **Symposium Adjourns**

TECHNICAL PROGRAM AT-A-GLANCE

Session Number	Session Title	Time	CE Credit Total/Type	Room
Sunday, July 29				
Workshop	Emerging Issues with Pathogens and Constituents of Emerging Concern (CECs) in Water Reuse and Surface Water Discharge	1:00pm-4:30pm	**0.3 CEUs**	Windsor A
Monday, July 30				
OGS	Opening General Session	8:30am-10:00pm	**1.5 GCHs**	Bellmont A
Session 1	Disinfection Fundamentals	10:45am–11:45am	**1.0 PDHs**	Bellmont B
Session 2	Wet Weather Disinfection – Current and Emerging Practices	10:45am–11:45am	**1.0 PDHs**	Bellmont A
Session 3	Innovative Disinfection Methods and Technology Trends	1:30pm–4:45pm	**2.5 PDHs**	Bellmont B
Session 4	Potable Reuse (DPR and IPR) - Safety, Water Quality Control, Demonstration	1:30pm–3:00pm	**1.5 PDHs**	Bellmont A
Session 5	Expanding Reuse Strategies and Markets	3:45pm–4:45pm	**1.0 PDHs**	Bellmont A
Tuesday, July 31				
Session 6	Disinfection Optimization and Control of Disinfection Byproducts (DBPs)	8:30am–11:45am	**2.5 PDHs**	Bellmont A
Session 7	New Water Reuse Approaches to Manage Tomorrow's Challenges	8:30am–10:00am	**1.5 PDHs**	Bellmont B
Session 8	Supporting Public Health with Water, Sanitation and Hygiene (WaSH) Projects	10:45am–11:45am	**1.0 PDHs**	Bellmont B
Session 9	Residuals, Sanitation and Hygiene: Concerns, Design and Application for Low Technology Processing	1:30pm–4:45pm	**2.5 PDHs**	Bellmont B
Session 10	Optimizing a Future for DPR – Design, Implementation and Operation of Robust Solutions	1:30pm–3:00pm	**1.5 PDHs**	Bellmont A
Session 11	Disinfection of WRRF Effluents for Surface Water Discharges	3:45pm–4:45pm	**1.0 PDHs**	Bellmont A

PRESENTATIONS AT-A-GLANCE

We encourage attendees to participate in presentations from among all sessions. To better plan your learning, see below and on the following pages a list of presentations offered during each time block.

Key:

Disinfection
Reuse
Both

Monday, July 30 - Morning		
	Bellmont B	**Bellmont A**
Session	**Session 01: Disinfection Fundamentals**	**Session 02: Sustainability as a Driver for Planning and Decision Making**
10:45 AM	Evaluation of Disinfection Alternatives to Meet Main Plant and CSO Permit Criteria	Intro to Disinfection, including recent WEF Disinfection Practices Survey
11:00 AM		Chlorination
11:15 AM	Bayonne, NJ Combined Sewer Overflow Treatment Demonstration Project: Disinfection Results	Ultra Violet
11:30 AM		Peracetic acid

Monday, July 30 - Afternoon		
Room	**Bellmont B**	**Bellmont A**
Session	**Session 03: Innovative Disinfection Methods and Technology Trends**	**Session 04: Potable Reuse (DPR and IPR) - Safety, Water Quality Control, Demonstration**
1:30 PM	Can MBR Reduce or Eliminate the Need for Disinfection in Water Reuse Projects	Are TOC and COD Limits Appropriate for Regulating Potable Reuse? Analyzing Organics through the Domestic Water Cycle: Results from WE&RF Project-15-04
2:00 PM	PAA, UV, and PAA/UV: Finding the Right "Bowl of Porridge" For Texas Type I Reuse Disinfection	Using Large Data Sets and Statistical Tools to Define "Failure" and Demonstrate the Safety of DPR
2:30 PM	Disinfection of Wastewater with PAA And UV Combined Treatment: A Pilot Study	A Framework for Regulating Raw Water Augmentation - Results from a Pathogen Benchmarking Study
Session	***Session 03, continued***	**Session 05: Expanding Reuse Strategies and Markets**
3:45 PM	Advanced Dose Control Strategy for Disinfection of Wastewater By Peracetic Acid: Pilot-Scale Validation And Performance	Stormwater Reuse: Aquifer Storage and Recovery Evaluation Using Municipal Residential Stormwater in the Pacific Northwest
4:15 PM	UV-C LED – Advances in Technology Foster Implementation	Telling the Story Of ONE WATER A Pint At A Time: What Four Years Of Drinking Beer Made From Wastewater Has Taught Us About The Language Of Water
Alternate 1	Re-Birth of Ozone Based IPR Treatment Trains and the Resulting Integrated Controls for an Efficient Process	
Alternate 2	Peracetic Acid Disinfection	

PRESENTATIONS AT-A-GLANCE

Tuesday, July 31 - Morning		
Room	**Bellmont A**	**Bellmont B**
Session	**Session 06: Disinfection Optimization and Control of Disinfection Byproducts (DBPs)**	**Session 07: New Water Reuse Approaches to Manage Tomorrow's Challenges**
8:30 AM	Disinfection By-Product (DBP) Formation and Control in Potable Reuse Applications	PureWaterSF: Direct Potable Reuse Demonstration in San Francisco
9:00 AM	Evaluation of Over 100 Disinfection Reactor Tracer Studies in Washington State	Quantification and Prediction of Degradation and Deactivation Kinetics of Antibiotic Resistance Genes During Disinfection/Oxidation Processes: From Bench-scale to Full-scale
9:30 AM	Ranitidine – A Potential Significant NDMA Precursor for Potable Reuse	Assessing Chemical Disinfectant Alternatives Using I-CT Dose-Response Kinetics for Multiple Microorganisms
Session	***Session 06 continued***	**Session 08: Supporting Public Health with Water, Sanitation and Hygiene (WaSH) Projects**
10:45 AM	Getting a Jump on Disinfection Byproducts: A Pro-Active Approach for Optimizing Reclaimed Water Disinfection	Managing Chlorinated Water for Ebola Disinfection
11:15 AM	Experiences and Successful Strategies for Reuse Water Chorine Residual Control	Increasing In-Home Water Availability in Remote Alaska's Homes with Water Reuse

PRESENTATIONS AT-A-GLANCE

Tuesday, July 31 - Afternoon		
Room	Bellmont B	Bellmont A
Session	**Session 09: Residuals, Sanitation and Hygiene: Concerns, Design and Application for Low Technology Processing**	**Session 10: Optimizing a Future for DPR – Design, Implementation and Operation of Robust Solutions**
1:30 PM	Sampling and Disinfection of Wastewater relative to Parasites and the Role of associated Surrogates, Indicators and Non-Vital Models	Accounting for Future Potable Reuse When Implementing Disinfection Solutions Today
2:00 PM	Effective Treatment of Biosolids and Manures – Multiple Additives/Factors for Disinfection, Stabilization and Product Quality	From Collection Systems to Tap: Resilience of Treatment Processes for Potable Reuse
2:30 PM	Low-Cost, Low-Tech Biosolids Treatment via combined Long-Tern Storage (Lagoon) and Air Drying: A Comparison of Two Pilot-Scale Studies	Design of a Managed Aquifer Recharge and Advanced Treatment Facilitygn of a Managed Aquifer Recharge and Advanced Treatment Research Center
Alternate 1		Planning for a Potable Reuse Future While Maximizing Recycled Water Today
Session	***Session 09, continued***	**Session 11: Disinfection of WRRF Effluents for Surface Water Discharges**
3:45 PM	Effectiveness of Tropical Lagoon Treatment for Disinfection of Domestic Septage	Aberdeen, WA Water Resource Recovery Facility Disinfection System Evaluation
4:15 PM	Usage of Low-Cost Bio Electro Technology to convert Class B Biosolids Processes (aerobic and anaerobic digestion) to Class A Disinfected and Stabilized Biosolids	Using Dynamic Dilution Methods to Derive More Accurate Effluent Total Residual Chlorine Limits for WWRFs

THE EXHIBITION

Exhibition Schedule

Networking events listed below are dedicated show hours.
Technical sessions will not take place during these times.

Monday, July 30	**10:00 am – 6:15 pm**	**Exhibits Open**
	10:00 am – 10:45 am	Networking Break
	11:45 am – 1:30 pm	Networking Luncheon
	3:00 pm – 3:45 pm	Networking Break
	4:45 pm – 6:15 pm	Networking Reception

Due to the proprietary nature of the displays, photography of displays and materials is forbidden without exhibitors' express permission.

EXHIBITORS

As of 7/11/2018

Canopus Water Technologies.. Booth 104
www.canopuswatertechnologies.com
Phone: 781-856-0476
7 Jenny's Hill Rd
Windham, NH 03087

Developer and manufacturer of non-chemical water disinfection products. UV, Ozone, and Plasma products with best in class sanitation performance for residential, industrial and commercial applications. Recently introduced – C-UV100, the best UV disinfection offering in the market based on state of the art UVC-LEDs and a unique reactor design.

CHEMetrics, Inc .... Booth 105
www.chemetrics.com
Phone: 800-356-3072
4295 Catlett Rd
Midland, VA 22728

CHEMetrics, Inc. manufactures water analysis test kits for testing impurities in process and product water and monitoring the efficacy of sterilization solutions. Test kits available for Peracetic Acid, Hydrogen Peroxide, Ozone and Chlorine.

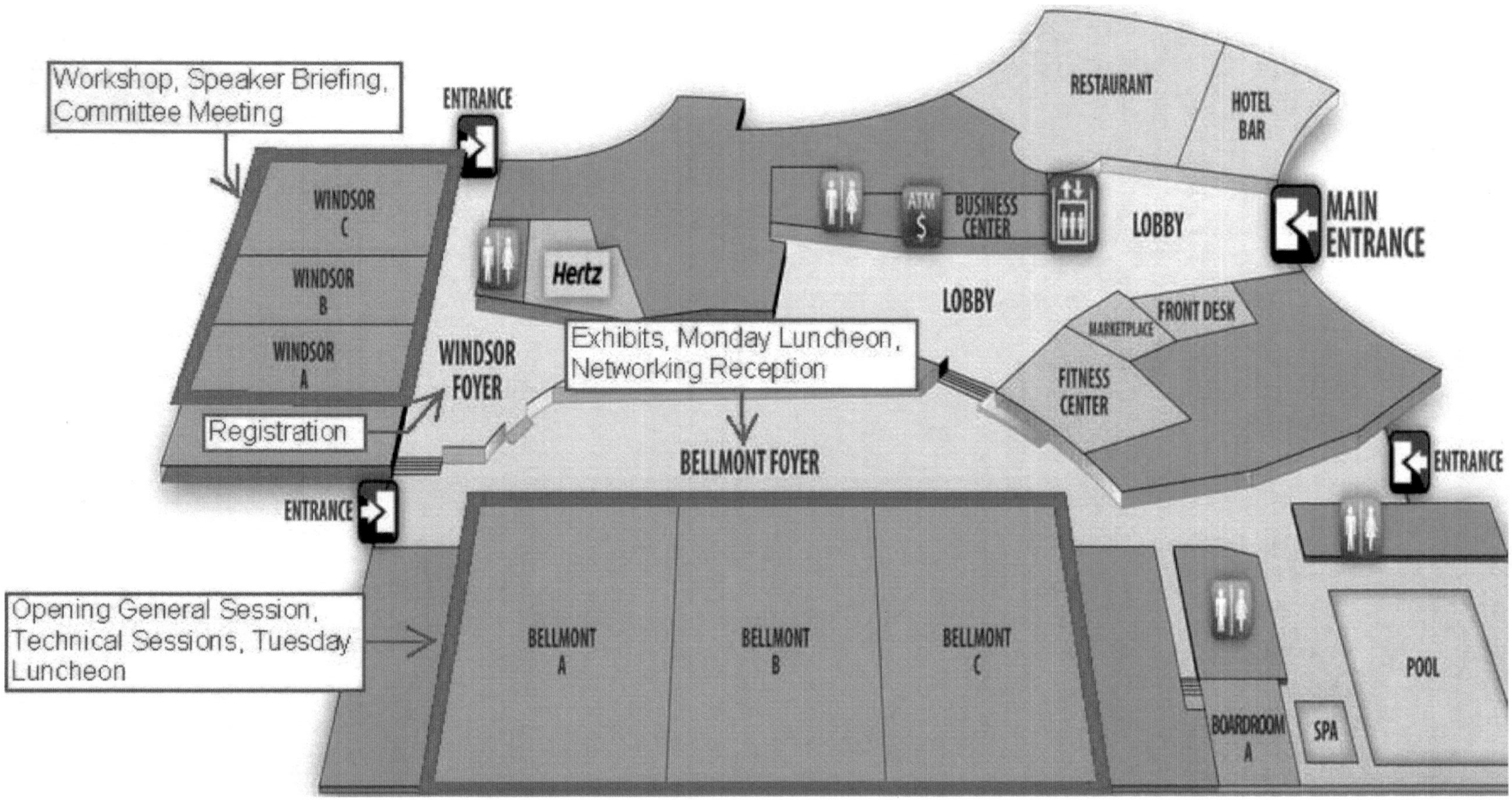

Workshop, Speaker Briefing, Committee Meeting
ENTRANCE
WINDSOR C
WINDSOR B
WINDSOR A
Registration
WINDSOR FOYER
ENTRANCE
Hertz
RESTAURANT
HOTEL BAR
ATM
BUSINESS CENTER
LOBBY
MAIN ENTRANCE
LOBBY
MARKETPLACE
FRONT DESK
FITNESS CENTER
Exhibits, Monday Luncheon, Networking Reception
BELLMONT FOYER
ENTRANCE
ENTRANCE
Opening General Session, Technical Sessions, Tuesday Luncheon
BELLMONT A
BELLMONT B
BELLMONT C
BOARDROOM A
SPA
POOL

PRESENTER AND MODERATOR DIRECTORY

Jennifer Becker, Michigan Tech University, **Presenter, 09**

Dwight Bowman, Cornell University Dept Of Micro & Imm, **Presenter, 09**

Cyndy Bratz, Tetra Tech, **Moderator, 11**

Leonard Casson, University of Pittsburgh, **Moderator, 09**

Jasmine Diaz, Water Systems Consulting, Inc., **Presenter, 10**

Aaron Dotson, University of Alaska Anchorage, **Presenter, 08**

Kaylee Dusek, Garver, **Presenter, 03**

Bertrand Dussert, Xylem, **Moderator, 03**

Robert English, Palintest, **Moderator, 01**

Ufuk Erdal, AECOM, **Presenter, 03**

Adam Faschan, Ardurra Group, **Presenter, 09**

Jennifer Field, Oregon State University, **Presenter, Workshop**

Val Frenkel, Greeley and Hansen, **Moderator, 06, 04**

Achal Garg, Metropolitan Sewer District, City Of Cincinnati, **Presenter, 03**

Daniel Gerrity, University of Nevada, **Presenter, Workshop**

Ajit Ghorpade, 980, **Moderator, 02**

Susan Glassmeyer, EPA, **Presenter, Workshop**

Josh Goldman-Torres, CDM Smith Inc, **Presenter, 01**
Steve Green, Xylem, **Presenter, 03**

Huan He, University of Washington, **Presenter, 07**

Stuart Humphries, Orthos Liquid Systems, Inc., **Moderator, 04**

Gary Hunter, Black & Veatch, **Presenter, 03**

Christopher Jepson, Van Cleef Engineering Associates, **Presenter, 03**

Mark Jockers, Clean Water Services, OR, **Presenter, 05**

Joel Nkiama Konde, Kinshasa School of Public Health, University of Kinshasa, **Presenter, 09**

June Leng, Jacobs, **Presenter, 10**

Bing Lin, EPCOR Water Services, **Presenter, 06**

James Lozier, CH2MHill, **Presenter, Workshop, 04**

Ting Lu, Clean Water Services, **Presenter, Workshop**

Dave MacNevin, Tetra Tech, **Presenter, 04**

Melanie Mann, Hazen and Sawyer, **Moderator, 10**

Erica Marti, University of Nevada, Las Vegas, **Moderator, 03, Presenter, 06**

Justin Mattingly, Water Research Foundation, **Moderator, 07, Presenter, 10**

PRESENTER AND MODERATOR DIRECTORY

Jason Melady, GSI Water Solutions, Inc., **Presenter, 05**

Andrea Netcher, Tetra Tech, **Presenter, 06**

Andrew Newbold, Hazen and Sawyer, **Presenter, 10**

Matthew Noesen, CH2M, **Moderator, Workshop, 02, 05**

Austa Parker, Carollo Engineers, **Presenter, 04**

Jurek Patoczka, Mott MacDonald, **Presenter, 02**

Patrick Regan, Evoqua WAter Technologies, **Moderator, 05**

Robert Reimers, Tulane University, **Presenter, 08, 09**

Jenny Reina, CH2M, **Moderator, Workshop, 07**

Caroline Russell, Carollo Engineers, **Presenter, 06**

Andrew Salveson, Carollo Engineers, **Presenter, Workshop, 07, Moderator, 10**

Siva Sarathy, Western University - Trojan Technologies, **Presenter, 03, 07**

Scott Schaefer, AE2S, **Moderator, 06, Presenter, 01**

Robert Sharp, Manhattan College, **Presenter, 02, 11**

Stanley Shmia, De Nora Water Technologies, **Presenter, 01**

Bill Sotirakos, Carollo Engineers, **Presenter, 01**

Keith Stewart, Gray & Osborne, Inc., **Presenter, 06**

Jay Swift, Gray and Osborne, **Moderator, Workshop, 01, Presenter, 11**

Michael Watts, Garver, **Moderator, 08**

CONFERENCE SCHEDULE AT-A-GLANCE

Sunday, July 29

12:00 PM – 4:30 PM	Registration
1:00 PM – 4:30 PM	Workshops A

Monday, July 30

7:30 AM – 4:45 PM	Registration
8:30 AM – 10:00 AM	Opening General Session
10:00 AM – 6:15 PM	Exhibit Hall Open
10:45 AM – 11:45 PM	Technical Sessions 1, 2
11:45 AM – 1:30 PM	Luncheon in Bellmont C
1:30 PM - 4:45 PM	Technical Session 3
1:30 PM - 3:00 PM	Technical Session 4
3:45 PM - 4:45 PM	Technical Session 5
4:45 PM – 6:15 PM	Networking Reception

Tuesday, July 31

8:00 AM – 5:00 PM	Registration
8:30 AM – 11:45 PM	Technical Session 6
8:30 AM – 10:00 AM	Technical Sessions 7
10:45 AM – 11:45 AM	Technical Sessions 8
11:45 AM – 1:30 PM	Luncheon in Bellmont C
1:30 PM – 4:45 PM	Technical Session 9
1:30 PM – 3:00 PM	Technical Session 10
3:45 PM – 4:45 PM	Technical Sessions 11

UPCOMING EDUCATION AND TRAINING EVENTS

WEFTEC 2018
September 29 - October 3, 2018
New Orleans, Louisiana
www.weftec.org

WEF Education and Training 2019 Events

Mark your calendars for our upcoming events.
More information coming shortly.

Forum 2019: James Barnard Research Conference on Emerging Themes in Biological Phosphorus Removal
January 2019, Austin, Texas

AWWA/WEF The Utility Management Conference 2019
March 5-8, 2019
Nashville, Tennessee
https://www.wef.org/utilitymanagement

Residuals and Biosolids 2019 Conference
May 7-10, 2019, Ft Lauderdale, Florida

Stormwater and Green Infrastructure 2019 Symposium
May 2019, Ft Lauderdale, Florida

Collection Systems 2019 Conference
June 4-7, 2019, Indianapolis, Indiana

Nutrient Removal and Recovery 2019 Symposium
July 2019, Midwest

WEF/AWWA Transformative Issues 2019 Symposium
August 2019, Washington, D.C.

NOTES

HDR

Evaluation of Disinfection Alternatives to Meet Main Plant and CSO Permit Criteria

R. Sharp[3*], L. Grieco[2], K. Mahoney[2], Xiao Lin[1] and S. Galst[1]

[1]Hazen and Sawyer, New York, NY
[2] New York City Department of Environmental Protection (NYCDEP)
[3]Manhattan College, NY
*Rsharp@hazenandsawyer.com

New York State is considering changing the regulated effluent pathogen indicator for wastewater and combined sewer overflow discharge permits from of Fecal Coliforms to *Enterococcus*. To prepare for such a change in regulations, a series of pathogen inactivation studies using chlorine and peracetic acid (PAA) were undertaken to determine the dose of each disinfectant needed to meet plant effluent and CSO disinfection permit criteria. Studies were carried out on four different plant effluents and two different CSO discharges. The results from these studies generally show that a higher dose of chlorine is needed to meet the proposed *Enterococcus* criteria compared to the current Fecal Coliform criteria, especially for plants that carry out Biological Nutrient Removal and have longer SRTs. In addition, a higher dose of PAA was required to meet both criteria compared to chlorine. This switch to *Enterococcus* presents several potential challenges for WRRFs. *Enterococcus* is widely considered more difficult to inactivate/kill using traditional disinfection methods compared to Fecal Coliforms (Noble, 2003) (Vilanova, 2003). In addition, CSO disinfection is being considered for a number of CSO facilities in the NYC system. The disinfection criteria are expected to include a 2-log kill of *Enterococcus* and a low TRC limit (<0.1 mg/l).

A series of batch pathogen inactivation and disinfectant decay studies were carried out on the plant effluent and CSO samples. The objective of the study was to determine if increased chlorine doses are needed to meet the new *Enterococcus* effluent criteria (35 cfu/100ml monthly geometric mean) compared to the current Fecal Coliform criteria (200 cfu/100ml). In addition, these studies investigated the effectiveness peracetic acid (PAA) as a disinfection alternative to meeting the new *Enterococcus* criteria in plant effluents and CSOs. Table 1 shows the current and new TRC limit for each plant as well as other associated permit criteria.

Table 1- New and Proposed Effluent Criteria for Each WRRF included in this Study

Plant	Type	Current TRC (mg/l)	New TRC (mg/l)	NH_4-N (mgN/L)	CN^- (Lbs/d)
Plant 1	BNR	1.0	0.07	13.2	2.1
Plant 2	BNR	1.0	0.05	1.03	0.92
Plant 3	2°	2.0	0.36	52.9	6.4
Plant 4	2°	2.0	0.26	33.9	11

BATCH PATHOGEN INACTIVATION AND DISINFECTANT DECAY STUDIES

The batch studies involved preparation of a series of 1000 ml well mixed batch reactors. Each batch reactor contained up to 800 ml of secondary effluent or CSO from one of the study facilities. The series of batch reactors were treated with varying doses of either chlorine (sodium hypochlorite) or PAA and were analyzed over time. The disinfectant doses varied among the different plant effluents, with some effluents requiring disinfectant doses greater than 5 mg/l, and CSO samples requiring up to 8 mg/l measured as Chemically Produced Oxidant (CPO – e.g. TRC as CPO or PAA as CPO). Testing of each sample at each dose was repeated a minimum of 4 times. Temperature and pH were monitored throughout each test and testing of all effluent and CSO samples was completed within 6 hours of sample collection. US EPA standard membrane filtration methods were used to enumerate *Enterococcus* and Fecal Coliforms. All other chemical analyses followed standard methods used in the wastewater industry (Standard Methods, 2012). The Ct values were calculated based on the residual disinfectant dose at the time of compliance with the specific fecal indicator bacteria criteria (C = TRC or PAA as mg/l CPO) multiplied by the contact or reaction time when compliance with the specific criteria was achieved during the test (t = minutes). Disinfection decay rates were determined by monitoring CPO using HACH PAO Amperometric Forward Titration and instantaneous CPO (chlorine or PAA) uptake and decay were correlated to various water quality criteria, including ammonia, COD and TSS.

RESULTS

Effluent Disinfection and Pathogen Inactivation: An example of the results from a series of chlorine and PAA inactivation tests for one of the plant effluent samples is shown in **Figure 1** and **Figure 2** respectively. **Figure 3** shows the inactivation dynamics of fecal coliforms and *Enterococcus* with both Chlorine and PAA disinfection. Typically, both PAA and chlorine rapidly inactivated both Fecal Coliform and *Enterococcus*, however, after approximately 20 minutes of contact time, the Fecal Coliforms level off, while the Enterococcus continued to decay at a relatively high rate.

CPO Decay Rates were measured for all effluent disinfection tests. Zeroth, first, and second order disinfectant decay rates were determined for each effluent sample at each disinfectant dose. The decay of PAA and Chlorine both has the best fit using a second order rate expression (Equation 1, highest R^2 value). **Table 2** shows the chlorine decay rates for each WWTP for specific chlorine doses. The decay rate for PAA was consistently lower than the decay rate for chlorine. For both chlorine and PAA the decay rate decreased as the applied dose increased, likely because more demand was realized in the early contact time at higher dose. However, at higher doses the instantaneous uptake increased with increased doe for both disinfectants. In addition, since second order rates predominated, it is apparent that the rate is a function of both time and the residual disinfectant concentration. The low PAA decay rates help to explain the persistence of the PAA residual in most plant effluent samples even after 30 minutes of decay.

Second Order Decay $$C(t) = \frac{C_0}{1+k \cdot C_0 \cdot t}$$ Equation 1

Figure 1 – Chlorine Disinfection: A) Fecal Coliform; B) Enterococcus inactivation; C) Chlorine Decay; D) Decay Characteristics at Chlorine dose of 1.0 mg/L

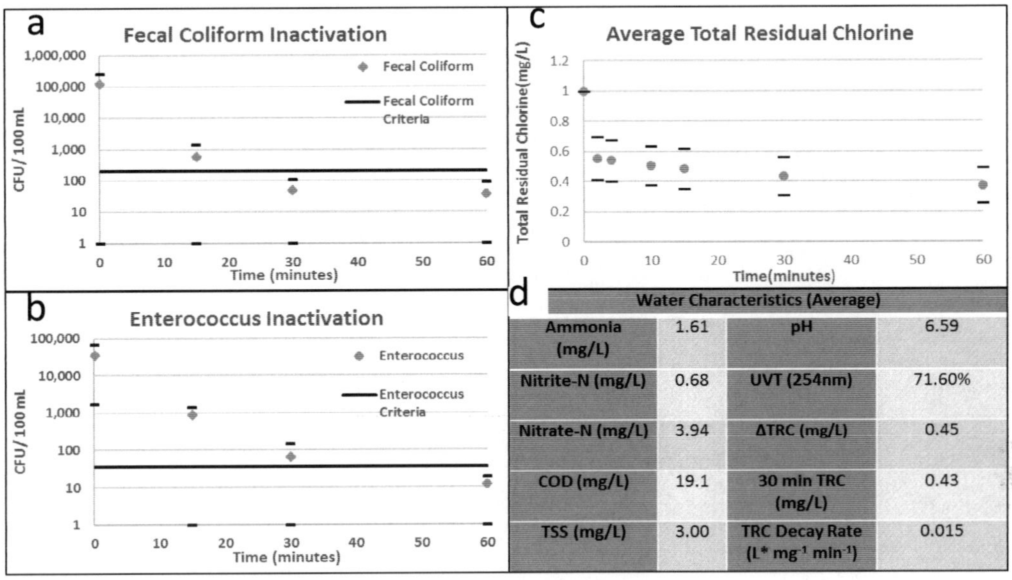

Water Characteristics (Average)			
Ammonia (mg/L)	1.61	pH	6.59
Nitrite-N (mg/L)	0.68	UVT (254nm)	71.60%
Nitrate-N (mg/L)	3.94	ΔTRC (mg/L)	0.45
COD (mg/L)	19.1	30 min TRC (mg/L)	0.43
TSS (mg/L)	3.00	TRC Decay Rate ($L^* mg^{-1} min^{-1}$)	0.015

Figure 2 – PAA Disinfection - A) Fecal Coliform; B) Enterococcus inactivation; C) PAA decay; D) Decay Characteristics at PAA dose of 2.0 mg/L

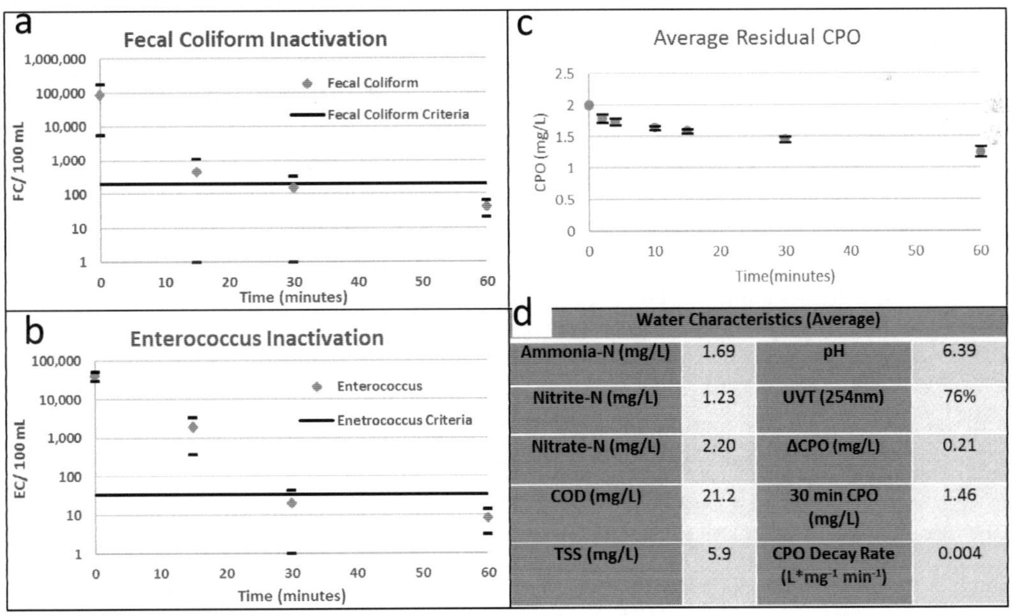

Water Characteristics (Average)			
Ammonia-N (mg/L)	1.69	pH	6.39
Nitrite-N (mg/L)	1.23	UVT (254nm)	76%
Nitrate-N (mg/L)	2.20	ΔCPO (mg/L)	0.21
COD (mg/L)	21.2	30 min CPO (mg/L)	1.46
TSS (mg/L)	5.9	CPO Decay Rate ($L^*mg^{-1} min^{-1}$)	0.004

Figure 3- Examples of *Enterococcus* and Fecal Coliform Inactivation Dynamics with Chlorine and PAA Disinfection

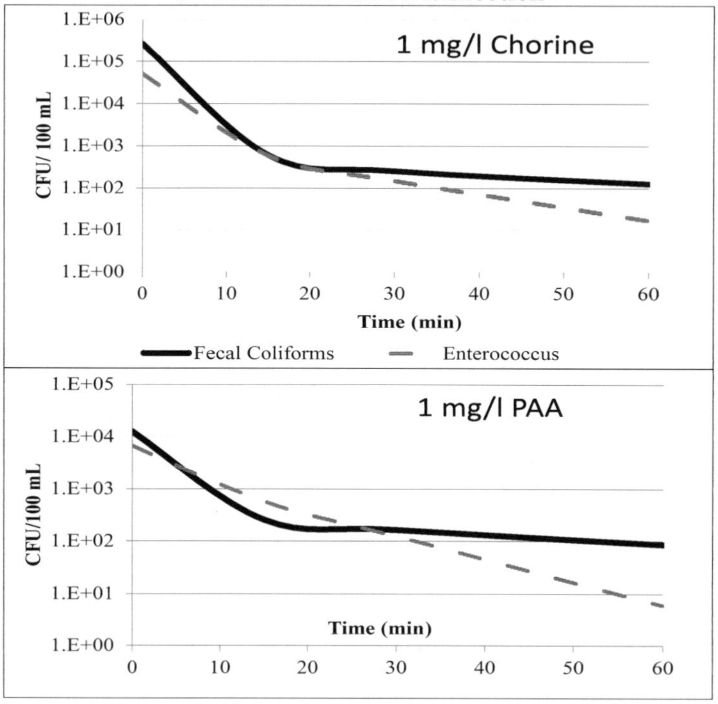

Table 2 - Disinfectant Decay Rate for Studied Applied Doses for Plant Effluents

Plant	Disinfectant	Dose (mg/L)	Decay Rate ($L*mg^{-1}min^{-1}$)
Plant 1 (BNR)	Chlorine	1.0	0.0087
		1.5	0.0046
	PAA	2.0	0.0033
		2.5	0.0022
Plant 2 (BNR)	Chlorine	1.0	0.015
		1.5	0.008
	PAA	2.0	0.004
Plant 3	Chlorine	0.75	0.012
		1.5	0.0057
	PAA	2.0	0.0064
		2.5	0.0028
Plant 4	Chlorine	2.5	0.0032
		3.0	0.0022
	PAA	6.0	0.0011
		7.0	0.0011

Figure 4 shows the relationship between chlorine decay rate and ammonia concentration. As shown in the figure, the higher chlorine decay rate was observed at low ammonia concentration, similar correlation was observed between ammonia concentration and initial chlorine demand. At low ammonia levels, chloramine will be oxidized into nitrous oxide and nitrogen, resulting in rapid decrease in TRC, therefore a higher decay rate, but also poor disinfection efficacy. Other wastewater quality parameters correlated poorly with the decay rate constants. This phenomenon may explain why lower chlorine doses can be more effective when ammonia levels are high: the key may be to keep the ammonia to chlorine ration above 5 and maintain a relatively high pH (>8).

Figure 4 - Chlorine Decay Rates as Function of Ammonia Concentration

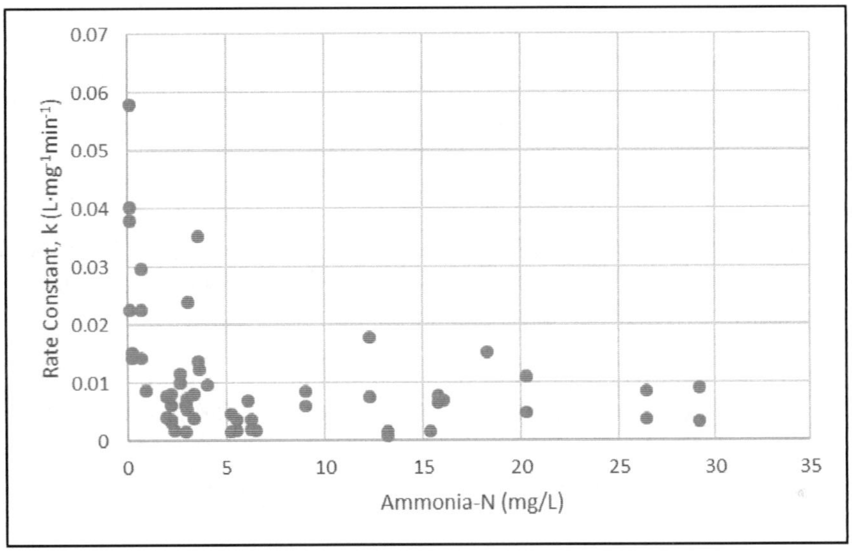

Simple correlations were performed to evaluate any possible relationship between different wastewater characteristics and PAA decay rate. Results show that most wastewater quality parameters correlated poorly with the PAA decay rate constants. **Figure 5** shows the relationship between PAA decay rate and applied dose. It appears that the decay rate decreases at higher applied dose, likely because most PAA demand is satisfied in the early stage (0-2 minutes) of the contact time at higher doses, and pathogen and inert substances contribute to the further decay of PAA (2-60 minutes contact time). However, as the PAA dose increased the instantaneous demand also increased. With limited demand remaining after the initial uptake in the effluent samples, the PAA residual remained high and did not dissipate appreciably after 60 minutes of contact time. Although PAA was an effective disinfectant, the PAA residual remained relatively high, with a low tailing decay rate in all of the effluent studies.

Figure 5 - PAA Decay Rate a Function of Applied Dose

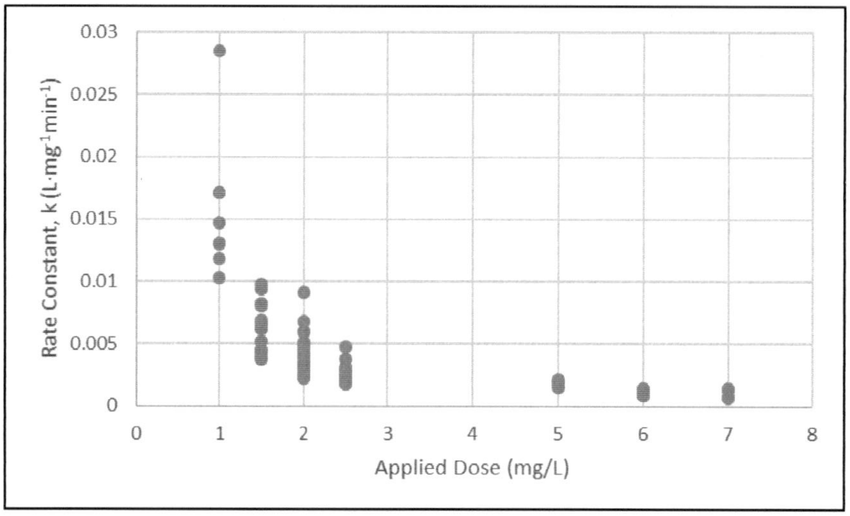

CSO Disinfection and Pathogen Inactivation: An example of the results from a series of chlorine and PAA inactivation tests for one set of CSO samples is shown in **Figure 6 (Chlorine dose of 7 mg/l)** and **Figure 7 (PAA dose of 7 mg/l)**.

Figure 6 – CSO Disinfection Characteristics: A) Fecal Coliform; B) Enterococcus inactivation; C) Chlorine Decay; D) Decay Characteristics at Chlorine dose of 7.0 mg/L

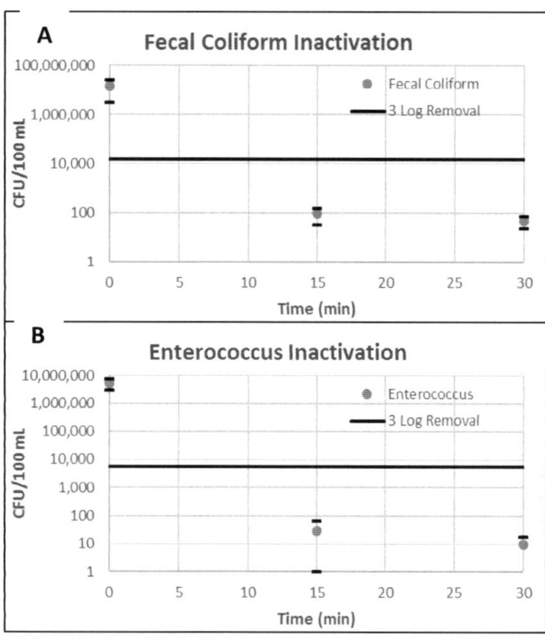

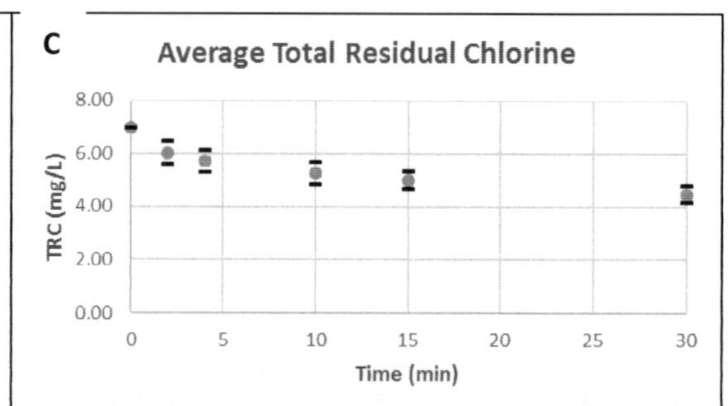

D

Water Characteristics (Average)			
Ammonia-N (mg/L)	17.79	pH	6.91
Nitrite-N (mg/L)	0.046	Flow (MGD)	117.3
Nitrate-N (mg/L)	0.63	ΔTRC (mg/L)	0.98
COD (mg/L)	120.1	30 min TRC (mg/L)	4.47
TSS (mg/L)	181.3	TRC Decay Rate (L/mg-min)	0.002

Figure 7 – CSO Disinfection Characteristics: A) Fecal Coliform; B) Enterococcus inactivation; C) Chlorine Decay; D) Decay Characteristics at PAA dose of 7.0 mg/L

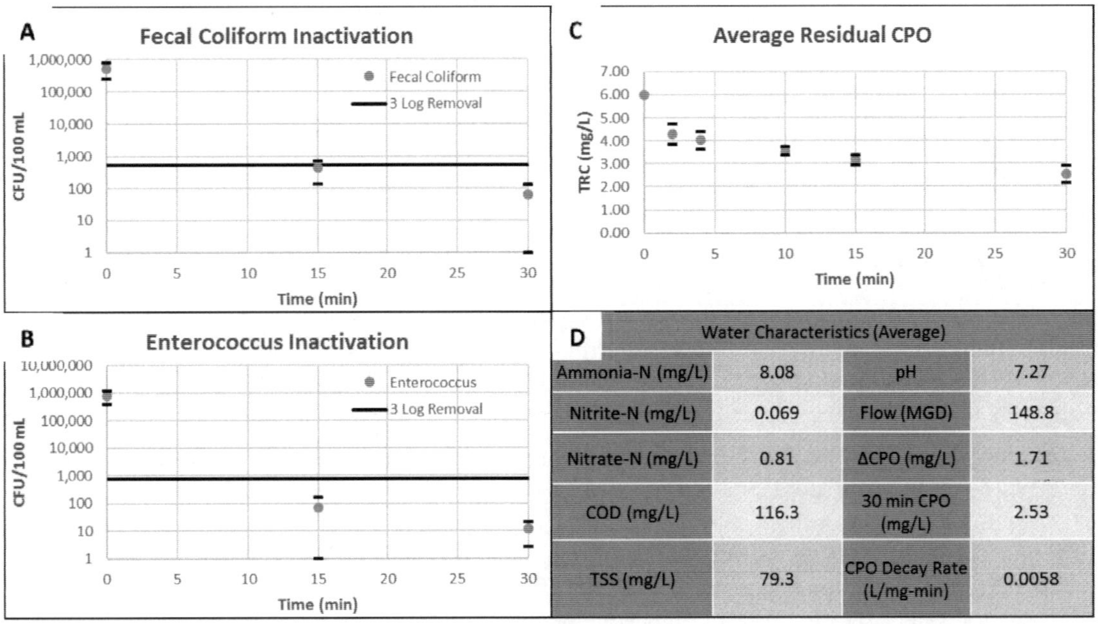

Table 3 shows the disinfection characteristics for the batch testing of two different sets of CSO samples. The CSO samples were dosed with either chlorine or PAA, and **Table 3** shows the time required to meet 2-log and 3-log inactivation of both *Enterococcus* and Fecal Coliform. In addition, **Table 3** shows the instantaneous uptake of TRC and PAA under each applied disinfectant dose along with the 2nd order decay rate constant. The higher the disinfectant dose the higher the initial uptake, however the decay rates were slower for the chlorine compared to PAA. **Table 4** shows the Ct (disinfection dose x contact time) to achieve 2-log and 3-log inactivation of Fecal Coliforms and Enterococcus in CSO samples. The results indicate that a higher dose of PAA or a longer contact time is required to meet the 2-log removal of both EC and FC compared to chlorine.

Table 3 - Contact time and TRC/PAA decay characteristics for CSO disinfection

Plant		Dose (mg/L)	Disinfection Time$_{2\text{-log}}$ (min)		Disinfection Time$_{3\text{-log}}$ (min)		ΔTRC/ PAA* (mg/L)	TRC/ PAA$_{30\text{ min}}$ (mg/L)	2° Decay Rate (L/mg-min)
			FC	EC	FC	EC			
CSO 1	Chlor	5	2.63	1.51	13.75	6.24	1.03	3.30	0.0017
		6	1.02	0.94	3.38	2.91	1.12	3.86	0.0019
	PAA	6	2.4	1.11	13	2.38	1.71	2.53	0.0058
CSO 2	Chlor	6	0.85	1.24	2.53	4.48	0.65	4.23	0.0018
		7	0.74	0.67	2.04	1.8	0.98	4.47	0.002
	PAA	14	1.31	0.73	4.83	2.0	4.31	5.01	0.0034

* Instantaneous chlorine/PAA demand

Table 4 - CTs for Two Log CSO Fecal Coliform and Enterococcus using Chlorine and PAA

Plant		Dose (mg/L)	$CT_{2\text{-log}}$ (mg/L-min)	
			FC	EC
CSO1	Chlor	5	10	6
		6	5	5
	PAA	6	10	5
CSO2	Chlor	6	5	7
		7	5	5
	PAA	14	13	8

The major findings from these studies are presented below. The complete results with a discussion on the implications of the results will be included in the presentation.

Summary of Results

- For BNR plants the EC criteria required a higher CT compared to FC, suggesting that either a higher chlorine dose or a longer contact time is needed to meet the EC criteria at BNR plants compared to FC.
- For secondary plants the FC criteria required a higher CT compared to EC, suggesting that current disinfection practices should be sufficient to meet the EC criteria because they have historically been successful in meeting the FC criteria.
- Typically the secondary plants had significantly higher fecal coliforms in the secondary effluent compared to the secondary effluent of the BNR plants.
- Compared to chlorine, a higher dose of PAA was required for all effluent samples to meet both the EC and FC criteria.
- The dynamics of EC and FC inactivation using chlorine shows that in general, FC and EC both die off quickly after the initial chlorine dose, however, FC numbers appear to level off after 15-20 minutes of contact time, while the EC numbers continue to decline, although at a slower rate compared to the initial rate.
- Instantaneous PAA and Chlorine uptake increased with increasing applied dose, however the decay rate after initial uptake decreased with increasing applied dose for both disinfectants, especially in the plant effluent disinfection studies where the applied doses were lower.
- PAA and chlorine both decay slowly after initial uptake, resulting in a relatively high residual after 30 minutes of contact time. At 2.0 mg/L PAA applied dose, 30 minutes residual PAA remained greater than 1.0 mg/L for all plant effluents.
- To meet the EC criteria using PAA, a higher Ct was needed compared to FC, indicating that EC was comparatively more difficult to inactivate with PAA. Initial PAA uptake was not significantly affected by water characteristics of the wastewater sample.
- Low levels of ammonia (< 2.0 mg/l) resulted in a very high initial demand of chlorine, indicating that plants which effective nitrification may be in danger of operating under breakpoint chlorination when ammonia levels are low, which adversely impacts disinfection efficacy.

- PAA was less effective at inactivation of both FC and EC compared to chlorine – a higher PAA dose and CT was needed to meet both FIB criteria compared to chlorine.
- CSO results indicate that a higher dose of PAA is required to meet the 2-log and 3-log removal of both EC and FC compared to chlorine.

REFERENCES

(EPA, 2011) U.S. EPA 2011. Recreational Water Quality Criteria. Office of Water 820-F-12-058. Web: https://www.epa.gov/sites/production/files/2015-10/documents/rwqc2012.pdf)

(Nobel, 2003) Noble RT[1], Moore DF, Leecaster MK, McGee CD, Weisberg SB. 2003 "Comparison of total coliform, fecal coliform, and enterococcus bacterial indicator response for ocean recreational water quality testing." *Water Res.* April, 37(7):1637-1643.

(Standard Methods, 2012) Standard Methods for the Examination of Water and Wastewater. American Water Works Association.

(Vilanova, 2003) Vilanova, X., Manero, A., Cerda`-Ce`llar, M. and Blanch, A.R. 2002. The effect of a sewage treatment plant effluent on the faecal coliforms and enterococci populations of the reception river waters. *Journal of Applied Microbiology* 92, 210–214.//

Bayonne, NJ Combined Sewer Overflow Treatment Demonstration Project: Disinfection Results

Jurek Patoczka* **and John Dening**

Mott MacDonald, Iselin, New Jersey
*email: jurek.patoczka@mottmac.com

ABSTRACT

A CSO pilot treatment program has been completed in Bayonne, New Jersey. The project focused on verifying performance of selected technologies for TSS removal and disinfection under field conditions. Following pretreatment for TSS removal in several alternative processes, the effluent was disinfected with peracetic acid (PAA) or ultraviolet (UV) light utilizing low-pressure and medium-pressure lamps. The test units had a design capacity of between 189 and 3,785 L/min (50 to 1,000 gpm) and were fed from an extensive CSO catchment area. PAA effectiveness was a function of the dose applied as normalized by COD for E. coli, fecal coliforms and Enterococci. Three (3) log reduction (deactivation) was achieved, on average, at 0.010 mg/L PAA dose per mg/L of COD present. Similar deactivation was achieved by UV dose of 25 mJ/cm^2 for low-pressure UV system and by 40 mJ/cm^2 for medium-pressure system.

KEYWORDS: Combined sewer overflow, CSO, peracetic acid, PAA disinfection, UV disinfection

INTRODUCTION

Bayonne Municipal Utilities Authority (BMUA), NJ, completed a Combined Sewer Overflow (CSO) Long Term Control Plan (LTCP) in 2006, as required by the New Jersey CSO Master General Permit. The LTCP evaluated a variety of technologies and methodologies for addressing combined sewer flows. Rapid treatment for TSS and disinfection at remote end-of-pipe facilities was included as a required element in the LTCP. Due to a lack of independent information, data from individual manufacturers was used as the basis for conceptual sizing of the required facilities. At the time a recommendation was made to validate selected manufacture's data prior to full scale implementation.

The BMUA retained Mott MacDonald as the Project Manager to develop, coordinate, and conduct this wet weather demonstration project and to publish the data for general use within the industry. The project was undertaken jointly by the BMUA, with grants from the NJDEP and EPA. A Technical Advisory Committee (TAC), and a Project Oversight Committee (POC) were formed, to review and comment on the means, methods, results, and conclusions of the project.

The goal and objective of the project was to develop scientifically valid performance data obtained under field conditions to evaluate the effectiveness of CSO treatment technologies and to gain an improved understanding of their potential use as satellite, end-of-pipe water treatment for CSO wet weather discharges. In addition to performance evaluation, aspects such as unit's reliability, scalability, anticipated capital and O&M costs, efficiency, and startup procedures were of interest.

TESTED TECHNOLOGIES

A solicitation for qualifications was published and various suppliers of wastewater and stormwater equipment responded offering hydrodynamic and gravimetric separators, filters, medium and low pressure ultraviolet disinfection devices and chemical disinfection units. The primary evaluation criteria for the units were; suitability for remote satellite facilities, documented performance, ease of operation, maintenance requirements, footprint and cost. The technologies selected for the demonstration project included the following existing available manufactured systems:

- Hydro International's Storm King® with Swirl Cleanse
- Terre Kleen TK-09
- WETTCO's Flex Filter™
- PeraGreen's INJEXX™ Per Acetic Acid (PAA) unit (Figure 1)
- Trojan's UV30000Plus™ (Figure 2)
- Aquionics Inline 250+W Medium Pressure UV (Figure 3)

Figure 1. Peracetic acid Peragreen INJEXX system on project site

Figure 2. Trojan UV3000 unit on project site

Figure 3. Aquionics UV 250+W unit on project site

The BMUA Oak Street Pumping Station was selected as the study location since its drainage area encompassed the entire City, the site provided adequate room, included a wet weather CSO discharge of up to 40 mgd, and provided consistent and extended CSO overflow periods.

The pilot facilities were laid out and designed to maximize flexibility in testing various configurations of TSS removal and disinfection units. Flow to each unit was controlled and measured so that it could be varied from storm to storm but held relative constant during any one storm. Figure 4 provides the aerial view of the completed pilot plant site, while Figure 5 shows the schematic layout of the pilot units and sampling locations.

Figure 4. Aerial view of the assembled pilot plant

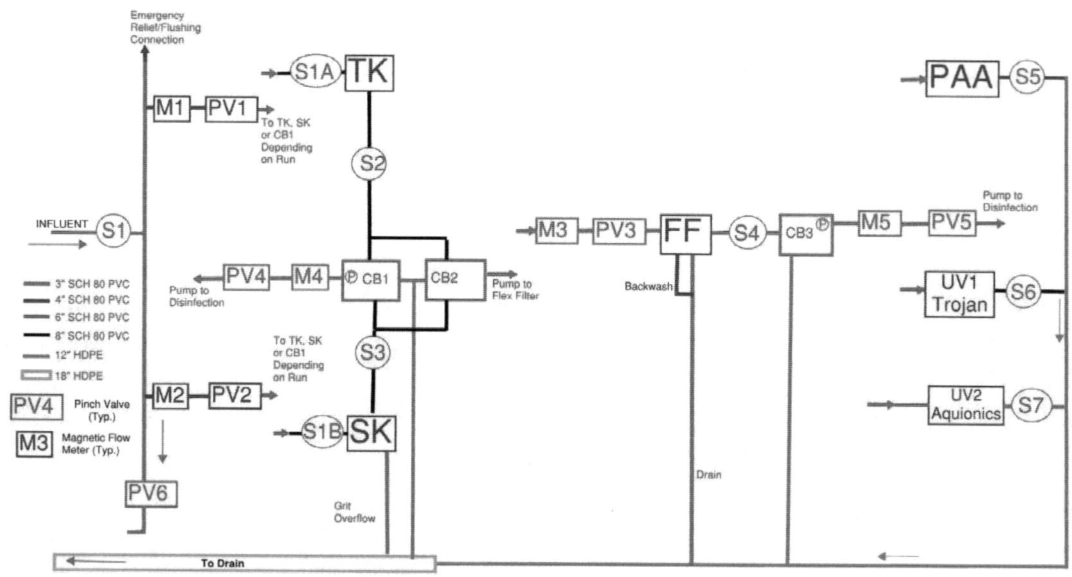

Figure 5. Schematics of the pilot plant layout and sampling locations

Testing Program

The project was initiated in the fall of 2014 with four (4) storm events (Test Runs) monitoring completed before a winter break. Testing resumed in the spring of 2015 but due to unusually dry weather and scheduling conflicts only 3 additional Test Runs were successfully completed by the end of September 2015. In order to complete the planned total of 9 sampling events, the program of "live" storm/overflow events was complemented by 2 events where regular BMUA sewage was diluted with groundwater to simulate the CSO discharge.

During each Test Run grab samples were obtained at 20 minute intervals for up to 4 hours at key influent and effluent locations and analyzed for appropriate water quality parameters. Analytical program included analysis (on selected samples, as appropriate) for the following parameters: TSS, VSS, soluble and total $CBOD_5$, COD, TOC, temperature, pH, turbidity and DO. Influent and effluent from disinfection units were analyzed for E. coli, fecal coliforms and Enterococci, as well as for UV transmittance and residual PAA, as applicable. Additionally, raw wastewater was tested at 60 minute intervals for settleable TSS and VSS utilizing the 60 minute Standard Methods gravimetric procedure.

Table 1 provides a summary of test equipment configuration and flows for each of the Test Runs. As can be seen, in all test runs the disinfection step was preceded by a TSS removal step(s). The sole exception was test run 7, where raw CSO influent was subjected to PAA disinfection.

Table 1. Summary of equipment set-up for individual Test Runs

Run No.	Date		Train 1			Train 2			Not Used	Comments
1	10/4/2014	Equipment	SK	FF	PAA	TK		AQ	TR	SK and TK screens plugged; FF backwashed once; influent screen plugged limiting flow to 50 gpm occasionally; PAA initially not fed, then overdosed
		Flow (gpm)	500	150	50	500		150		
2	10/16/2014	Equipment	TK	FF	AQ	SK	TR		PAA	SK screen manually cleaned; TK fed w/o screen; FF fed with a pump with a screen which was periodically manually cleaned; FF backwashed twice
		Flow (gpm)	300	150	120	300	150			
3	10/22/2014	Equipment	TK	AQ		SK	FF	PAA	Trojan	Limited sampling for PAA; TK and SK functioned w/o plugging; FF fed with a pump with a screen which was periodically manually cleaned; no backwash on FF;
		Flow (gpm)	600	160		400	100	40		
4	11/6/2014	Equipment	TK	PAA		SK	FF	TR	AQ	SK screen was plugging and was periodically manually cleaned; TK water level overtopped internal weir; FF operated with a pump w/o screen, backwashed 4 times for ~ 35 minutes; PAA dose erratic due to oversized pump
		Flow (gpm)	400	50		600	150	130		
5	7/30/2015	Equipment	TK	FF	TR	SK	PAA		AQ	SK screen was plugging and was periodically manually cleaned; TK water level overtopped internal weir; FF operated backwashed 3 times for ~ 35 minutes; PAA off for first sample.
		Flow (gpm)	550-375	150	130	450-300	100			
6	8/11/2015	Equipment	TK	PAA		SK	FF	AQ	TR	SK screen was plugging and was periodically manually cleaned; FF operated backwashed 3 times for ~ 35 minutes; PAA Tank volume restored to 300 gal; AQ "Water too hot" alarm
		Flow (gpm)	300-200	105		300-200	160-140	140-100		
7	9/10/2015	Equipment	PAA			SK	TR		FF, AQ	SK screen was plugging and was periodically manually cleaned; FF not working; PAA influent samples taken from CB-3
		Flow (gpm)	100			400-345	150			
8	10/15/2015	Equipment	SK	FF	AQ	SK	PAA		TK, TR	Wedge wire screen on SK, plugged and was periodically cleaned; FF backwashed 2 times for ~ 35 minutes; Could not achieve PAA residual
		Flow (gpm)	475-710	150	130	475-710	100			
9	10/27/2015	Equipment	TK	FF	TR	TK	PAA		SK, AQ	Unusual "sludge blanket" on FF filter media, Could not achieve PAA residual
		Flow (gpm)	900-850	145	140-100	900-850	61-105			

DISINFECTION WITH PAA

Overview of Testing Procedures

A 12% solution of PAA (Proxitane WW-12) with a specific density of 1.11 g/mL was used during all runs. The volume of the PAA reactor tank initially provided by the supplier was 1,136 L (300 gallons). It was temporarily reduced to 568 L (150 gallons) by internal overflow modifications for Test Runs 3 and 4, before being restored to 1,136 L for the Test Run 5 and all subsequent Runs. The design flow rate of wastewater varied between 95 and 189 L/min (25 to 50 gpm) for the initial Test Runs, before being set at 189 L/min for Run 5 and all subsequent Runs. The resulting design hydraulic retention time (HRT) for all sampling events is provided in Table 2 and it varied from 3 to 6 minutes, with HRT standardized at 3 minutes after the Run 4. The wastewater flow rate to the PAA unit fluctuated to some degree during each Test Run in response to hydraulic head and other factors, and these variations were taken into account during the analysis of individual data sets.

PAA solution was initially pumped into the PAA reactor system by a dosing pump with 22.7 L/day (6 gpd) flow rate. The initial results indicated that the resulting dose was too high and that it was difficult to adjust the pump dosing rate to smaller flow rate with a stable output. Consequently, starting with the Test Run 5, with a smaller, 11.4 L/day (3 gpd) dosing pump was utilized. During the Test Runs the dosing pump settings (stroke and speed) were adjusted in response to the results of the PAA residual field measurement. The objective was to maintain the PAA residual in the range 1 to 2 mg/L, but this was difficult to accomplish in real time, with limited ability to conduct frequent grab sampling and measurements for adjustments.

Summary of Conditions During Test Runs

A summary of the operating conditions for all Test Runs with PAA is provided in Table 2, which lists information such as the pretreatment unit used, average PAA dose and measured residual, HRT and average water quality parameters, including feed (or influent) concentration of pathogen indicators. The average performance results in terms of log reduction of pathogen indicators are also provided. Due to large variability in the wastewater quality and in the PAA dose delivered within some of the Test Runs, these average performance data for individual Test Runs are listed for general information and are not further discussed or correlated. The subsequent analysis focuses on data sets from individual sampling events on corresponding samples collected during all Test Runs.

Table 2. Summary of the operating conditions for all Test Runs with peracetic acid

Storm #	PAA								Geo. Mean in Feed			Average Log Reduction		
	Source of Influent	Design Flow, gpm	Average PAA Dose Applied, mg/L	Average PAA Residual Measured, mg/L	Design Hydraulic Retention Time, min	Average TSS in Influent to PAA, mg/L	Average Soluble $CBOD_5$ in Influent to PAA, mg/L	Average COD in Influent to PAA, mg/L	E. Coli	Fecal coliform	Enterococci	E. Coli	Fecal coliform	Enterococci
1	FF	50	6.9	1.50	6	128	6.3	321	835,970	5,527,250	303,650	2.46	1.48	2.26
3	FF	40	0.56	1.05	4	12.1	28.3	113	688,520	3,041,255	748,488	3.78	0.73	0.91
4	TK	20	1.7	1.13	7.5	417	20.7	417	1,976,300	3,312,149	1,230,198	1.06	0.62	0.88
5	SK	50	1.7	1.82	3	225	22.1	364	1,176,657	8,318,359	370,081	1.87	2.15	1.07
6	TK	100	2.8	1.09	3	235	13.1	350	1,537,320	1,246,712	502,264	2.96	2.61	2.29
7	Raw	100	2	0.92	3	137	30.4	254	2,518,877	27,478,039	1,303,941	2.83	2.45	2.24
8	SK	100	2.8	0.00	3	90.3	53.8	312	4,339,314	2,991,547	1,327,225	0.18	0.72	-0.08
9	TK	100	2.8	0.18	3	113	48.2	342	4,428,413	1,949,520	1,375,294	0.65	0.25	-0.11

NOTE: For calculating Log Reduction for pathogen indicators, results reported as "less then" were interpreted as 1/2 of the detection level.

During the simulated Test Runs 8 and 9 influent wastewater was generated by mixing groundwater from the underground tankage at the site (former primary clarifiers) with raw wastewater at an approximate ratio of 1:1. Unfortunately, during these Tests Runs no measurable PAA residual was achieved, despite the fact that the PAA feed pump was operating at full capacity. Accordingly, little or no reduction in density of the bacteria during these Runs was observed across the PAA unit (Table 2).

A possible explanation of this lack of PAA residual is an accelerated degradation of PAA caused by high salinity. Such effects were reported by Liu *et al.* (2014), where 1% and 3% sea water solutions were found to significantly accelerate degradation of PAA solutions, resulting in half-life times of 30 to 60 minutes. In tests on undiluted seawater Howarth (2003) found half-life of PAA to be 12 to 30 minutes, depending on the initial PAA concentration. Since the pilot test site is adjacent to coastline, groundwater in the underground tanks could have been contaminated with intruding seawater. During the Test Run 9 conductivity of the simulated wastewater was measured at 4.2 uS/cm, which is consistent with about 10% contribution of seawater. Subsequent test of the groundwater from the underground tanks confirmed that contamination, with the TDS measured there at 4,630 mg/L. Even though the contact time in the Test Runs 8 and 9 was only 3 minutes, the high salinity in our simulated wastewater was likely contributing to the accelerated decay of the PAA and the lack of residual.

The relatively high soluble $CBOD_5$ in the simulated wastewater (Table 1) could have been another factor in the lack of a measured PAA residual in Test Runs 8 and 9.

In any case, due to the lack of PAA residual and meaningful log removal, the results of PAA disinfection tests for the Test Runs 8 and 9 were excluded from further analysis.

Table 3 provides a brief summary of pathogen indicator data from all valid individual sampling events in the Test Runs 1 through 7 indicating that the average log reduction was in the range 1.7 to 2.3 logs for all three indicators. In contrast, in HDR (2014) study on wet weather primary effluent it was found that higher PAA and chlorine residual is required to inactivate E. coli and Enterococci to its potential regulatory limits than is required for fecal coliform. The specific log reductions were not identified in that study, but from the graphical presentation of the data it appears that at about 3 mg/L PAA residual and 15 min. contact time the log reduction for fecal coliform averaged 2.5 to 3, while it was approximately 2 for E. coli and Enterococci. Similarly, in WERF (2005) study on wet weather plant influent (corresponding to CSO) chlorine and chlorine dioxide were most effective against fecal coliform and least effective against E. coli, although the overall removals were better than in the current study.

Table 3. Summary of Pathogen Indicator Data for PAA Tests

Pathogen Indicator	Initial Count Range (cfu/100 mL)	Average Log Reduction
E. coli	5.2E+05 to 4.9E+06	2.3
Fecal coliform	6.0E+05 to 5.5E+07	2.0
Enterococcus	4.0E+04 to 2.1E+06	1.7

The lower removals demonstrated during the current program are likely related to a relatively low PAA dose applied (Table 2), which generally was under 3 mg/L (targeting 1 to 2 mg/L residual) and shorter contact time. This compares to order of magnitude larger dose of chlorine in the WERF (2005) study (8.5 to 28 mg/L range).

Discussion of the Results

Initially, all valid data sets from individual sampling events during Test Runs 1 through 7 were analyzed to determine correlation between disinfection effectiveness and the applied PAA dose or PAA residual, including data normalized by contact time. The obtained correlations were not satisfactory.

Subsequently, the applied PAA dose was normalized with respect to COD by dividing the dose by COD measured in the corresponding wastewater sample. The normalized dose correlates quite well with the log reduction for all three pathogen indicators (Figures 6 through 8). The only significant outliers are 3 data points for fecal coliforms for the Test Run 1, where a number of interferences and out of range results were reported by the laboratory.

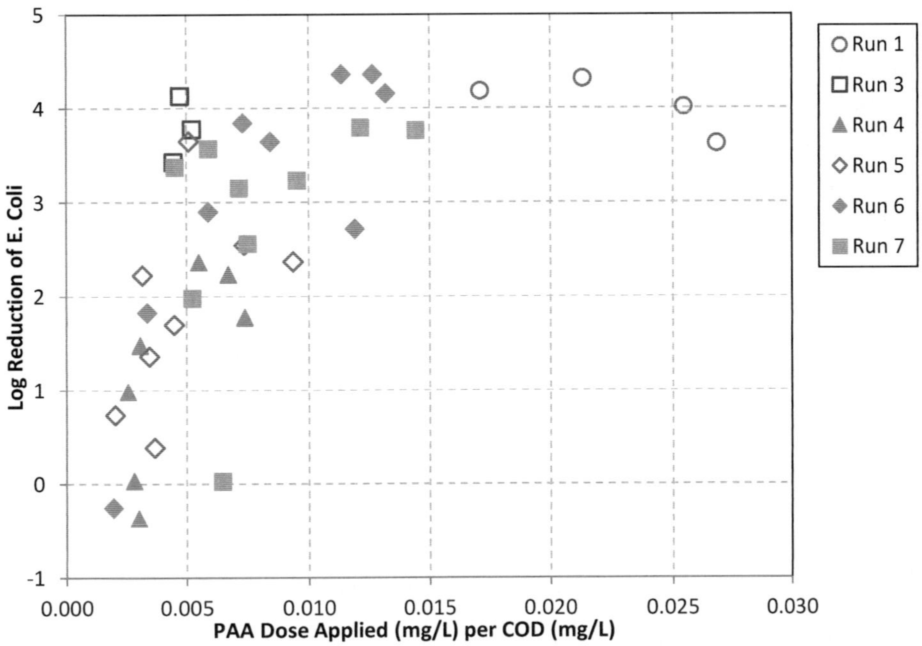

Figure 6. PAA Dose per COD vs Log Reduction of E. Coli

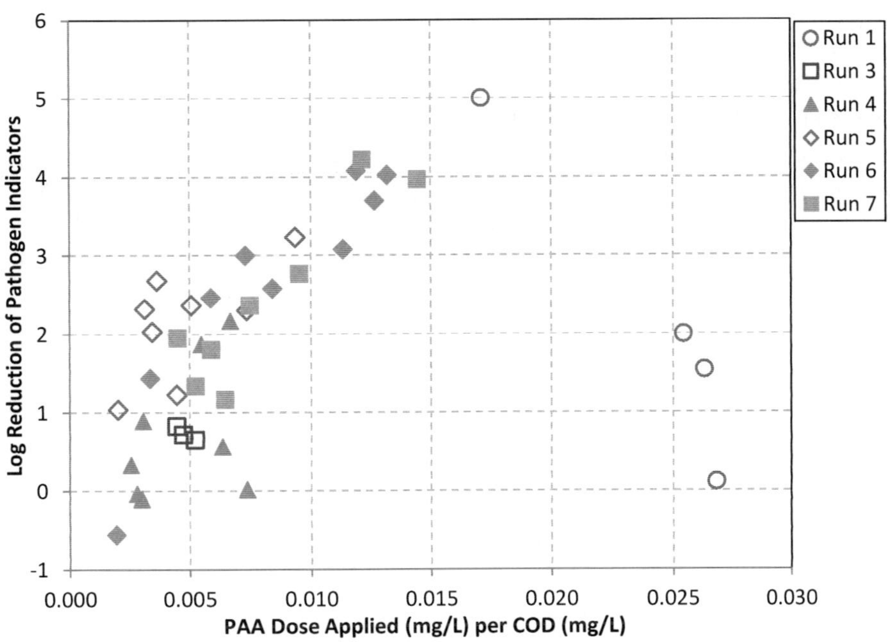

Figure 7. PAA Dose per COD vs Log Reduction of fecal coliforms

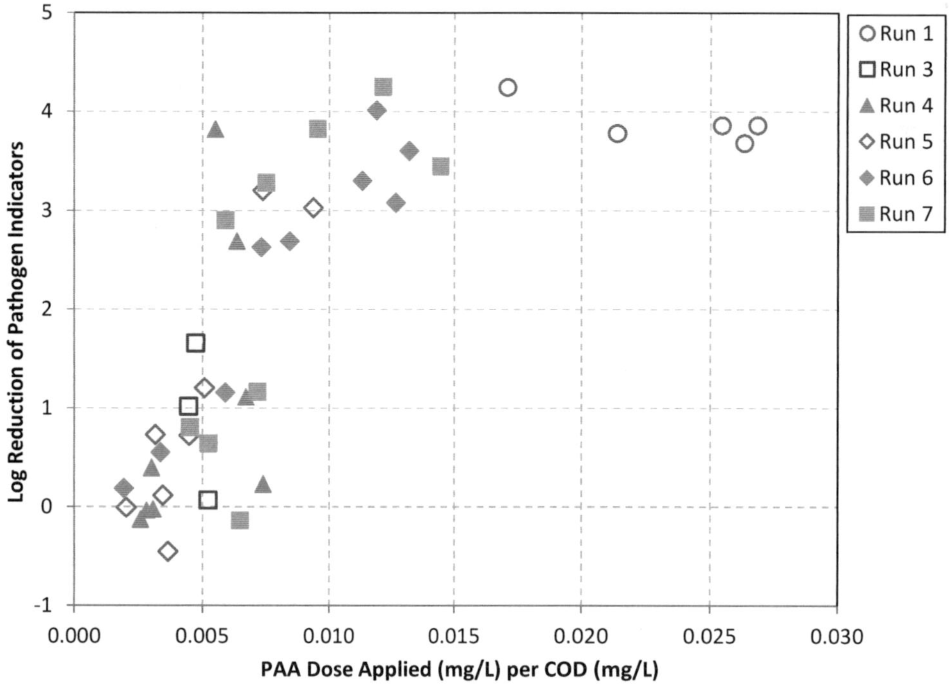

Figure 8. PAA Dose per COD vs Log Reduction of Enterococci

Figure 9 provides data from all Test Runs consolidated for each of the 3 pathogen indicators, with the exception of the 3 outlying data points for fecal coliforms. Best fit logarithmic regressions lines shown on this graph indicate a very good fit, with better than 99% confidence level (considering value of the R^2 and number of data points).

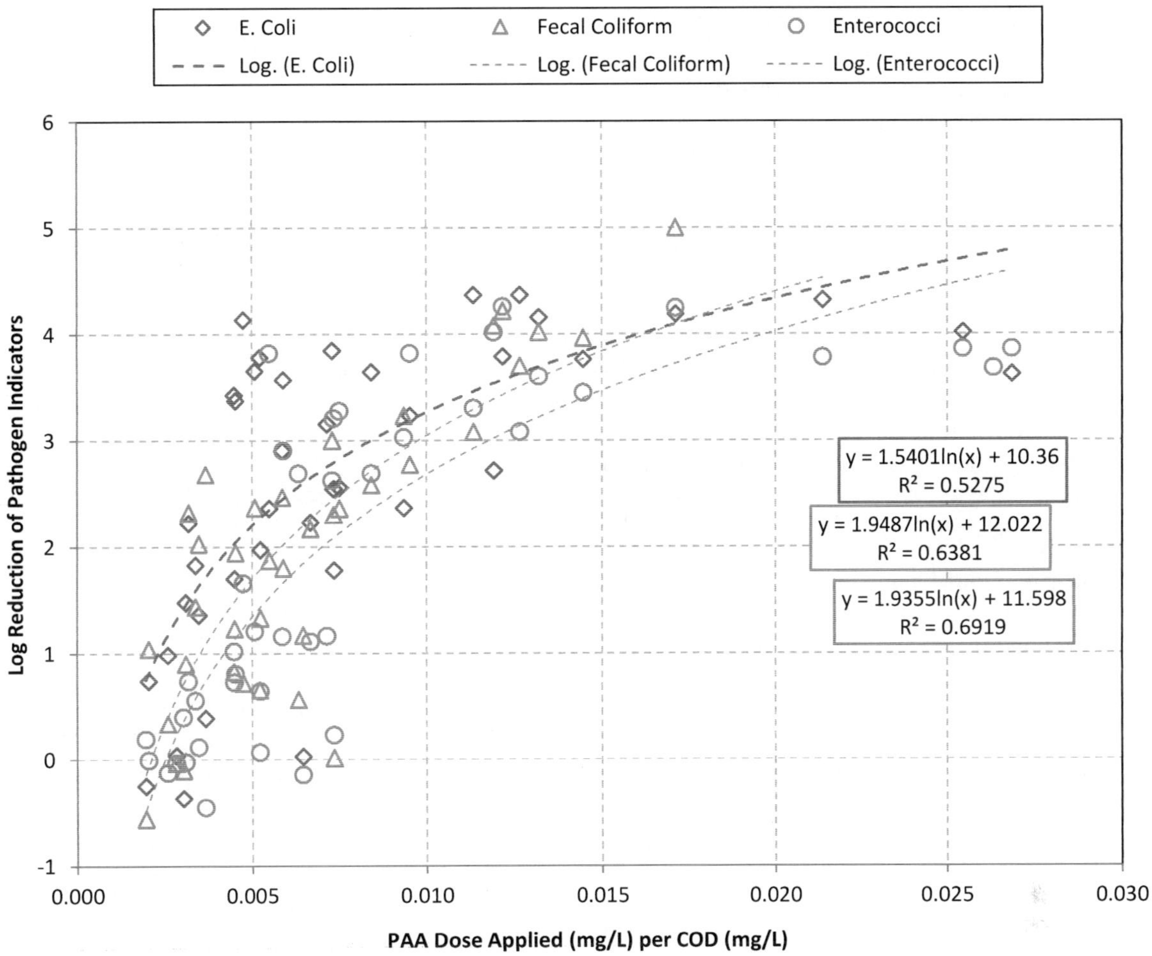

Figure 9. PAA Dose per COD vs Log Reduction of Pathogen Indicators

From Figure 9 it could be inferred that PAA dose of 0.01 mg/L of PAA per mg/L of COD typically results in 3 log reduction of fecal coliforms, with slightly higher effectiveness for E. coli and slightly lower for Enterococci. Increasing the relative dose to above 0.015 mg/L of PAA per mg/L of COD increased log reduction to 4, although data in that range are too limited to allow for a firm conclusion. Further increase of the PAA dose appeared to have limited effect on further increasing reduction of the bacterial densities, based on very limited data.

ULTRAVIOLET LIGHT DISINFECTION

Overview of Testing Procedures

Two UV disinfection units were tested during the demonstration project: a low-pressure, high-intensity Trojan UV3000Plus model, and a medium-pressure, high-intensity Aquionics UV 250+W model. The units supplied by the manufacturer were rated as follows:

- Trojan: maximum hydraulic capacity of 946 L/min (250 gpm). No further specifications, such as acceptable UV transmittance (UVT) range, were provided,
- Aquionics: flow range 379 – 1,136 L/min (100 - 300 gpm) at 45 to 65% UVT.

In the case of the Aquionics unit the specified flow range was designed to provide a minimum UV dose of 30 mJ/cm^2 under the above listed conditions. The 65% is a typical minimum value of UVT found in secondary effluents and is a standard design value for UV disinfections systems aimed at three to four log reduction of pathogen indicators (HydroQual, 2006). The unit's flow rating decreased to 397 L/min (100 gpm) at UVT values of 45% in order to achieve the same UV dose.

Summary of Conditions During Test Runs

A summary of the operating conditions for all Test Runs utilizing either UV disinfection unit is provided in Table 4. Table 4 lists information such as the pretreatment unit used, average water quality parameters, including average count of pathogen indicators in the influent. The average performance results in terms of log reduction of pathogen indicators are also provided. The water quality parameters were measured in the effluent from the upstream TSS removal unit. Due to large variability in the wastewater quality during the individual the Test Runs, these average performance data are listed for general information only and are not further discussed or correlated. The subsequent discussion and analysis of the results is based on sets of individual data collected during all Test Runs.

Table 4. Summary of conditions during UV disinfection Test Runs

Storm #	UV Disinfection											
	Source of Influent	UV Unit Type (Trojan or Aqu-ionics)	Design Flow, gpm	Trans-mittance %	Average TSS in Influent to UV, mg/L	Average Soluble $CBOD_5$ in Influent to UV, mg/L	Geo. Mean in Feed			Average Log Reduction		
							E. Coli	Fecal coliform	Entero-cocci	E. Coli	Fecal coliform	Entero-cocci
1	TK	A	200	16.8	199	9.9	784,244	7,316,329	636,498	1.25	1.35	1.15
2	SK	T	150	44.8	196	6.9	493,504	1,934,295	364,299	2.31	2.87	1.91
	FF	A	120	62.8	9.9	5.1	300,370	726,924	235,048	2.27	2.48	1.91
3	TK	A	160	27.0	97	31	889,599	5,664,802	886,950	1.51	1.83	1.14
4	FF	T	130	27.1	31	16.2	1,281,190	1,602,734	742,545	2.29	2.30	1.76
5	FF	T	130	39.6	29.2	21.9	522,495	3,944,017	287,200	2.27	3.40	1.98
6	FF	A	130	40.1	27.8	13.4	1,150,762	907,888	342,454	1.89	1.52	1.07
7	SK	T	150	23.8	118	28.4	2,468,098	21,815,203	1,262,797	1.42	1.98	1.27
8	FF	A	130	25.9	28	41.3	3,393,289	1,455,492	1,019,468	0.94	1.38	0.73
9	FF	T	130	27.0	37.9	39	3,425,122	2,288,616	1,349,306	1.23	1.53	1.27

NOTE: For calculating Log Reduction for pathogen indicators, results reported as "less then" were interpreted as 1/2 of the detection level.

Discussion of the Results

Table 5 provides a summary of pathogen indicator data from all valid individual UV sampling events. From these data it is apparent that UV was most effective in inactivation of fecal coliforms, while least effective in inactivation of Enterococci. These results are consistent with the data reported in WERF (2005) study on wet weather plant influent (corresponding to CSO), although the overall removals were better than reported below. This is undoubtedly due to the much lower UV dose applied in the current study, as discussed below.

Table 5. Summary of Pathogen Indicator Data for UV Tests

Pathogen Indicator	UV Unit	Initial Count Range (cfu/100 mL)	Average Log Reduction
E. coli	Trojan	1.2E+05 to 4.6E+06	1.9
	Aquionics	8.0E+04 to 5.3E+06	1.5
Fecal coliform	Trojan	5.0E+05 to 3.9E+07	2.4
	Aquionics	1.2E+05 to 2.8E+07	1.7
Enterococcus	Trojan	1.2E+05 to 2.1E+06	1.6
	Aquionics	6.4E+04 to 1.7E+06	1.2

During each individual sampling event involving UV disinfection units (typically every 20 minutes), the value of the UVT of incoming wastewater was measured. These discrete UVT values, together with the actual wastewater flow at the time of sampling, were provided to the UV units' manufacturers. Based on this information and results of validation tests for their equipment, the units' suppliers calculated the irradiation power delivered during each sampling event. For several sampling events with particularly low transmittance, the calculated dose is an approximation, as the parameters were outside of the validated range.

Figure 10 illustrates log reduction of E. coli recorded in all individual samples during all Test Runs as a function of the calculated UV dose, separately for Trojan and Aquionics units. Figures 11 and 12 provide the same information for fecal coliforms and Enterococci, respectively.

Inspection of Figures 10, 11 and 12 indicate an expected trend of increasing log reduction of pathogen indicators as UV dose increases. Despite the ostensibly low values of correlation coefficients (R^2) shown on these figures, the correlations (forced through the origin) are statistically significant at 99% confidence level for 4 out of 6 data sets, with the remaining 2 being at 95% level. This is due to the relatively large number of data points available for these correlations.

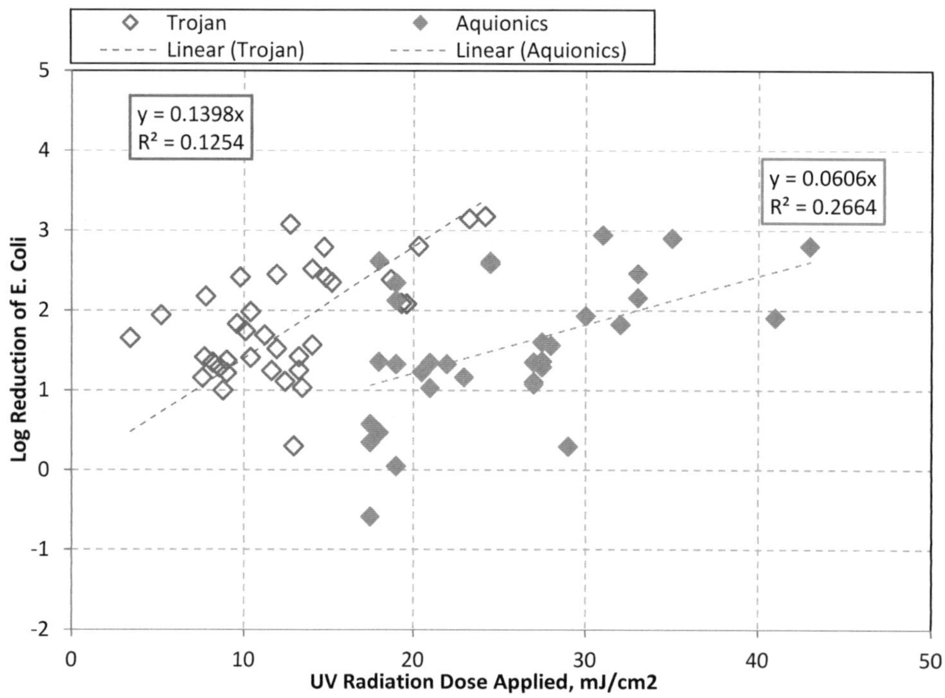

Figure 10. UV dose applied vs log reduction of E. Coli

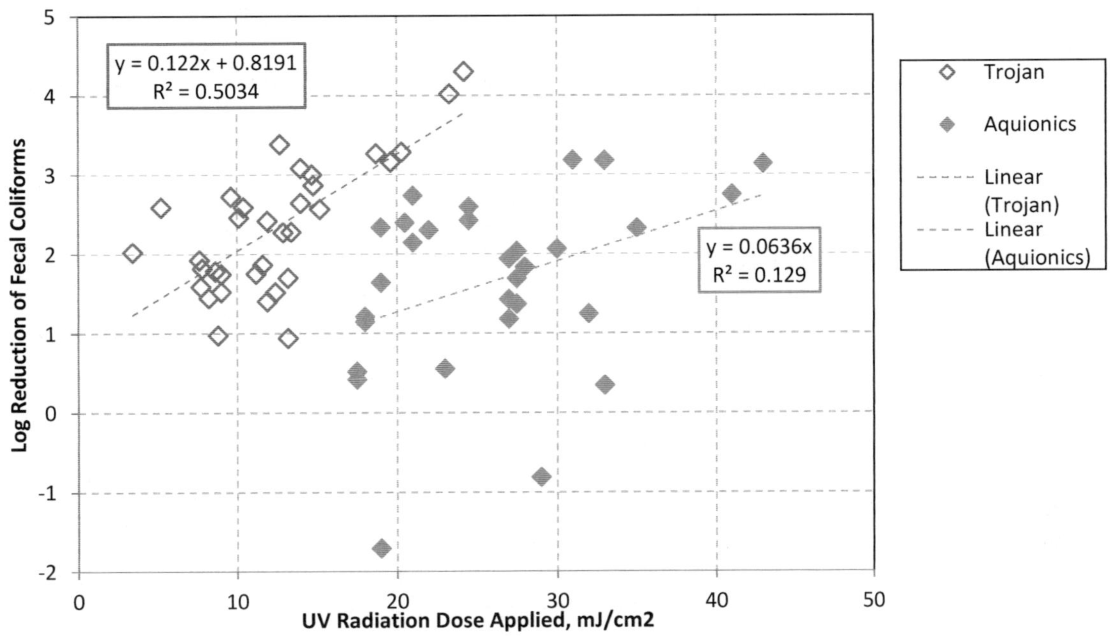

Figure 11. UV dose applied vs log reduction of fecal coliforms

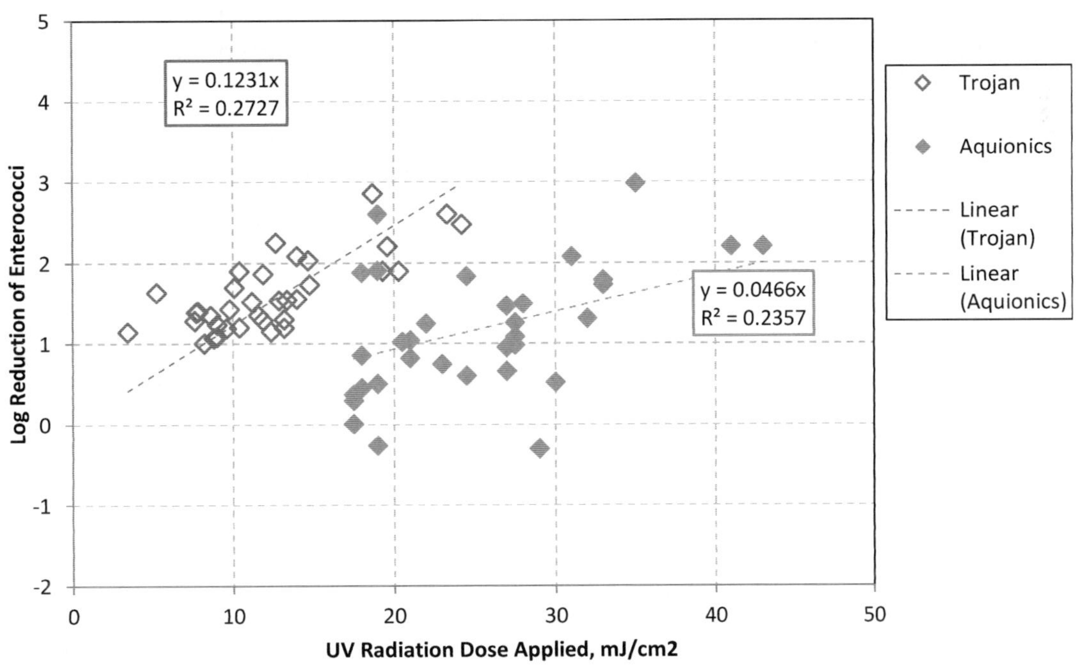

Figure 12. UV dose applied vs log reduction of Enterococci

Attempts to further improve the resulting correlations by normalizing the UV dose by COD were not successful. It is assumed that this is because the organic strength of the wastewater, as measured by COD, is already factored into the UV dose calculation by independent measurement of the transmittance.

Inspection of Figures 10, 11 and 12 indicates that the Trojan UV unit performed better than the Aquionics UV unit at the same calculated UV dose. It is not clear if the disparity in performance is a result of systemic difference in the validation procedure and effective dose calculation performed by different manufacturers, or if it is also related to the lower efficiency in generation of UV in the germicidal range by the polychromatic medium pressure lamps as compared to the relatively monochromatic low pressure lamps.

The most obvious observation is that the relatively low log reduction of bacterial densities achieved by the UV units is the inadequate UV dose caused by frequently very low transmittance of the wastewater. This can be most readily inspected on Figure 13, where the expected, strong relationship between the wastewater TSS and transmittance is evident. Figure 14 presents the same data grouped by the Test Run. The transmittance ranged from single digits to 60%, with majority clustered in the 20 to 50% range. These low transmittance values are consistent with expectations. For example, transmittance of primary effluent is quoted to be in 20 to 50% range in Metcalf & Eddy/AEOCOM, (2014). HDR (2014) report on wet weather primary effluent found the UTV values to be somewhat higher, in the range from 40 to 60%.

It is clear that the flow rating of the supplied UV units was suitable for a typical, secondary effluent application, without taking into account the expected, significantly worse quality of the CSO effluent. As a result, the applied dose for the Trojan UV unit never exceeded 25 mJ/cm^2 and was below 45 mJ/cm^2 for the Aquionics unit. This is contrasted with much higher effective UV dose applied during the wet weather tests reported by WERF (2005) report, when it ranged from 65 to 220 mJ/cm^2.

Subsequent Figures indicate that correlation between UVT and total $CBOD_5$ (Figure 15) and COD (Figure 16), are even better than with TSS, attesting to the contribution of soluble organics to the UV absorbance.

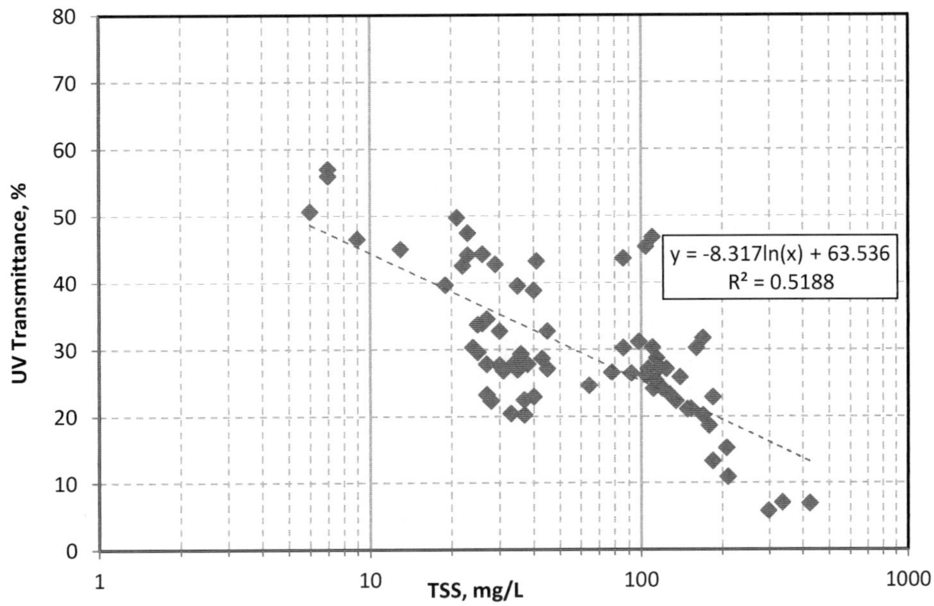

Figure 13. Effect of TSS on UV transmittance

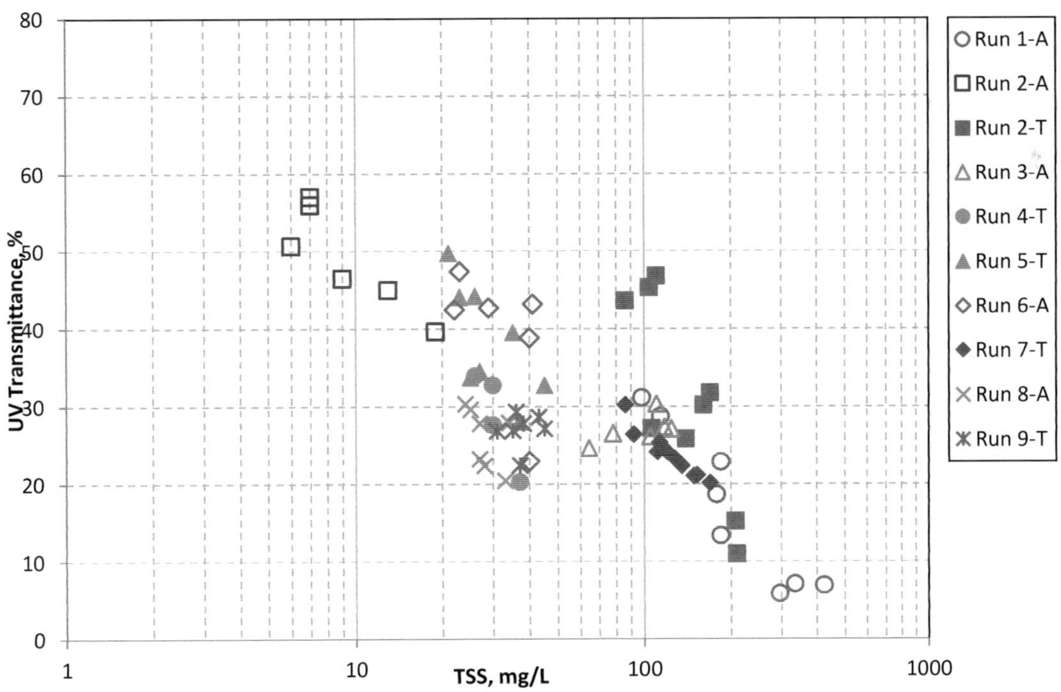

Figure 14. Effect of TSS on UV transmittance – grouped by Test Run

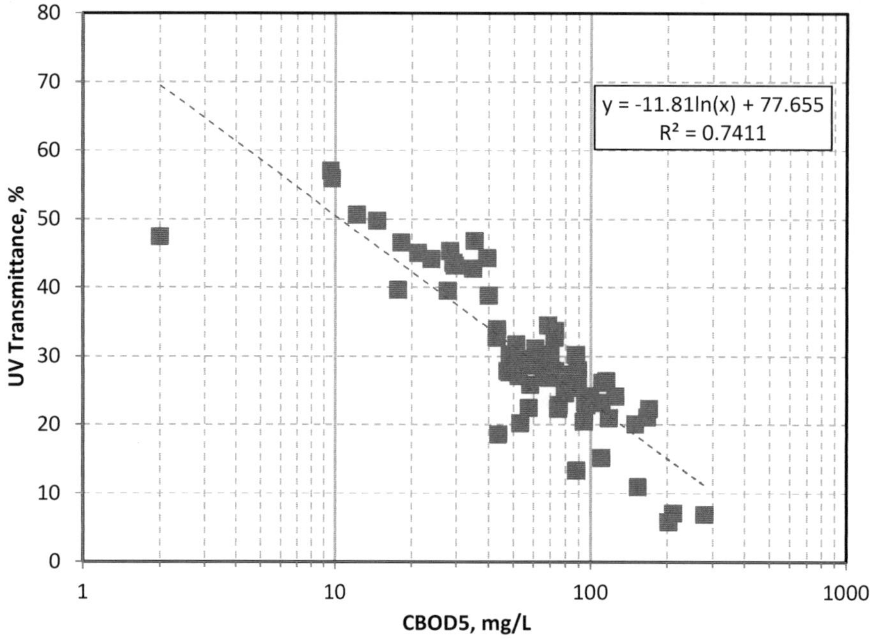

Figure 15. Effect of $CBOD_5$ on UV transmittance

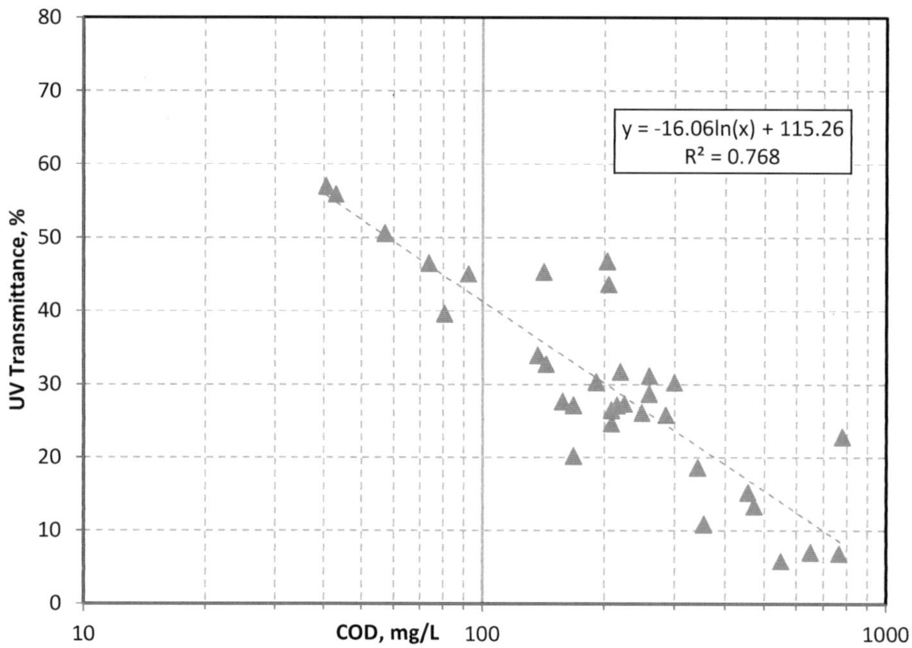

Figure 16. Effect of COD on UV transmittance

CONCLUSIONS AND RECOMMENDATIONS

PAA Disinfection Tests:

1. The most important finding from the PAA pilot study was definition of a predictive relationship between the applied dose of PAA per mg/L of COD present in the wastewater and the log reduction of pathogen indicators. PAA dose of 0.01 mg/L of PAA per mg/L of COD predicted 3 log reduction of fecal coliforms, with slightly higher effectiveness for E. coli and slightly lower for Enterococci.
2. Increasing the relative dose to above 0.015 mg/L of PAA per mg/L of COD increased log reduction to 4. Further increase of the PAA dose appeared to have limited effect on further increasing reduction of the bacterial densities, although data in that range are too limited to allow for a firm conclusion.
3. The PAA contact time and dose applied in most of the Test Runs were relatively low; nevertheless a removal of 99% (or two log) reduction of the pathogen indicator organisms was documented, on average. Higher applied dose, as modified by COD concentration, may be needed to satisfy the disinfection requirements and guidelines of many States and the Federal government.
4. Should applicability of the relationships discussed under items 1 and 2 above be confirmed at other locations, it would be desirable to adjust the PAA application rate based on both wastewater flow and organic strength. The organic strength could potentially be measured in real time by a surrogate parameter such as TOC, but this could be practical only at large sites. Alternatively, a typical COD profile of CSO discharge could be developed based on historical data and PAA dose adjusted based on that profile and instantaneous flow. Lacking this, the only available strategy to accomplish significant disinfection would be to apply a pre-set PAA dose effective at the high end of the possible COD concentrations. This, however, will result in potentially significant residual PAA concentration, which could be toxic to the aquatic life.
5. The issue of toxicity of residual PAA to aquatic life in the receiving stream is a significant issue in selection of appropriate disinfection strategy. The key consideration is to balance impact of un-disinfected, or partially disinfected, CSO discharges on the designated water uses with potential toxicity of the residual PAA to the aquatic life.
6. Use of PAA in satellite CSO locations could be complicated by a need for on-site storage of large volumes of the chemical, which requires secondary containment and appropriate safety measures.

UV Disinfection Tests:

1. The UV units tested exhibited the expected effectiveness commensurable with the modest UV dose applied, as limited by frequently very low UV transmittance of the CSO wastewater
2. The pathogen deactivation results follow the expected relationship with the applied UV irradiation dose.
3. Trojan UV3000Plus unit using low-pressure lamps required approximately 25 mJ/cm^2 irradiation energy input to achieve 3 log deactivation of pathogen indicators, on average.
4. Aquionics 250+W unit using medium-pressure lamps required approximately 40 mJ/cm^2 irradiation energy input to achieve 3 log deactivation of pathogen indicators, on average.

5. Design flow of UV equipment, when used in a "dirty water" application, must be significantly lowered (de-rated) to account for poor transmittance of the CSO wastewater treated (in the absence of adequate pre-treatment to increase UVT).
6. Wastewater transmittance showed an expected, strong correlation with water quality parameters such as TSS, $CBOD_5$ and COD.

REFERENCES

HDR (2014), *ALCOSAN Disinfection Assessment, Part II: Pilot Disinfection Study of ALCOSAN Primary Effluent,* (unpublished report).

Howarth, J. (2003) Decay Kinetics of Peroxyacetic Acid (PAA) and Hydrogen Peroxide in a Variety of Water Matrices (unpublished report by Enviro Tech Chemical Services), Modesto, California.

HydroQual (2006) *Technical Memorandum: Evaluation of UV Disinfection Process for City of Elizabeth, NJ Combined Sewer Overflow Long Term Plan*, (unpublished report).

Kitis M (2004) Disinfection of wastewater with peracetic acid: a review. *Environment International*; **30**, (47-55).

Liu, D., Steinberg C.E.W., Straus, D. L., Pedersen L. F., Meinelt, T., (2014) Salinity, dissolved organic carbon and water hardness affect peracetic acid (PAA) degradation in aqueous solutions, *Aquacultural Engineering*, **60** (35-40).

Metcalf & Eddy/AECOM (2014) *Wastewater Engineering; Treatment and Resource Recovery*, McGraw-Hill Education, New York, New York.

WERF (Water Environment Research Foundation) (2005) *Identifying Technologies and Communicating the benefits and Risks of Disinfecting Wet Weather Flows*; Alexandria, Virginia

Can MBR Reduce or Eliminate the Need for Disinfection in Water Reuse Projects

Ufuk G. Erdal, Zeynep Erdal and Beverley Stinson
AECOM

ABSTRACT

Membrane Bio Reactors (MBRs) combine activated sludge biological treatment with an integrated membrane filtration system to provide biological treatment and suspended solids removal. MBR uses a low-pressure membrane filtration system (e.g., microfiltration [MF] or ultrafiltration [UF]) and eliminates the need for both solids separation and tertiary filtration for production of high quality product water. Sampling of MBR permeate at operating reuse facilities frequently demonstrates that no coliform bacteria are present because of excellent rejection of bacteria through the membranes. In addition, virus studies conducted at pilot and full-scale MBR water reuse facilities also indicated substantial virus removal (0.5-log to as high as 6 log removal) depending upon virus types used, pore sizes of membranes (MF vs. UF) and characteristics of membranes utilized in MBR systems as well as MBR sludge characteristics and design and operating conditions. Given that, many plant operators and managers ultimately ask why stringent disinfection requirements are needed since reclaimed water pathogen limits are mainly met through MBR alone. In this paper, pathogen removal mechanisms in MBR systems are provided. In addition, pathogen removal results of the two full scale MBR facilities for varying operating conditions (i.e. clean membranes following sodium hypochlorite and citric acid maintenance clean, at increased fluxes, increased and reduced SRT) were presented to show how efficient MBR system to remove bacteria and other pathogenic microorganisms.

KEY WORDS

Direct integrity testing, MBR, MF/UF, pathogen credit, coliform, virus, LRV

BACKGROUND AND OBJECTIVES

Membrane bioreactors (MBRs) combine conventional activated sludge treatment with a membrane filtration that provides high quality water and eliminate the need for secondary clarification. Elimination of secondary clarifiers allows MBR systems to operate much higher mixed liquor suspended solids concentration than conventional activated sludge systems which greatly reduce footprint requirement for MBR systems. For a given activated sludge basin volume, MBRs can be operated at longer SRTs than CAS systems which further enhance removal of bulk (COD, BOD) and trace organics (TOrC, CECs).

As in other low pressure membrane systems (MF/UF) used in tertiary filtration applications, MBR can produce solids free effluent with relatively low silt density index values (<3) which allows direct coupling of MBR with RO in reuse applications for TDS control and industrial reuse applications. Although a limited reference exists, MBR has also been coupled with RO in potable reuse projects (Gibbsland, Australia) to augment drinking water supplies.

Indirect potable reuse (IPR) projects via groundwater injections in California require implementation of full advanced treatment consisting of RO and advanced oxidation to meet pathogen, TOC, nitrogen, trace organic other requirements specified in the most recent ground water recharge regulations. In many cases, MF/UF is also used as part of the full advanced treatment. The main functions of MF/UF in the process train are to greatly reduce solids and pathogen concentrations and provide adequate pretreatment for downstream RO process. In drinking water and indirect potable reuse applications, the State Water Resources Control Board Division of Drinking Water awards up to 4 log credit for Cryptosporidium and Giardia for MF/UF systems by passing a daily direct integrity test. For IPR via groundwater recharge, a total of 12, 10 and 10 log removal is needed for viruses, Crypto and Giardia that includes credits gained thru retention of water in ground.

Similarly MBR system do not get pathogen credits for non potable reuse applications in California and many other states despite sampling of MBR permeate at operating reuse facilities frequently demonstrates that no coliform bacteria and protozoa are present because of excellent rejection of bacteria and protozoa through the membranes. Since coliform bacteria (0.6-1.2 μm in diameter by 2-3 μm in length), and protozoa (i.e., Giardia, Crypto; 8-14 μm and 4-7 μm, respectively) are larger than the absolute pore size of the membranes used in MBR systems (typically 0.1-0.5 μm), no coliform bacteria and protozoa are expected in the permeate streams of the intact membranes (Zhang and Farahbakhsh, 2007). Size exclusion was found to be the dominant mechanism for the removal of bacteria and protozoa by the membrane filtration (Zhang and Farahbakihsh, 2007).

Viruses (i.e., enteric viruses, coliphages) are usually smaller than absolute pore sizes used in membranes used in MBR systems. However virus studies conducted at MBR facilities also indicated substantial virus removal (0.5-log to as high as 6 log removal) depending upon virus types used in challenge studies, pore sizes of membranes (MF vs. UF) and characteristics of membranes utilized in MBR systems as well as MBR design and operating conditions (Madaeni et al., 1995, Ueda and Horan, 2000). Unlike solids and bacteria removal, virus removal in MBR systems depends on many factors. The literature studies suggested that indigenous viruses can be removed by MBR systems via the following mechanisms (Shang et al., 2005):

1. biological completion/predation (predator organisms such as protozoa and fungi could consume small organisms)
2. direct removal via size exclusion by membranes (molecules larger than membrane pore sizes will be rejected regardless of their surface properties (i.e., charge, polarity)
3. absorption into the biomass and sequential removal through filtration
4. absorption into biofilm/cake layer and removal through backwash and/or chemical cleaning
5. pore blocking (large molecules such as carbohydrates, polysaccharides and MLSS flocs block the pores of the membranes and restrict passage of small viruses
6. reduction of the effective pore size due to biofilm growth, gel, scale cake layer formation on the membrane surface and pores.

Within the six mechanisms above, the later five have been considered as the main mechanisms for the removal of viruses in the MBR systems (Shang et al., 2005). MBR flocs are subjected to erosion because of a higher mixed liquor suspended solid (MLSS) concentration and increased shear (Schwarz et al., 2006). Yi et al. (2006) showed that measured median particle diameter was much smaller in MBR systems (10 μM) than conventional activated sludge (CAS) systems (120 μM) operated under identical conditions. Smaller flocs observed in MBR systems increase the exposed surface area and enhance adsorption and retain of small particles and pathogens by the MLSS (Water Environment Federation, 2011).

While direct integrity testing vastly relies on size exclusion it does not account for additional removal mechanisms (defined above) involved in MBR systems. The objectives of this paper are to:

- show if on-line turbidity readings can be used to reliably evaluate membrane integrity and get pathogen credit based on turbidity readings
- present what other approaches can be used to assess membrane integrity in MBR systems

METHODS

Pathogen challenge studies were conducted in two phases. In Phase I of the study, the candidate viruses that are quantifiable in MBR feed were identified at two full-scale MBR facilities located in Nevada (City of Henderson Water Reclamation Facility (WRF) and City of North Las Vegas WRF). The MBR permeate from City of Henderson WRF was disinfected via UV disinfection whereas chlorine disinfection is used at North Las Vegas WRF to disinfect MBR effluent. Both facilities are designed and operated for biological nitrogen removal mode to meet total nitrogen limit of no higher than 10 mg/L in regular basis. In each facility utilizes GE Zenon 500 d membranes with nominal and absolute pore sizes of 0.04 and 0.1 micron, respectively.

The tested viruses included Human Adenovirus (HAdV), Norovirus G1, Norovirus G2, MS-2 Coliphage, Somatic Coliphage, Enterovirus, Hepatitus A and Rotovirus. The periodic sampling from tested MBR systems indicated that MBR influents were free of Polio virus. Therefore, no sampling was performed for Polio virus.

The aim of the Phase II of the study was to determine pathogen (virus and coliform) removal capabilities of MBR systems under five different operating scenarios:

(1) Sampling for normal operating conditions (design SRT, design flux and design UV dose (Facility 1) and chlorine dose and CT (Facility 2)
(2) Sampling immediately after sodium hypochlorite maintenance clean
(3) Sampling immediately after citric acid maintenance clean
(4) Operating at increased membrane flux (25% increase on the design value)
(5) Normal operating conditions with lowered UV dose (40 mJ/cm2; Facility 1) and lowered chlorine dose and CT (0.5 mg/L and CT of 150 mg-min/L)

Phase II also covered sampling for fecal and total coliform and monitoring the key plant operating parameters such as solids retention time (SRT), wastewater temperature, flux, transmembrane pressure, filtrate turbidity, mixed liquor suspended solids concentration.

The sampling locations are presented in Figure 1.

Figure 1. Sample Collection Locations from City of Henderson Water Reclamation Facility (top) and City of North Las Vegas Water Reclamation Facility (bottom)

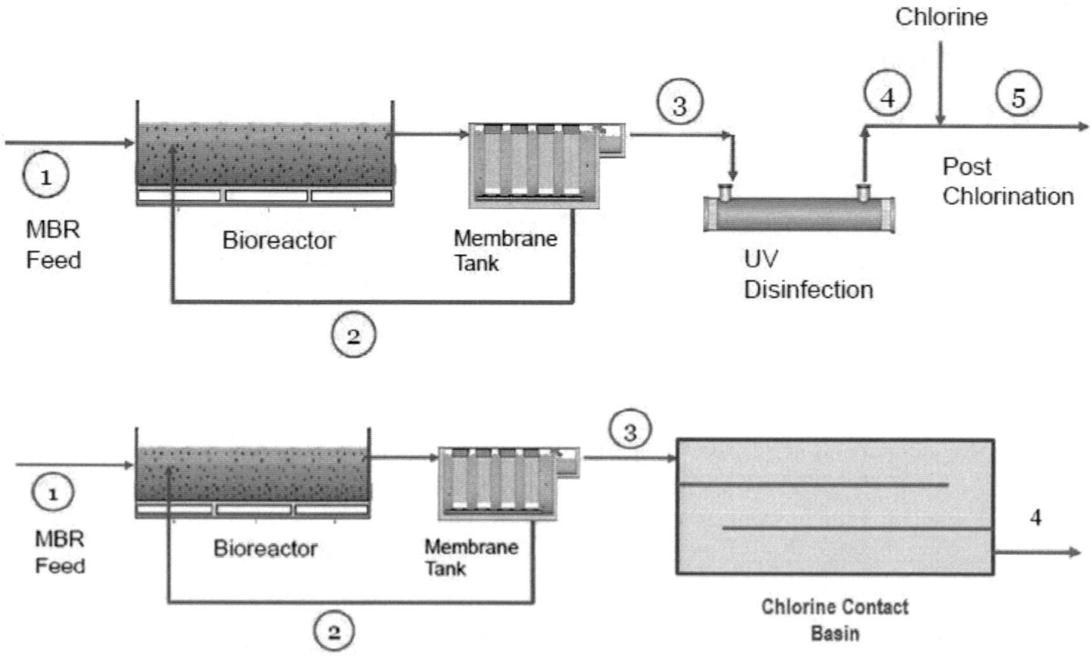

At City of Henderson WRF, with the exception of the Coliphage analysis, all pathogen analysis including coliform was performed by Southern Nevada Water Authority. Coliphage analysis was performed at the Biowir Laboratories. Low solids containing virus analysis (i.e., MBR permeate, UV disinfection) was performed per EPA Method 1602 whereas high solids containing samples (i.e., bioreactor influent, RAS) was performed per Double Layer Adams Method. At City of North Las Vegas WRF, with the exception of fecal and total coliform, all virus analysis was performed at the Biowir Laboratories.

RESULTS

Initial Screening (Phase I)

Enterovirus, Hepatitis A, Rotovirus were not detected in the influent and any sampled locations. They were excluded in Phase II of the study. Human Adenovirus (HAdV), Norovirus GI, Norovirus GII, MS-2 and Somatic Coliphage were detected in the MBR feed with a varying concentration. Along with total and fecal coliform, HAdV, Norovirus GI and GII, Coliphages were identified as the indicator organisms and included in Testing Phase of the study

Testing Phase (Phase II)

Without disinfection, the MBR systems were very effective for complete retention of Human Adenovirus (HAdV), Norovirus GI, Norovirus GII, total and fecal coliform in any tested conditions at both testing locations. Since these microorganisms were non-detectable in MBR permeate streams, the actual upper limit of log removal values (LRVs) achieved in MBR could not be determined. The minimum LRVs observed was between 1.94-6.23 for viruses and 5.95-7.31 for coliforms. Tables 1 and 2 present pathogen concentrations in MBR feed, permeate and calculated log removal values (LRVs) for MBR facility 1 and 2, respectively.

Table 1. Pathogen Concentrations in MBR Feed, Permeate and Calculated Log Removal Values at Full City of Henderson WRF

	Size, nm	MBR Feed, pfu/100 mL	MBR Permeate, pfu/100 mL	Log Removal Value
HAdV	70-110	2,016-37,572, 256	ND (<1)	>3.30->7.57
Norovirus GI	35-42	100-1,038,037	ND (<1)	>2.00->6.02
Norovirus GII	35-42	1,900-197,974	ND (<1)	>3.28->5.30
MS2 Coliphage	24-26	1,800-4,400	ND(<1)-21	1.93->3.64
Somatic Coliphage	30-95	1,500-6,200	ND(<1)-7	2.33->3.79
Total Coliform	~1,000	12,000,000-17,000,000	ND (<2)	>6.78->6.93
Fecal Coliform	~1,000	3,000,000-11,000,000	ND (<2)	>6.18->6.74

Table 2. Pathogen Concentrations in MBR Feed, Permeate and Calculated Log Removal Values at North City Water Reclamation Facility

	Size, nm	MBR Feed, pfu/100 mL	MBR Permeate, pfu/100 mL	Log Removal Value
HAdV	70-110	56,640-474,720	ND (<1)	>4.75->5.68
Norovirus GI	35-42	3,650-11,240	ND (<1)	>3.56->4.05
Norovirus GII	35-42	1,900-197,974	ND (<1)	>3.47->3.80
MS2 Coliphage	24-26	1,800-4,400	ND(<1)-21	1.99->3.51
Somatic Coliphage	30-95	1,500-6,200	ND(<1)-7	2.57->3.72
Total Coliform	~1,000	8,000,000-10,000,000	ND (<2)	>6.60->6.70
Fecal Coliform	~1,000	3,800,000-7,000,000	ND (<2)	>6.28->6.54

Membranes used in each MBR system have a nominal pore size of 0.04 micron (absolute pore size of 0.1 micron) which is larger than the mean diameter of the MS-2 (0.024 micron). MS-2 Coliphage is a commonly used surrogate in virus challenge studies because size and surface characteristics of MS-2 Coliphage are very similar to that of Poliovirus. The MBR membranes in each facility were very effective for complete retention of MS-2 (the smallest virus tested) under normal operating conditions (i.e. fouled membrane conditions) thereby indicating that sieving (size exclusion) is not the sole pathogen rejection mechanisms while dynamic cake layer formation, gel formation, pore blocking are playing major roles for the removal of small pathogenic organisms in fouled membranes.

Lower concentrations of MS-2 and Somatic Coliphages were observed in MBR permeate streams immediately following hypochlorite and citric acid maintenance cleans in each facility. This indicates that small viruses can cross the membranes because pore blocking, cake and gel layer formation mechanisms are not established and therefore these mechanisms have a little effect for the retention of viruses when membranes are clean.

Detectable MS-2 concentrations in MBR permeate streams under clean membrane conditions enabled accurate determination of MS-2 log removal values (LRVs) across MBR membranes. Figure 2 presents MS-2 log removal values as function of MBR permeate turbidity for Henderson and North Las Vegas WRFs, respectively. The data in Figure reflect the results obtained immediately after sodium hypochlorite and citric acid maintenance clean at each WRF. The high R2 values (>0.85) of the regression lines indicate that there is a good correlation between MBR permeate turbidity and permeate MS-2 concentration. Therefore, based on limited data generated from each WRF, it may be concluded that MBR permeate turbidity values may be used as surrogate for assessing membrane integrity in MBR systems and establishing pathogen credits for MBR systems.

Figure 2. Observed MS-2 LRVs as Function of MBR Permeate Turbidity Values for Cleaned Membranes

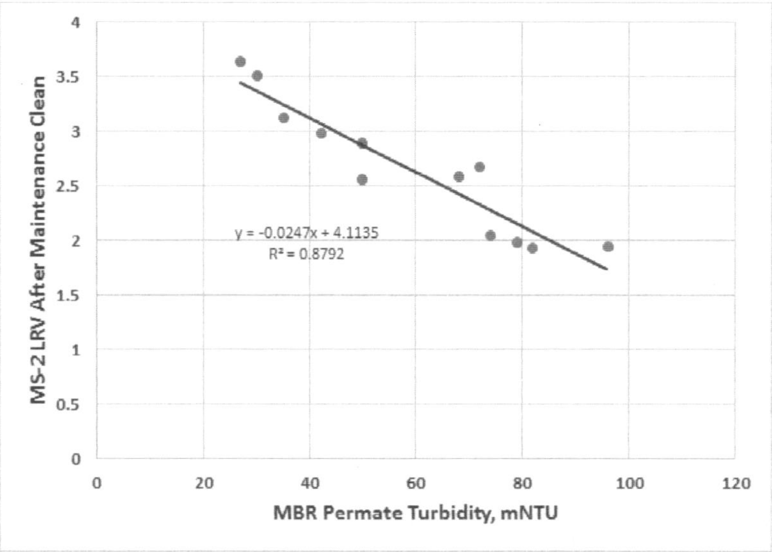

DISCUSSION AND CONCLUSIONS

Complete rejection of MS-2 (average size of 0.024 micron) by the tested MBR membranes (nominal and absolute pore sizes of 0.04 and 0.1 microns, respectively) under several operating conditions (membranes were moderately fouled) indicated that multiple removal mechanisms may be contributing for the removal of pathogen in MBR systems including:

1. biological completion/predation
2. absorption into the biomass and other pathogens
3. direct removal via size exclusion by membranes
4. absorption into biofilm/cake layer and removal through backwash and/or chemical cleaning
5. pore blocking
6. reduction of the effective pore size due to biofilm growth, gel, scale cake layer formation on the membrane surface and pores.

Direct Integrity Testing (DIT) used in drinking water applications has limitations and shortcomings and therefore it may not be directly applicable to MBR systems to assess membrane integrity and get pathogen credits based on DIT results. These limitations and shortcomings may include:

- Historically MBR manufacturers are not used to providing pressure decay testing. DIT test pressure is relatively high and cannot be applied to flat sheet MBR membranes. In addition, DIT test pressure also exceeds pressure requirements of the most of the MBR hollow fiber products (3-5 psi).
- Lack of correlation between PDT and LRV in MBR; due to the action of mechanisms (i.e. pore blocking, cake and gel layer formation, absorption by MLSS, sludge wasting, etc.) other than pure size exclusion.
- DIT conducted per Enhanced Surface Water Treatment Rule (ESWTR) and USEPA Membrane Guidance Manual (USEPA, 2005) to warrant pathogen credit for drinking water membrane filtration applications relies on a volumetric concentration factor (VCF) of 1 in most cases (dead end filtration without recirculation). Volumetric concentration factor is defined as the ratio between membrane feed and raw water total suspended solids concentrations. In wastewater applications, raw wastewater typically contains total suspended solids of 150-300 mg/L whereas membrane feed may contain up to 10,000-12,000 mg/L suspended solids which yield a VCF of 40-60. The CVF values in MBR systems will be much higher than the values used in drinking water applications (typically 1). Because VCF is in the denominator of the LRV formula, a VCF of 60 compared to 1, for example, can reduce Crypto LRV by 1.78 in MBR applications even though the same membranes are used in MBR and drinking water applications.

The limited data in this study indicate that permeate turbidity values may be used to assess MBR integrity. The high R2 values (>0.85) of the regression lines obtained in this study may indicate that MBR permeate virus concentrations are well correlated with the MBR permeate turbidity values and therefore permeate turbidity values may be used to assess membrane integrity and in establishing pathogen credits for MBR systems. However, more comprehensive studies must be done to further assess membrane pathogen removals under intact and compromised membrane conditions (i.e. cut membrane fibers) as well as varying operating and treatment process upset conditions (i.e. wet weather, sludge de-flocculation, sludge bulking) and validate if turbidity

- is a suitable parameter that can accurately correlate permeate pathogen concentrations
- can be reliably used to assess membrane integrity

- can be approved by California State Water Board to warrant pathogen credit for MBR systems

Other potential methods/approaches to assess membrane integrity in MBRs include:

- Real-Time Detection via Multi-Angle Light Scattering MALS
- On-line Particle Counting
- Membrane Integrity Sensor
- Real-Time Detection via ATP Production Luminultra

On-line particle counting has been used in drinking water applications with moderate success. Relatively intensive calibration requirements and lower accuracy of the equipment due to interferences (i.e. air bubble), it has been widely used. All other methods are relatively new and their performances and accuracies yet to be proven.

More comprehensive data will be completed during pilot testing to further assess membrane pathogen rejections under intact and compromised membrane conditions and if MBR permeate turbidity is a viable approach to understand membrane integrity and used it to warrant pathogen credits toward MBR systems. The pilot testing results will be incorporated into the final manuscript.

ACKNOWLEDGEMENTS

The authors sincerely thank to City of Henderson and North City staff for sharing and inclusion of pathogen data collected during full-scale testing.

REFERENCES

California Department of Public Health. Regulations Related to Recycled Water. June 18, 2014 (Revisions effective on 6/18/14).

Fact Sheet - Long Term 2 Enhanced Surface Water Treatment Rule. United States Environmental Protection Agency EPA 815-F-05-009. December 2005.

Madaeni, S.S., Fane, A.G., Grohmann, G.S., 1995. Virus removal from water and wastewater using membranes. Journal of Membrane Science 102, 65–75.

Schwarz, A. O.; Rittmann, B. E.; Crawford, G. V.; Klein, A. M.; Daigger, G. T. 2006. Critical Review on the Effects of Mixed Liquor Suspended Solids on Membrane Bioreactor Operation. *Sep. Sci. Technol.*, **41**, 1489.

Ueda, T., Horan, N.J., 2000. Fate of indigenous bacteriophage in a membrane bioreactor. Water Res. 34, 2151–2159.

United States Environmental Protection Agency 2005. Membrane Filtration Guidance Manual. United States Environmental Protection Agency. EPA 815-R-06-009. November 2005.

Water Environment Federation. 2011. Membrane Bioreactors. Water Environment Federation Manual of Practice 36. 2011. Mc Graw Hill. Newyork, NY.

Yi, T; Harper, W. F., Jr; Holbrook, D.; Love, N. G. 2006. Role of Particle Size and Ammonium Oxidation in Removal of 17-Ethinyl Estradiol in Bioreactors. *J. Environ. Eng.* (Rcston, Virginia), Nov, 1527.

Zhang, K., Farahbakhsh, K., 2007. Removal of native coliphages and coliform bacteria from municipal wastewater by various wastewater treatment processes: implications to water reuse. Water Res. 41, 2816–2824.

PAA, UV, and PAA/UV: Finding the right "bowl of porridge" for Texas Type I reuse disinfection

AUTHORS:
Kaylee Dusek, EIT (Garver)
Michael Watts, PhD, PE (Garver)
Matthew Jalbert, PE (TRA)

ABSTRACT:
In recent years, utilities have been faced with the difficult challenge of eliminating the need for storage of hazardous gases on-site, and the onerous risk mitigation measures that accompany their usage. The Trinity River Authority's (TRA) Central Regional Wastewater System (CRWS) is no exception. Not only is CRWS confronted with the challenge of eliminating risk exposure to these hazardous chemicals, but is also faced with the challenge of incorporating innovative alternative disinfection measures for future reuse.

CRWS began operation in December of 1959. The initial rated capacity was 30 million gallons per day (MGD) and has since expanded to 162 MGD average daily flow with a 405 MGD peak 2-hour capacity to serve twenty-one contracting parities of approximately 1.4 million people in the Dallas/Fort Worth area. The next permit is anticipated to be 189 MGD average daily flow and 550 MGD peak 2-hour which will put even more demand on their disinfection needs.

To mitigate issues associated with chlorine disinfection, CRWS has evaluated two (2) innovative alternative disinfection technologies – ultraviolet (UV) and peracetic acid (PAA) disinfection, or a cost-effective combination of both. Based on previous evaluations at CRWS, two (2) advanced low-pressure-Hg, High Output UV technologies were considered for capital improvement planning purposes. In addition, a non-contact UV system (where treated water flows through UV transparent tubes, surrounded by lamps) was evaluated to offer operations and maintenance advantages over traditional UV disinfection technologies. As regional economics and the viability of PAA disinfection have improved, bulk PAA treatment was examined while accounting for the impact of ongoing solids improvements projects at CRWS and anticipated future discharge permits while providing opportunities for expanded reuse with each of the alternative disinfection scenarios. The plant's current reuse portfolio includes the augmentation of nearby recreational lakes and impoundments with disinfected effluent.

A 20-year, total-lifecycle cost analysis was performed, with a Monte-Carlo framework to account for the risk of future price fluctuations in power, chemicals, and labor, to assess each disinfection alternative for the plant. Dosing for each disinfectant was based on extensive laboratory, pilot, and full-scale testing performed by the Trinity River Authority.

PAA as the primary disinfectant at CRWS, represented the least expensive alternative for the twenty-year project period with or without quenching considered. However, these total project costs are minimally influenced by total annual operation and maintenance (O&M) costs. Therefore, when evaluating the cost breakdown of fixed, chemical and power associated O&M costs for the first year of operation for each of the evaluated alternatives it demonstrated that UV offers a lower annual O&M cost for CRWS as compared to PAA.

Non-economic factor analysis confirmed the results of life cycle cost analysis. A significant contributing factor to the recommendation of future disinfection with UV was the observed residuals in PAA treated effluents at CRWS. Notably, these residuals included substances observed to contribute to increased effluent BOD (at a ratio of ~0.9:1 BOD-to-PAA-dosed on average in CRWS effluent dosed with PAA), as determined through on-site testing.

In this presentation, the results of this disinfection optimization study for the largest water resource recovery facility in Texas will be presented. The outcomes of a complete life cycle, cost-optimization model for average-day, and peak-flow disinfection with UV and/or PAA will have both State, Regional, and National implications. For wastewater disinfection planning, this innovative approach for adapting to evolving disinfection needs will aid both TRA and other utilities.

Disinfection of Wastewater with Peracetic Acid (PAA) and UV Combined Treatment: A Pilot Study

Achal Garg[1], Vasudevan Namboodiri[2], Bruce Smith[1], Abdulaziz Al-Anazi[1], Brindha Murugesan[1], and Tyler Bowman[1]

[1]Metropolitan Sewer District of Greater Cincinnati, Cincinnati, OH 45204 and [2]US Environmental Protection Agency, Cincinnati, OH 45220

***Corresponding author: Achal Garg; email: achal.garg@cincinnati-OH.gov**

Abstract
UV irradiation is one of the most commonly used wastewater disinfection methods. Significant advantages of using the UV-based disinfection are that it leaves no residual and does not form any disinfection by products (DBPs) in the treated wastewater. However, the high-energy cost is the major downside of the UV-based disinfection. For the past three years, Metropolitan Sewer District of Greater Cincinnati (MSD) has been investigating the use of peracetic acid (PAA) alone and in combination with UV to increase the disinfection efficiency and reduce the cost of disinfection treatment. Here we are reporting the results from a full-scale plant-level pilot study we conducted at MSD's Muddy Creek Treatment Plant (MCTP) to investigate the impact of PAA pre-treatment on UV disinfection and the rate of microbial inactivation. It was observed that pre-treating secondary effluent with low doses of PAA results in an increased UV efficiency, which can significantly increase the rate of microbial inactivation when compared with individual treatments with PAA or UV separately. In our studies, the combined PAA-UV treatment achieved significantly greater log reduction in fecal coliform, and E. coli counts. The results from this plant-level pilot study validates our previously conducted lab and side-stream pilot studies that a PAA-UV sequential treatment is more effective than the UV or PAA treatments separately. The disinfection efficiency was directly proportional to the dose of PAA during the pre-treatment of secondary effluent.

Key Words: Peracetic acid, UV irradiation, Disinfection, Effluent, Wastewater, Combined PAA-UV treatment, Pilot study

Introduction
Disinfection is an essential component of wastewater treatment to eliminate the disease-causing organisms such as bacteria, viruses, and parasites. It is required to protect public health and the environment. Various methods and technologies are applied to disinfect the final effluent before releasing it to the environment. These methods include ultraviolet (UV) irradiation (Carmineo et al., 1994; Lazarova et al., 1998; Kolch, 2000), ozone treatment (Lazarova et al., 1998; Andreottola et al., 1996), and the use of various chlorine derivatives (Hajenian & Butler, 1980; Zanetti et al., 1996; Legnani et al., 1996). Among these, the chlorine derivatives are the most common disinfectants because of their wide spectrum disinfection efficiency and low cost. However, recent research has evoked concerns that chlorination can promote the formation of toxic, mutagenic, and carcinogenic disinfection by-products (DBPs) in the treated water. Harmful DBPs can negatively influence the water quality and the environment (Madoni et al., 1998; Dell'Erba et al. 2007; Kauppinen et al. 2012; Veschetti et al. 2003). These concerns have led to the search and evaluation of new disinfectants such as peracetic acid (PAA). In the recent years, many municipalities are developing and adopting environmental-friendly approaches and disinfection treatment strategies (Amin et al., 2013; Lester et al., 2016; Dunkin et al., 2017; Luukkonen & Pekhonen, 2017). These approaches are

aimed to achieve several goals such as protecting the environment through improved effluent treatment practices, eliminating toxic treatment byproducts, and reducing treatment cost.

Peracetic acid (PAA) is a strong oxidizing compound with a wide spectrum of antimicrobial/biocidal properties similar to liquid chlorine or NaOCl. It is a clear and colorless liquid commercially available at a concentration of 12% to 15% in an equilibrium mixture of acetic acid, hydrogen peroxide and water (Block 2001; Alsari et al., 1992; Alsari et al., 1993; Gehr et. al., 2003):

$$\underset{\text{Acetic Acid}}{CH_3COOH} + \underset{\text{Hydrogen Peroxide}}{H_2O_2} \leftrightarrow \underset{\text{Peracetic Acid}}{CH_3COOOH} + \underset{\text{Water}}{H_2O}$$

PAA is considered to be the primary component responsible for the disinfection action in the solution (Baldry, 1983; Baldry and French, 1989; Kitis, 2004; Lubello et al., 2002).

It has been widely used in the food, beverage, medical, and pharmaceutical industries for over 20 years (Orth, 1998; Kitis, 2004; Rasimus et al., 2011). Because of its strong antimicrobial properties, PAA has been getting a lot of attention as a wastewater disinfectant to replace chlorine in the recent years (Lefevre et al., 1992; Baldry et al., 1995; Sanchez-Ruiz et al., 1996; Stampi et al., 2001, 2002; Wagner et al., 2002). It has been reported that PAA and sodium hypochlorite have demonstrated similar antimicrobial activities against *E. coli*, fecal coliform, and total coliform (Baldry et al., 1991; Veschetti et al., 2003) but PAA holds multiple advantages over sodium hypochlorite as a disinfectant for wastewater effluent. These advantages include: need for lower doses, lower residuals, faster disintegration, and absence of disinfection byproducts (DPBs) in the treated effluent (Booth and Lester, 1995; Liberti and Notarnicola, 1999; Monarca et al., 2000; Kitis 2004; Veschetti et al., 2003; Crebelli et al., 2005; Koivunen & Heinonen-Tanski, 2005; Antonelli et al., 2013).

Combining UV with an oxidizing agent, such as PAA, H_2O_2, and ozone (O_3), results into an Advanced Oxidative Process (AOP) (Lubello et al., 2002; Koivunen and Heinonen- Tanski, 2005; Caretti and Lubello, 2003, Lester et al., 2016). When oxidants are exposed to UV irradiation, a photolytic reaction takes place which leads to the formation of highly reactive hydroxyl (°OH) radicals (Ceretti & Lubello, 2003). In the photolysis process, the photons interrupt the O-O bond in the PAA molecule forming the hydroxyl radical (°OH) which reacts vigorously with biological, organic and inorganic matters. Formation of hydroxyl radicals is considered as the key for the disinfection effect of PAA+UV combination treatment.

$$\text{CH3COOOH} + UV \xrightarrow{hv} {}^{\circ}\text{OH} + \text{CH3CO}_2{}^{\circ} \xrightarrow{\text{+Pollutants}} CO_2 + H_2O + \text{Inorganic ions}$$

Commercial PAA contains significant amount of hydrogen peroxide (H_2O_2) in the mixture. The presence of H_2O_2 maintains PAA in an equilibrium and also produces additional hydroxyl (°OH) radicals when exposed to UV irradiation. The formation of additional °OH radicals by photolysis of H_2O_2 is a contributing factor to the synergism of combined PAA+UV irradiation treatment.

This study had two goals: 1. validate findings from our previously conducted studies in the laboratory and in a side-stream pilot study with a full-scale plant-level pilot study, and 2. evaluate potential synergism between PAA pre-treatment and UV on wastewater disinfection in the field study. Our previous laboratory and side-stream studies found a significant increase in the microbial inactivation in effluent when the PAA treatment preceded UV irradiation. These findings are consistent with several published reports (Lubello & Cerettti, 2000; Lubello et al, 2000).

Materials and Methods

Materials: A pilot study was conducted at the Muddy Creek wastewater treatment plant (MCTP) operated by the Metropolitan Sewer District of Greater Cincinnati (MSD), Ohio, USA. On an average day, MCTP treats about 15 million gallons of wastewater (15 MGD). The plant treats wastewater using the activated sludge process and is fitted with a UV disinfection system (Trojan Technologies, Inc.) for the disinfection of secondary effluent.

Peracetic acid (15%), marketed under the commercial name of VigorOX WWT II, was obtained from PeroxyChem, Philadelphia, USA. *E. coli* and fecal coliform broth, micro filters, sampling bottles, and microbiological petri dishes were obtained from Thermo Fisher Scientific, USA. Buffered water with magnesium chloride was from Hardy Diagnostics.

Methods:

Experimental Set-Up

For this pilot study, three sampling locations (L1 - L3) were identified to collect secondary effluent samples before and after the treatment (Fig. 1). All samples were grab samples. Location L-1 was a concrete chamber where the raw secondary effluent from two clarifiers was mixed. This location (L1) provided untreated control samples of raw effluent to estimate the baseline of the water parameters prior to PAA of UV treatments.

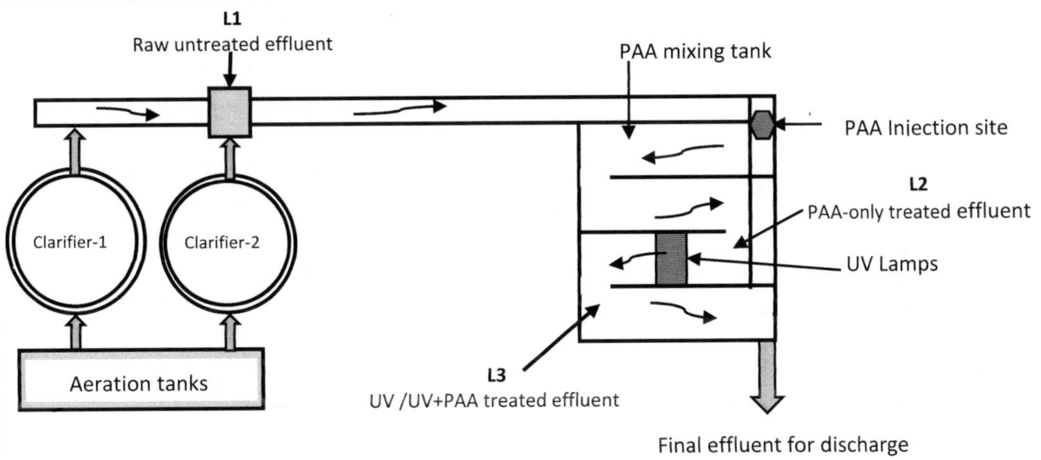

Fig. 1: Schematic of the wastewater effluent treatment with PAA and UV combination

The samples treated with PAA-only were collected at location "L2". Various doses of PAA ranging from 0.75 to 2 mg/l was injected at the beginning of the mixing tank that provided approximately 17-20 minutes of contact time depending on the flow rate. Following PAA treatment, the effluent is passed over the UV lamps for final disinfection. The final sampling site, L3, was located at the end of the plant and was used to collect samples to determine the effect of UV-only disinfection or PAA-UV combined treatment.

PAA and UV treatments: The UV treatment system at Muddy Creek treatment plant, Cincinnati, is comprised of two banks. Each bank contains six rows (modules) of 8-UV lamps each. Thus, each bank has 48 UV lamps. Both UV banks work independent of each other and can be controlled separately. As described in Table-1, five combinations of PAA and UV treatment were tested to evaluate the treatment efficiency individually or in combination. The treatment protocols were: PAA alone, 50% UV alone (41 mJ/cm^2 dose), 100% UV alone (89 mJ/cm^2 dose), PAA+41 mJ/cm^2 UV, and PAA+89 mJ/cm^2 UV. Various concentrations of PAA, ranging between 0.75 and 2 mg/L were tested alone or in combination with UV. For the treatment purpose, when all UV lamps of

Table 1

Treatment	Sampling location	Treatment conditions
Control	L1	Untreated raw effluent
PAA only	L2	Various doses of PAA
UV only (41 mJ/cm^2 dose)	L3	No PAA pre-treatment; Only one UV bank turned on
UV only (89 mJ/cm^2 dose)	L3	No PAA pre-treatment; Both UV banks turned on
PAA+UV (41 mJ/cm^2 dose)	L3	Effluent pre-treated with PAA followed by UV irradiation. Only one UV bank turned on
PAA+UV (89 mJ/cm^2 dose)	L3	Effluent pre-treated with PAA followed by UV irradiation. Both UV banks turned on

both banks were turned on, it was considered 100% UV exposure. The 100% UV dose was equal to an average of 89 mJ/cm^2. However, the UV confluence varied depending on the flow rate and level of suspended solids at the time. To achieve the 50% UV confluence/dose, one UV bank was turned off which on average yielded about 41 mJ/cm^2 UV dose. This experimental design allowed an understanding of the relative contribution of each of the treatment components in the combined disinfection system. As described in Table-1, the samples were collected at locations L1 for control (no treatment), L2 for PAA alone and at L3 for either UV only or combination of UV and PAA.

Wastewater characterization:
The effluent samples were analyzed for E. coli, fecal coliform, total suspended solids (TSS), pH, chemical oxygen demand (COD) and PAA residuals. For microbial analysis of E. coli and fecal coliform the treated and untreated secondary effluent samples were collected in 100 ml sterile plastic bottles containing 10 mg sodium thiosulfate (Thermo Fisher Scientific, Cat# 05-719-361) to neutralize residual PAA and H_2O_2 instantaneously. The fecal coliform and *E. coli* were analyzed using a membrane filter method. The plates were incubated for 24 ± 2 hours in a 44.5 ± 0.2°C for fecal coliform and in a 35 ± 0.5ºC water bath for *E. coli.* Chemical oxygen demand (COD), total suspended solids (TSS) and pH were analyzed using the procedures as described in "Standard Methods for the Water and Wastewater, 23rd Ed., 2017).

PAA Residual Measurement:
The PAA residual was measured at the end of plant location (L3) using the Single Analyte Meter (SAM) (ChemTrics, Midland, VA, USA, Cat. # I-2020) and self-filling ampules (ChemTrics, Midland, VA, USA, Cat. # K-7913) on grab samples. PAA residual levels were measured within 5 minutes of collecting the sample.

Results:

Full Scale Plant-level Pilot Study:

Effect of Wastewater Characteristics on PAA Treatment: Attempts were made to assess the effect of water parameters such as COD and TSS on PAA treatment efficiency. No significant correlation was found between the levels of above mentioned water parameters and PAA disinfection efficiency to inactivate E. coli and fecal coliform. Additionally, we did not notice any impact of PAA treatment (0.75 to 2.0 mg/L doses) on either COD or TSS.

Table-2 (values are arithmetic mean; mg/L)

	Control	PAA only
COD	23	20
TSS	1.15	1.12

Wastewater Disinfection with PAA and UV Individual Treatments:
The disinfection efficiencies of PAA and UV were determined individually and in various combinations by measuring inactivation of E. coli and fecal coliform after the treatment. As expected, the UV fluence of 89 mJ/cm^2 was sufficient to inactivate 99% (2-log reduction) fecal coliform and E. coli. At this dose, there was an average 2.2-log reduction in fecal coliform and 2.4-log reduction in the E. coli concentration. The average number of fecal coliform and E. coli in the final effluent treated with 89 mJ/cm^2 was less than 50 CFU/100 ml. In comparison, when one UV bank was turned off, reducing the UV dose by almost 50% to 41 mJ/cm^2, the inactivation rate of E. coli dropped from 2.4 to 1.8-logs and for fecal coliform from 2.2 to 1.6 log reduction (Fig. 3 and 5). A 50% reduction in the UV fluence or 41mJ/cm^2 UV dose was insufficient to achieve the inactivation level needed to meet the plant's NPDES permit requirement for fecal coliform and E. coli.

Disinfection with PAA-only treatment was found to be highly dose-dependent. Four doses of PAA ranging from 0.75 to 2.0 mg/l were tested independently or in combination with UV. With PAA-only treatment, the inactivation ranged from 0.6 log (0.75 mg/l dose) to 1.5 log (2.0 mg/l dose) for E. coli and 0.8 (0.75 mg/l dose) to 1.8 log (2.0 mg/l dose) in case of fecal coliform. Doses from 0.75 to 1.5 were found insufficient to meet the NPDES permit of the plant. However, at 2 mg/l PAA, the number of fecal coliform was reduced below the permit requirement of 126 E. coli CFU/100 ml (Fig. 2 and 4). Thus, 2 mg/l PAA dose or 89 mJ/cm^2 dose of UV were able to provide sufficient disinfection individually to meet the plant's permit microbial requirements.

The most interesting data came when PAA and UV treatments were combined. When the secondary effluent was pre-treated with PAA followed by UV, an additive effect on disinfection efficiency was observed. We did not observe synergism in the PAA-UV combined treatment in any treatment combination. The efficiency of combined PAA and UV treatment depended on the doses of both PAA and UV. However, every combination of PAA and UV that we tested during this study yielded better disinfection or microbial inactivation than the individual treatments with either PAA or UV because of additive effect of the treatment. For instance, 0.75 mg/l PAA and 41 mJ/cm^2 of UV fluence individually yielded about 0.8- and 1.6-log reduction in fecal coliform respectively. However, when 0.75 mg/l PAA treatment was followed by 41 mJ/cm^2 UV treatment, we achieved an average of 1.9 log reduction in fecal coliform counts. Similarly, at 1.0 mg/l PAA and 41 mJ/cm^2 dose, there was 2.5-log reduction in both E. coli and fecal coliform, which was equal to the sum of 0.8-log reduction with PAA and 1.6-log reduction with 41 mJ/cm^2 UV separately. When both banks were turned on (100% UV efficiency; 89 mJ/cm^2 dose), a 2.2 to 2.4 log reduction was observed in fecal coliform and E. coli with UV-only treatment reducing the population of these microbes more than 99%. This level of disinfection virtually eliminated all fecal coliform and E. coli bacteria in the final effluent. The additive effect of PAA on UV treatment was more prominent at 1 mg/l or higher doses. When 1 mg/l dose of PAA was combined with 89 mJ/cm^2 UV fluence, the disinfection efficiency of the combined treatment was more than 3-logs (i.e., 99. 9%) (Fig. 3 and 5).

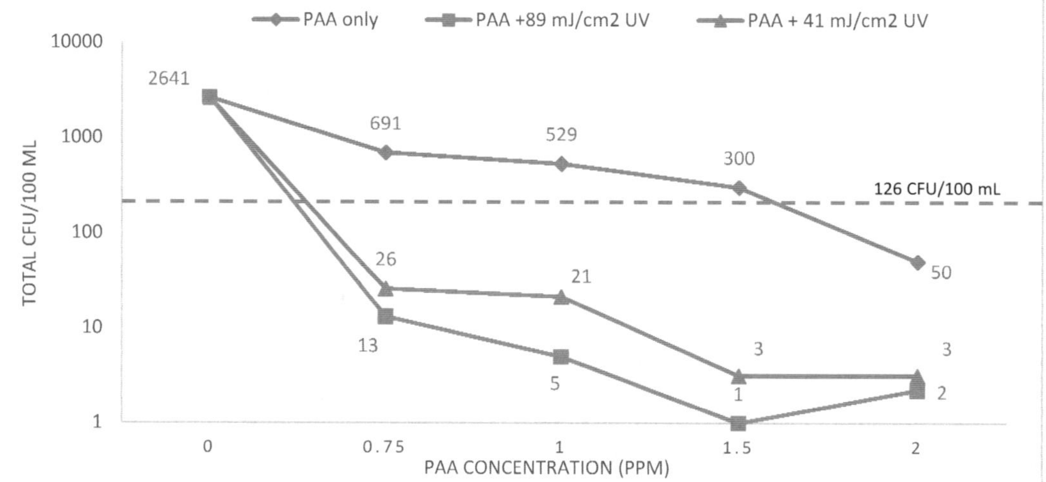

Fig 2- Effect of PAA and UV combined treatment on E. coli inactivation

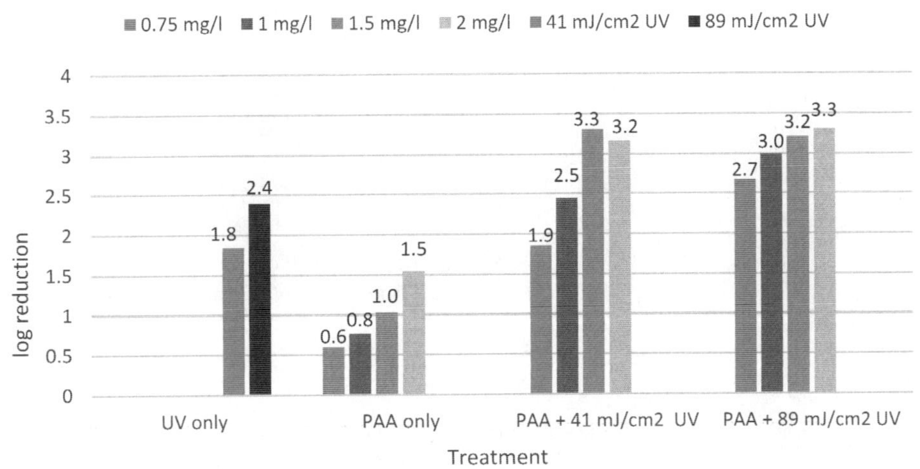

Fig 3. Log reduction of E.coli after UV, PAA and PAA+UV treatments

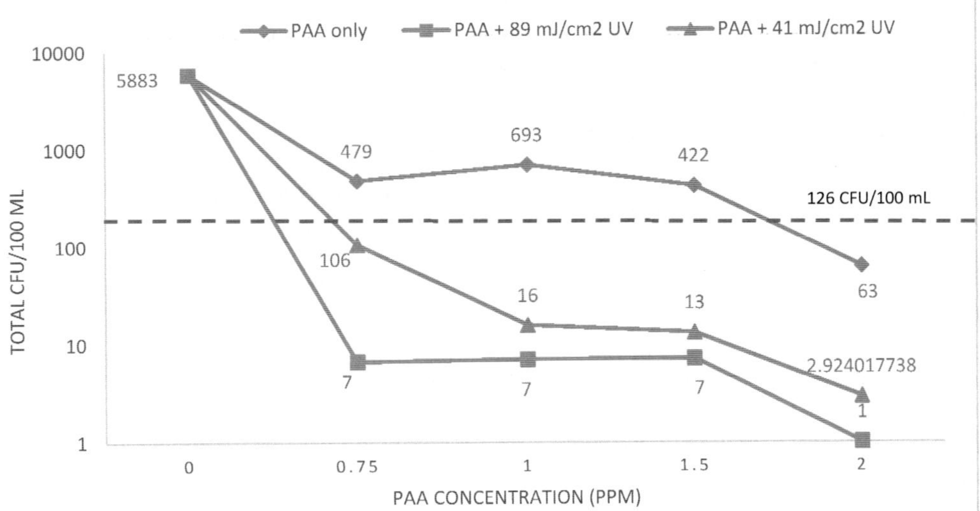

Fig 4- Effect of PAA and UV combined treatment of fecal coliform inactivation

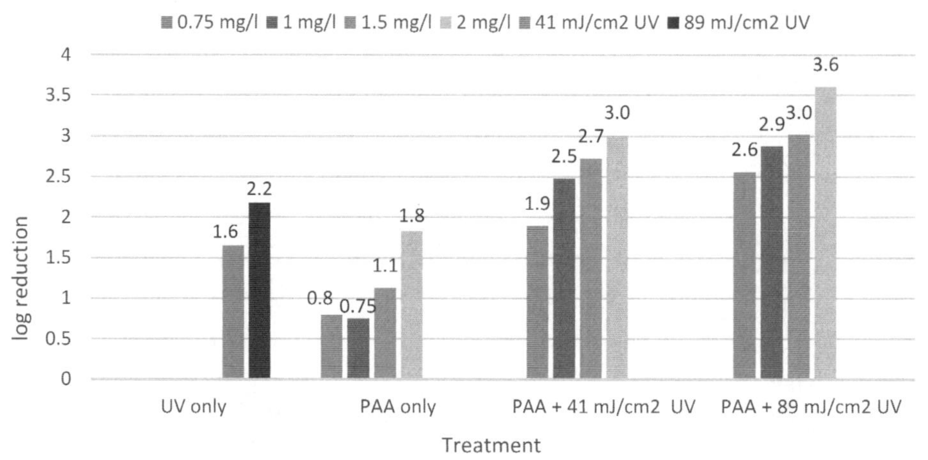

Fig 5- Log reduction of fecal coliform after UV, PAA or PAA+UV treatments

Residual peracetic acid: The PAA residual were measured at location L3, just prior to discharge. After approximately 20 to 23 minutes of contact time, traces of PAA were detected in the final effluent. As shown in Table-3, the residual level depended on the initial dosing. Up to 1 mg/L dosing, on an average 0.27 mg/L PAA residual was detected after about 20 minutes. At 1.5 mg/L and 2.0 dosing, the average residual levels were 0.48 and 0.78 mg/L respectively (Table-3).

Table-3	
PAA Residual (Arithmetic mean; mg/L)	
Dose	Residual
0.75	0.27
1	0.27
1.5	0.48
2	0.78

Discussion:

Prior to this full-scale pilot study, we conducted extensive benchtop and side-stream pilot studies to investigate the potential synergism between PAA and UV treatments. Our benchtop as well as pilot studies provided evidence of increased disinfection efficiency or additive benefic when PAA and UV are combined to treat wastewater effluent. When only PAA was used for treatment, the disinfection efficiency appeared to be affected by both concentration and contact time suggesting that a low dose of the disinfectant with prolonged contact time can achieve the same degree of disinfection as the higher dose with shorter contact time. This observation was consistent with previously published reports (De Luca et al., 2008; Koivunen & Heinonen-Tansski, 2005; Wagner et al., 2002).

Commercially available peracetic acid (PAA) comes as a mixture of PAA and hydrogen peroxide (H_2O_2) in a stable equilibrium state. Both PAA and H_2O_2 are known to possess antimicrobial activity though PAA, being an organic peroxide, is a much stronger antimicrobial. PAA can penetrate the bacterial cell wall more efficiently than H_2O_2 and can cause more extensive damage to bacteria's cellular contents such as enzymes and transport system compared to H_2O_2 (Koivunen & Heinonen-Tanski, 2005). Although the disinfection efficiency of H_2O_2 is significantly lower, it may still contribute to the PAA's total disinfection efficiency by producing additional hydroxyl radicals ($^{\cdot}OH$) in the presence of UV. Hydroxyl radicals are highly reactive and quickly degrade organic and biological material. Formation of hydroxyl radicals is considered the primary factor for PAA's disinfection activities.

It has been suggested that PAA disinfection is a two-step process. A fast and initial reaction between PAA and biological matter, including the microbes, followed by inactivation at a slower rate (Koivumen & Heinonen-Tanski, 2005; Luukkonen et al., 2014**;** Antonelli et al., 2006; Falsanisi et al., 2006). It is believed that free-floating microbes are inactivated promptly by PAA in the first phase (Tyrrell et al., 1995; Liberti et al., 2000). Although the precise mechanism of microbial inactivation in the second or slow phase is not clearly understood, it is proposed the presence of suspended solids in the effluent appear to provide shelter to microbes from PAA thus requiring either a higher PAA dose or longer contact time (Koivumen & Heinonen-Tanski, 2005). This further suggests microbial inactivation is a direct function of both PAA concentration and contact time.

Just like PAA, UV is also a strong wastewater disinfectant. Its broad spectrum of disinfection properties, high rates of disinfectant efficiency at shorter contact times, little to no disinfection byproduct (DBP) formation, and ease of use are some of the benefits associated with UV (Koivunen & Heinonen-Tanski, 2005; Nurizzo et al., 2005). However, combining PAA with UV in a sequential treatment has shown significant improvement in the microbial disinfection efficiency over individual UV or PAA treatments (Koivunen & Heinonen-Tanski, 2005).

Caretti & Lubello (2003) have reported that when UV irradiation is applied before PAA to the effluent (UV-PAA combination), there is no synergistic benefit from the UV-PAA combined treatment. However, when the treatment conditions are reversed and PAA is applied before UV, the treatment efficiency is increased significantly (Caretti & Lubello, 2003). Our studies were conducted by treating the effluent with PAA prior to UV irradiation in a PAA-UV combination. Every PAA and UV combination that was tested in this study resulted in additive increase in the efficiency of PAA and UV treatments. However, no synergistic benefit was observed. The disinfection efficiency increased when either the PAA dose or the UV dose was increased. This increased efficiency can be attributed to the increased formation of hydroxyl radicals and photolysis of PAA and H_2O_2 (Lubello et al., 2002; Caretti & Lubello, 2003). Our data show that 1.0 mg/l PAA when applied individually yielded up to 0.8-log reduction in E. coli and fecal coliform and a fluence of 41 mJ/cm^2 UV achieved a 1.6-1.8 log reduction in the microbes. However, when the effluent was pretreated with 1 mg/l PAA dose followed by 41 mJ/cm^2 UV dose, the combined treatment achieved a 2.5 log reduction. Similar additive effect was observed with higher PAA and UV doses. UV and PAA inactivate microbes by different mechanisms. UV irradiation attack the genetic material, i.e., DNA, of the bacteria while PAA kills the bacteria by causing damage to its cell wall and its critical enzymes by removing electrons from susceptible chemical groups or the cellular components (Melly et al., 2002; Linley et al., 2012; Leggett et al., 2015; Finnegan et al., 2010). Thus, when the bacteria are exposed to a low dose of PAA, it may cripple the enzyme system that repairs the cellular damage. The inability of bacteria to repair cellular and DNA damage leads to its destruction when the bacteria is irradiated subsequently with UV. When both PAA and UV are applied together, extensive damage to cellular contents caused by highly reactive hydroxyl radicals and nuclear damage caused by UV can overwhelm the bacteria's repair mechanisms leading to its inactivation and subsequent death.

The levels of PAA residuals in the treated wastewater is important measurement to understand the PAA degradation kinetics and the role other water parameters can play in this process. We observed quick degradation of PAA in our studies. It has been reported that PAA breaks down in 2 phases: initial, almost instantaneous decrease in concentration followed by subsequent slow decomposition of residual (Antonelli et al., 2006; Falsanisi et al., 2006). Several factors can affect the initial degradation of PAA such as particulates, reduced organic species, iron, manganese and microorganisms (Falsanisi et al., 2006). Small residues of H_2O_2 are constantly present in water bodies as H_2O_2 is generated in the surface water layers by sunlight – initiated through photoredox reactions involving organic compounds (e.g. humic or fulvic acids), dissolved oxygen and trace metals (e.g. Fe or Mn) (Stumm and Morgan, 1996; Scully et al., 1995). Small amounts of PAA and hydrogen peroxide are, however, considered harmless to an aquatic environment and have been suggested to be beneficial in reducing the amounts of biofilms in effluent discharge pipes (Koivunen and Heinonen-Tanski., 2005).

Using the combined PAA-UV disinfection treatment over individual disinfectant (i.e., PAA or UV) has two-fold advantages. One, it can potentially reduce the disinfection cost by requiring lower dose of UV and thus saving the energy cost. Second, it can improve the bacteriostatic effect of UV disinfection by broadening the range of microbial disinfection. For PAA/UV combination treatment to be successful, it is important that the PAA is introduced prior to UV treatment so the reactive oxidative and hydroxyl radicals can be formed during the UV irradiation. By combining the disinfection powers of PAA and UV, an extensive and permanent damage can be caused to a much wider range of bacteria potentially at a lower cost. Additional studies are currently in progress to investigate the potential economic impact of the PAA-UV combined disinfection on treatment and disinfection of wastewater at Muddy Creek Treatment plant.

Conclusions: Peracetic acid (PAA) is a potent oxidizing agent with a wide spectrum antimicrobial activity. UV is a very effective wastewater disinfectant. When used individually, both PAA and UV need high doses to achieve full inactivation of fecal coliform and *E. coli* bacteria. However, pretreating secondary effluent with a low dose PAA prior to UV irradiation has additive disinfection effect and significantly reduce the UV dose needed for disinfection.

Acknowledgements: Our sincere thanks to Bill Beyer, Treatment Supervisor, Tim Hauck, Plant Supervisor and Muddy Creek plant operators and support staff for their crucial support to this study. The chemical pump was provided by Pelton Environmental Services, Cincinnati, Ohio, at no cost for this study.

Disclaimer: *This document has been prepared in accordance with the City of Cincinnati policy. Any mention of trade names, commercial entities, or commercial products does not constitute endorsement or recommendation for use. The findings and conclusions in this report have not been formally disseminated by the U.S. EPA and should not be construed to represent any agency determination or policy.*

References:

1. Alasri A, Roques C, Michel G, Cabassud C, Aptel P., 1992. Bactericidal properties of peracetic acid and hydrogen peroxide, alone and in combination, and chlorine and formaldehyde against bacterial water strains. Can J Microbiol 38:635–642.
2. Alasri A, Valverde M, Roques C, Michel G, Cabassud C, Aptel P., 1993. Sporocidal properties of peracetic acid and hydrogen peroxide, alone and in combination, in comparison with chlorine and formaldehyde for ultra-filtration membrane disinfection. Can J Microbiol 39: 52–60.
3. Amin, MM., Hashemi, H., Bovini, AM and Hung, YT (2013). A review on wastewater disinfection. Int J Env Health Eng, 2, 22-30
4. Andreottola, G., Bertola, P., Ziglio, G. (1996) Indagine sperimentale comparata sulla disinfezione di acque reflue urbane con ozono e acido peracetico. Ingegneria Ambientale 25, 200–206.
5. Antonelli, M., Rossi, S., Mezzanotte, V., and Nurizzo, C. (2006). Secondary effluent disinfection: PAA long-term efficiency. *Environ. Sci. Technol.*, 40, 4771–4775.
6. Antonelli M., Turolla, A., Mezzanotte, V., Nuizzo, C., 2013, Peracetic acid for secondary effluent disinfection: a comprehensive performance evaluation. Water Sci. Technol., 68, 2638-2644.
7. Baldry M. (1983) The bactericidal, fungicidal and sporicidal properties of hydrogen peroxide and peracetic acid, J Applied Micro., 54, 417-423
8. Baldry M. and French M.S. Disinfection of Sewage Effluent with Peracetic Acid. Water Science & Technology 21(3):203-206
9. Baldry M., French MS, Slater, D., 1991. The activity of Peracetic acid on sewage indicator, bacteria and viruses. Water Sci. Technol., 23, 353-357.
10. Baldry MGC, Cavadore A, French MS, Massa G, Rodrigues LM, Schirch PFT, et al., 1995, Effluent disinfection in warm climates with peracetic acid. Water Sci Technol, 31, 161–164.
11. Block SS. 2001. Peroxygen compounds, p 185–204. In Block SS (ed), Disinfection, sterilization and preservation. Lippincott Williams & Wilkins, Philadelphia, PA.
12. Booth, R.A. & Lester, J.N. (1995). The potential formation of halogenated byproducts during peracetic acid treatment of final sewage effluent. Water Res. 29, 1793–1801.
13. Caretti, C. and Lubello, C. 2003. Wastewater disinfection with PAA and UV combination treatment: A pilot study. Water Research, 37, 2365-2371.
14. Carmineo, D., Contini, E., Di Marino, R., Donadio, F., Liberti, L., Ranieri, E. (1994) Wastewater disinfection by UV at Trani municipal plant. Water Sci. Technol 30, 125–132.

15. Crebelli, R. Conti, L., Monarca, S., Feretti, D., Zerbini, I., Zani, C., Veschetti, E., Cutilli, D., Ottaviani, M., 2005, Genotoxicity of the disinfection byproducts resulting from Peracetic acid or hypochlorite disinfected sewage wastewater. Water Res., 39, 1105-1113.
16. De Luca, G., Sacchetti, R., Zanetti, F. Comparative study on the efficiency of peracetic acid and chlorine dioxide at low doses in the disinfection of urban wastewaters. Ann Agri Environ Med 2008; 15: 217–24.
17. Dell'Erba, A., Falsanisi, D., Liberti, L., Notrnicola, M., Santoro, D., 2007, Disinfection by-products formation during wastewater disinfection with peracetic acid. Desalination 215, 177–186.
18. Dunkin, N., Weng, S., Schwab, KJ, McQuarrie, J. and Bell, K. (2017). Comparativ inactivation of murine norovirus and MS2 bacteriophage by peracetic acid and monochloramine in municipal secondary wastewater effluent. Environ. Sci. Technol., 51, 2972-2981.
19. Falsanisi, D., Gehr, R., Santoro, D., Dell'Erba, A., Notarnicola, M., Liberti, L., 2006, Kinetics of PAA demand and its implications on disinfection of wastewaters. Water Qual. Res., J. Can., 41, 398-409.
20. Finnegan, M., Linley, E., Denyer, SP., McDonnell, G., Simons, C., Maillard, J., 2010, Mode of action of hydrogen peroxide and other oxidizing agents: differences between liquid and gas forms. J. Antimicorb Chemother, 65, 2108-2015.
21. Gehr, R., Wagner, M., Veerasubramanian, P., Payment, P., 2003, Disinfection efficiency of peracetic acid, UV and ozone after enhanced primary treatment of municipal wastewater. Water Res. 37, 4573–4586.
22. Hajenian, H. & Butler, M. (1980) Inactivation of f2 coliphage in municipal effluent by the use of various disinfectants. Journal of Hygiene (Lond.) 84, 247–255.
23. Kauppinen, A., Ikonen, J., Pursiainen, A., Pitkänen, T., Miettinen, T., 2012, Decontamination of a drinking water pipeline system contaminated with adenovirus and Escherichia coli utilizing peracetic acid and chlorine. J. Water Health, 10, 406-418.
24. Kitis, M., 2004, Disinfection of wastewater with Peracetic acid: a review. Environ. Int., 30, 44-55.
25. Koivunen, J. and Heinonen-Tanski, H., 2005, Inactivation of enteric microorganisms with chemical disinfectants, UV irradiation and combined chemical/UV treatments. Water Res. 39, 1519–1526.
26. Kolch, A., 2000, Sewage disinfection towards protection of drinking water. Schriftenr ver Wasseries Boden Lufthyg, 105, 331-336.
27. Lazarova, V., Janex, ML, Fiksdal, L., Oberg, C., Bercina, I., Pommepuy, M., 1998, Advanced wastewater disinfection technologies: Short and long term efficiency. Water Sci Technol., 38, 109-117.
28. Lefevre, F., Audic, JM, Ferrand, F., 1992, Peracetic acid disinfection of sewage effluents discharged off secondary coastal seawater. Water Sci Technol., 25, 155-164.
29. Leggett MJ, Schwarz JS, Burke PA, McDonnell G, Denyer SP, Maillard J-Y. 2015. Resistance to and killing by the sporicidal microbicide Peracetic acid. J Antimicrob Chemother 70:773–779.
30. Legnani, P.P., Leoni, E., Baraldi, M., Pinelli, G., Bisbini, P. (1996) Evaluation of disinfection treatment systems for municipal wastewater reclamation and re-use. Zentralblatt für Hygiene und Umweltmedizin 198, 552–566.

31. Lester, Y., Aga, DS., Love, NG, Singh, RR. Morrissey, I. and Linden, KG (2016) Water Environ. Res., 88, 1085-1093.
32. Liberti and Notarnicola, 1999, Advanced treatment and disinfection for municipal wastewater reuse in agriculture. Water Sci Technol, 40:235–245.
33. Liberti, LO., Lopez A, Notarnicola M, Barnea N, Pedahzur R, Fattal B. 2000, Comparison of advanced disinfecting methods for municipal wastewater reuse in agriculture. Water Sci Technol, 42, 215–220.
34. Linley E, Denyer SP, McDonnell G, Simons C, Maillard J-Y. 2012. Use of hydrogen peroxide as a biocide: new consideration of its mechanisms of biocidal action. J Antimicrob Chemother 67:1589–1596.
35. Lubello, C. and Caretti, C. (2000). Wastewater disinfection with Peracetic acid and UV. In: Proceedings of International Symposium on sanitary and Environmental Engineering, Trento 18–23.
36. Lubello,C., Caretti, C. and R. Gori (2000). Comparison between PAA/UV and H2O2/UV disinfectionfor wastewater reuse. Water Science & Technology Water Supply 2:205-212
37. Luukkonen, T., Teeriniemi, J., Prokkola, H., Ramo, J., Lassi, U., 2014, Chemical aspects of Peracetic acid based wastewater disinfection. Water SA, 40, 73-80.
38. Luukkonen, T. and Pehkonen, SO (2017) Peracids in water treatment: A critical review. Critical Rev. Env Sci and Technol., 47, 1-39.
39. Madoni, P., Gorbi, G., Taje, E., 1998, Toxic effects of chemical disinfection of wastewater on freshwater ciliates. Acta Protozoologica, 37, 221-225.
40. Melly E, Cowan AE, Setlow P. 2002. Studies on the mechanism of killing of Bacillus subtilis spores by hydrogen peroxide. J Appl Microbiol 93:316–325.
41. Monarca, S., Feretti, D., Collivignarelli, L., Zerbini, I., Bertanza, G., et al., 2000, The influence of different disinfectants on mutagenicity and toxicity of urban wastewater. Water Res., 34, 4261-4269.
42. Nurizzo, C., Antonelli, M., Profaizer, M., Romele, L., 2005, Byproducts in surface and reclaimed water disinfected with various agents. Desalination, 176, 241-253.
43. Orth, R., 1998. The importance of disinfection for the hygiene in the dairy and beverage production. In. Biodeterior. Biodegrad., 41, 201-208
44. Rasimus, S., Kolari., M., Rita, H., Hoonstra, D. and Salkinoja-Salonen, M., 2011, Biofilm-forming bacteria with varying tolerance to peracetic acid from a paper machine. J. Ind. Microbiol. Biotechnol., 38, 1379-1390.
45. Sanchez-Ruiz, C., Martinez-Royano, S., Tejero-Mozon, I., 1996, An evaluation of the efficiency and impact of raw wastewater disinfection with peracetic acid prior to ocean discharge, Water Sci. Technol., 32, 159–166.
46. Scully, N, Mcqueen, D., Lean, D. and Cooper, W. (1995) Photochemical formation of hydrogen peroxide in lakes: Effect of dissolved organic carbon and ultraviolet radiation. Can J. Fish Aquatic Sci., 52 2675-2681.
47. Stampi, S., Luca, G., Zanetti, F., 2001, Evaluation of the efficiency of Peracetic acid in the disinfection of sewage effluents. J Appl. Microbiol., 91, 833-838.
48. Stampi, S., De Luca G., Onorato., M., Ambrogiani, E., Zanetti, F., 2002, Peracetic acid as an alternative wastewater disinfectant to chlorine dioxide. J. Appl. Microbiol., 93, 721-731.

49. Stumm, W. and Morgan, J.J. (1996) Aquatic Chemistry, Chemical Equilibria and Rates in Natural Waters. 3rd Edition, John Wiley & Sons, Inc., New York.
50. Standard Methods for the Examination of Water and Wastewater. 23nd Ed. 2017, APHA, AWWA, WPCF, Washington, DC, USA.
51. Tyrrell, S., Ripley, SP., Watkins WD., 1995, Inactivation of bacterial and viral indicators in secondary sewage effluent, using chlorine and ozone.
52. Veschetti, E., Cutillli, D., Bonadonna, L., Briancesco, R., Martini, C., Cecchini, G., Anastasi, P., Ottaviani, M., 2003, Pilot-plant comparative study off Peracetic acid and sodium hypochlorite wastewater disinfection. Water Res., 37, 78-94.
53. Wagner, M. Brumelis, D. Gehr, R. 2002. Disinfection of Wastewater by Hydrogen Peroxide of Peracetic Acid: Development of Procedures for Measurement of Residual Disinfectant and Application to a Physico-chemically Treated Municipal Effluent. Water Environ. Res., 74 1 33-50.
54. Zanetti, F., De Stampi, S., Luca, G., Varoli, O., Tonelli, E. (1996) Comparative disinfection of secondary-treatment sewage with chlorine dioxide and bromine chloride. Zentralblatt für Hygiene Umvelmedizine 198, 567–579.

Advanced dose control strategy for disinfection of wastewater by peracetic acid: Pilot-scale validation and performance

Kyriakos Manoli[1,2], Siva Sarathy[1,2], Paris Neofotistos[3], Domenico Santoro[1,2]

[1]Trojan Technologies
[2]University of Western Ontario
[3]USP Technologies

ABSTRACT

An innovative dose control strategy for peracetic acid (PAA) disinfection of wastewater has been developed. The proposed dose-pacing control was benchmarked against a conventional control strategy, i.e. flow-pacing control. The controls were compared in a pilot-scale continuous-flow PAA disinfection of secondary effluent wastewater. Results showed that dose-pacing control outperformed flow-pacing control in terms of maintaining disinfection performance and reducing PAA usage.

KEYWORDS

Disinfection, Dose, Control, Peracetic acid, Wastewater

INTRODUCTION

Disinfection of wastewater is critical to protect the environment and human health. Insufficient disinfection may cause waterborne diseases with negative consequences for society (Xiao et al., 2018). UV-based as well as chemical-based disinfection processes are commonly used for water and wastewater disinfection.

The following aspects should be considered when designing a disinfection process according to the integrated disinfection design framework (IDDF): a) microbial disinfection/ disinfection by-products, b) demand/decay of disinfectant, and c) hydraulics of the reactor (Bellamy et al., 1998). Understanding of water quality and treatment requirement in terms of dose, i.e. the target ICT dose (mg·min/L) in the case of chemical disinfection or target UV dose (mJ/cm^2) for UV disinfection, is important for the development of advanced control strategies. For peracetic acid (PAA) disinfection, worth noting is the definition of ICT dose as the area under the chemical disinfectant decay curve introduced in Santoro et al. (2015).

A low disinfection performance due to under-dosing of a disinfectant may result in violation of regulatory limits, while over-dosing results in increased operational cost (higher amount of disinfectant is used) and possible violation of regulatory chemical residual limits. Moreover, over-dosing of certain chemical disinfectants can contribute to the formation of toxic disinfection by-products. Therefore, disinfection control strategies are essential in order to avoid under-treatment and over-treatment. Nowadays, the most common chemical disinfection control strategies target a constant concentration of the disinfectant using feed-forward flow-pacing control (Kobylinski et al., 2006). These strategies do not maintain constant ICT dose, hence the disinfection performance varies with water quantity and quality.

Particularly, in wastewater disinfection, the issue of advanced ICT dose monitoring, optimization, and control is even more obvious because wastewater flow and quality variations occur frequently and sometimes unexpectedly because of many reasons such as the variation of diurnal flow, wet-weather events, upstream process upsets, and industrial discharge. Under such circumstances, conventional control methodologies such as flow-pacing would be inadequate as they are unable to respond appropriately to maintain sufficient disinfection during these events. The wastewater flow and quality variation makes ICT-dose pacing control a necessity, considering the side reactions between the chemical disinfectant and water constituents that occur during a disinfection process. In addition, the use of innovative strategies for ICT-dose pacing control would become essential in order to properly assess the feasibility in cost and disinfection performance of novel disinfectants, such as PAA, where the unit price of the disinfectant is typically higher than conventional disinfectants such as chlorine or electricity in the case of UV disinfection. Recently, an innovative ICT-dose pacing control strategy has been developed (Sarathy et al., 2017).

In this study, an innovative chemical dosing control technology for chemical disinfection of wastewater (e.g. PAA), which is based on controlling the ICT dose (ICT-dose pacing), has been evaluated and compared with the commonly used flow-pacing strategy which is based on controlling the concentration of the chemical disinfectant. The ICT-dose pacing strategy has been demonstrated and validated under realistic pilot-scale experimental conditions, and the benefit of the different chemical dosing strategies has been quantified.

METHODOLOGY

Pilot studies were carried out in 3035-L chemical contact basin (CCB) located in a wastewater treatment plant employing activated sludge-based secondary treatment. This enabled the use of real-time secondary effluent wastewater as the influent of the CCB. PAA was dosed upstream of the pilot-scale serpentine CCB consisting of four passes (Figure 1). A PLC programmed with flow-pacing logic as well as the innovative ICT-dose pacing logic was used to control the PAA dosing pump. PAA probes were installed at the inlet and outlet of the CCB in order to monitor PAA demand and decay online, and incorporate water quality changes into the ICT-dose pacing control strategy (Figure 1). The flow of secondary effluent wastewater has been adjusted accordingly to simulate relevant water flow variation during a day resulting in hydraulic retention times (HRTs) from 14 to 51 minutes. Inlet and outlet viable fecal coliforms (FC) were measured to monitor disinfection performance.

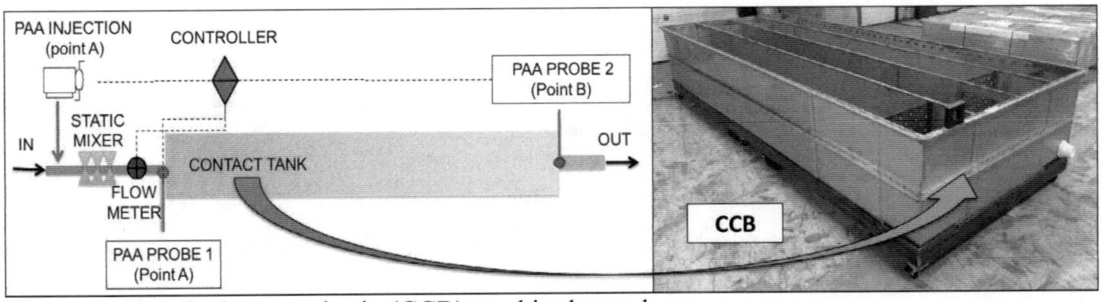

Figure 1. Chemical contact basin (CCB) used in the study.

RESULTS

Initially, a flow-pacing experiment was carried out. The flow ranged from 1.0 L/s (overnight) to 3.5 L/s (morning and evening peak flows). Results are presented in Figures 2 and 3. Because the concentration of PAA at the injection point is fixed to a setpoint concentration (Figure 3), the ICT dose varies during the day, resulting in high ICT dose at low flow (i.e. longer HRTs, greater disinfectant decay), and low ICT dose at high flows (i.e. shorter HRTs, lower disinfectant decay) (Figure 2). This variation in ICT dose is confirmed through the observation that concentration of viable FC at the outlet is not consistent. That is, the FC counts increased with increasing flow since the ICT dose was inadequate. The reduction in the ICT dose is an immediate indication of a potential disinfection permit violation (Figure 2). In contrast, ICT-dose pacing was able to achieve closely the setpoint ICT dose (16 mg·min/L) through the numerous flow changes thereby ensuring consistent disinfection performance (Figures 4 and 5). The small fluctuation of the outlet counts of FC is due to the variation of the FC counts at the inlet (Figure 4).

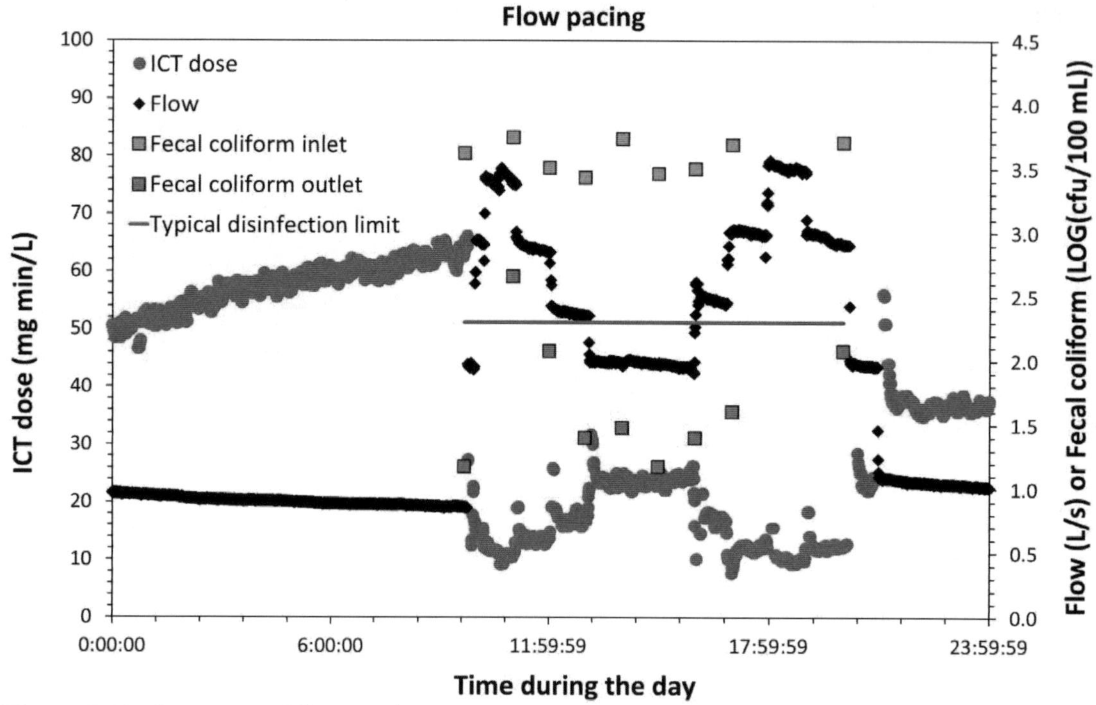

Figure 2. Performance of flow-pacing control strategy in pilot test.

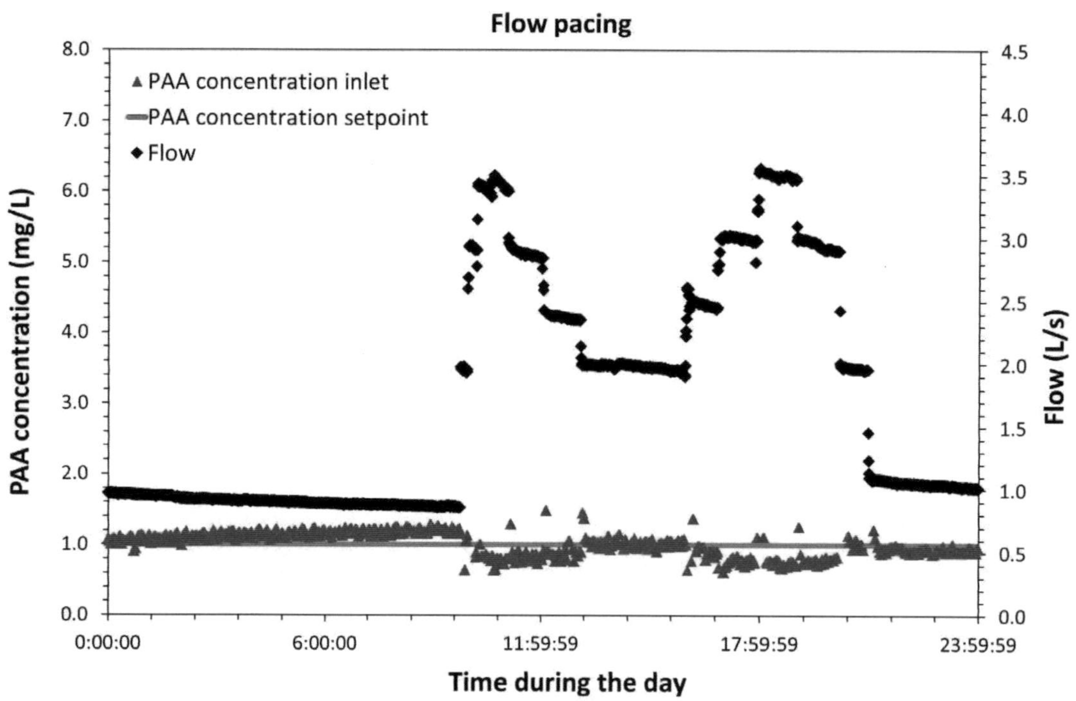

Figure 3. Inlet and setpoint concentration of PAA for flow-pacing strategy in pilot test.

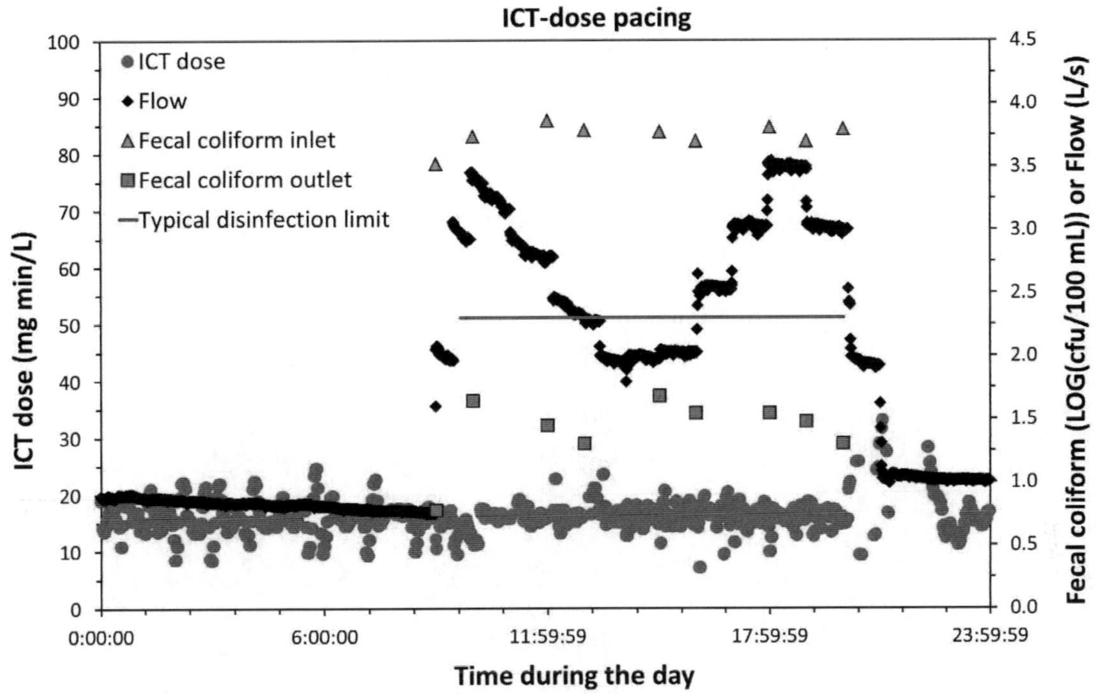

Figure 4. Performance of ICT-dose pacing control strategy in pilot test.

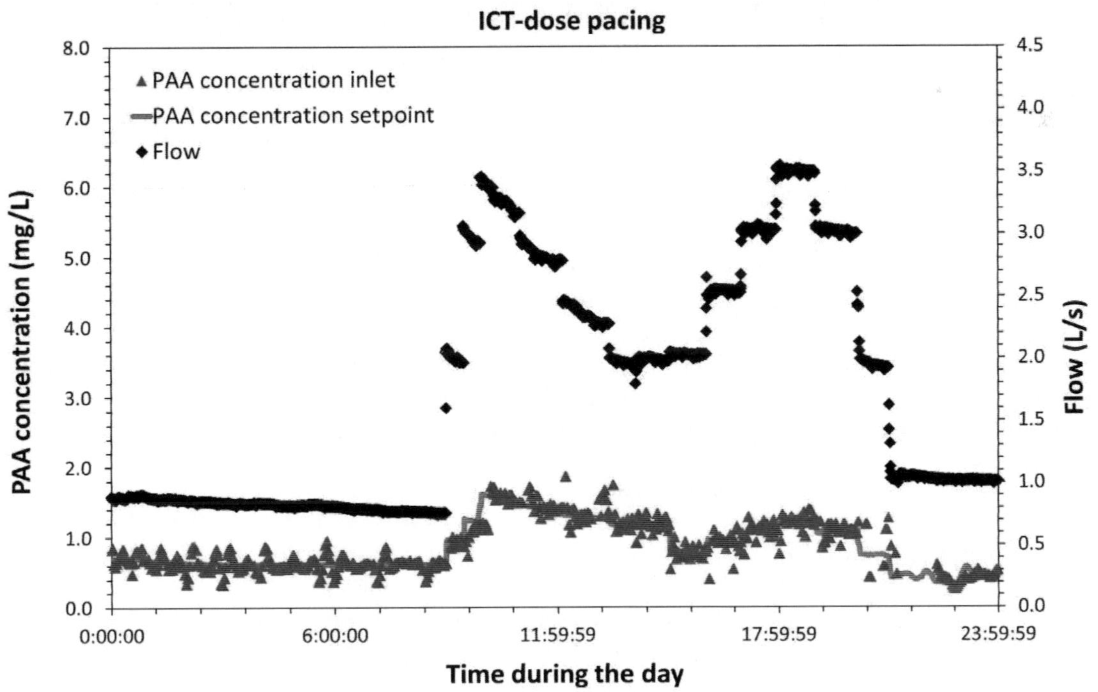

Figure 5. Inlet and setpoint concentration of PAA for ICT-dose pacing strategy in pilot test.

Importantly, data analysis show that flow-pacing operation required twice the amount of PAA than ICT-dose pacing, in order to maintain a minimum ICT dose of 16 mg·min/L and ensure consistently meeting disinfection limits. Figure 6 shows the kg PAA required to treat 1000 m^3 of secondary effluent wastewater, for three different days of operation.

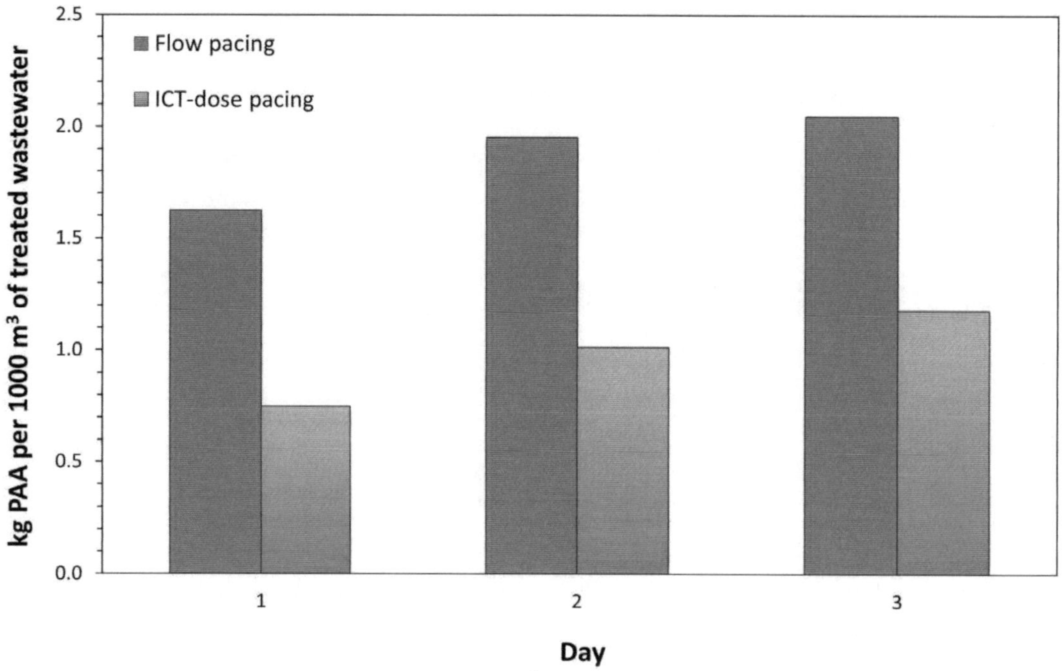

Figure 6. Amount of PAA required per 1000 m^3 of treated secondary effluent wastewater.

CONCLUSIONS

In this study, the innovative (patent pending) ICT-dose-based chemical disinfection control strategy has been validated under realistic conditions. Results from pilot-scale experiments show clearly the benefit of the novel ICT-dose pacing over the conventional flow-pacing control. The ICT-dose pacing control strategy that has been demonstrated and evaluated herein may be applicable to other chemical disinfection/oxidation processes such as chlorination and ozonation of wastewater.

REFERENCES

Bellamy WD, Finch GR, Haas CN (1998) Integrated disinfection design framework. AWWA Research Foundation and AWWA.

Kobylinski EA, Hunter GL, Shaw AR (2006) On-line control strategies for disinfection systems: Success and failure. WEFTEC 2006.

Santoro D, Crapulli F, Raisee M, Raspa G, Haas CN (2015) Nondeterministic computational fluid dynamics modeling of Escherichia coli inactivation by peracetic acid in municipal wastewater contact tanks. Environmental Science & Technology, 49, 7265−7275.

Sarathy S, Santoro D, Gaida D, Canali A, Turolla A, Antonelli M, Lawryshyn Y, Walton J, Neofotistos P (2017) This is how we do it – Dose based control for wastewater disinfection with peracetic acid. WEFTEC 2017.

Xiao S, Hu S, Zhang Y, Zhao X, Pan W (2018) Influence of sewage treatment plant effluent discharge into multipurpose river on its water quality: A quantitative health risk assessment of Cryptosporidium and Giardia. Environmental Pollution, 233, 797−805.

UV-C LED – ADVANCES IN TECHNOLOGY FOSTER IMPLEMENTATION

Oliver Lawal[1], Gary Hunter[2], and Ray Ehrhard[3]
[1]AquiSense Technologies, 4400 Olympic Blvd, Erlanger, KY, USA;
[2]Black & Veatch, Kansas City, MO, USA
[3]Washington University, St. Louis, MO

ABSTRACT

In the wastewater industry, energy and operation costs are major factors in selection of a disinfection technology. Ultraviolet Light Emitting Diode (UV-C LED) systems based on a solid-state technology, generally tends to operate with low electrical consumption and less relative space to conventional technology. A comparison of LED UV to UV systems applied to wastewater indicating that there was large (over 50 percent) cost savings for this technology when applied to wastewater systems. This comparison addressed the differences in lamp technology as well as the construction and operating costs for each system.

UV LED Technology currently is on a rapid development scale. How UV LED units were produced 2 years ago has dramatically changed increasing the output and decreasing the production costs. With the rapid changes that are occurring, the question exists: Where LED UV will be in 2 years (short- term as well as in 10 and 20 years)?

A summary will be provided of the advances of LED UV, focusing on its ability to reduce microbial transfer across several applications. There are often limits in place that prevent fully utilizing newer technology. When capital and operating costs come down, availability of products increases and new possibilities open up to apply technology. UV LED technology can be seen to follow the same path. This technology has been used for applications such as sensors and low-flow water treatment products but not much more. It is the premise that advancements in output power and efficiency of UV-C LEDs will see it grow into new markets – including city level water treatment.

INTRODUCTION

Light emitting diodes (LEDs) are semiconductor devices that produce photons based on an input current. Depending on the composition of the LED, it can produce photons at wavelengths anywhere from infrared, visible, and since the turn of the century (Figure 1) in the deep UV (UV-C) range. The introduction of UV-C or germicidal LEDs that emit 240 to 280 nm will have a profound impact on the water, air and surface disinfection markets as society continues to seek more reliable, efficient, and environmentally friendly light sources for disinfection applications.

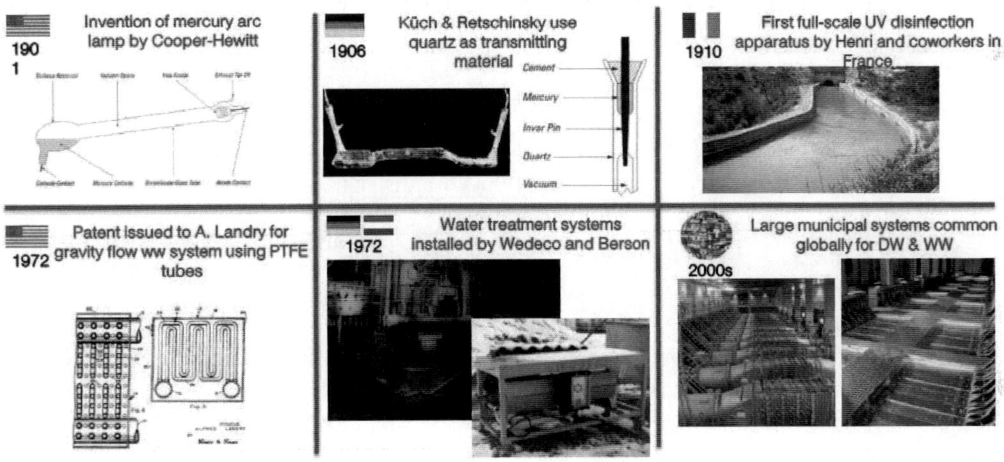

Source: AquiSense Technologies

Figure 1 – History of UV systems

Whether UV-C LEDs will eventually replace all traditional mercury-vapor based UV lamps for conventional disinfection applications is yet to be seen (Table 1) ; however, it is clear they are already enabling the creation of entirely new disinfection applications due to a number of different characteristics, including:

1. Mercury-free – Conventional UV lamps contains up to several hundred milligrams of mercury in a liquid or amalgam form. However, UV LEDs are mercury-free and contain

only tiny elements of metals held within a crystalline structure that has no ability to leach in case of breakage or disposal.

2. Compact footprint – High-power-density UV-C LEDs and advanced controls allow for a much smaller footprint compared to traditional UV sources and their electronic drivers. Unlike gas discharge lamps, UV LEDs can be switched on and off without any warm-up times. This enhances power savings and leads to prolonged lamp replacement intervals in batch process applications.
3. Unlimited cycling – Lamp life is not impacted by on/ off cycles, allowing for unlimited lamp cycling.
4. Temperature independent – LED photon emission is from a different surface as the heat emission meaning they can be engineered not to transfer heat into the water, thus reducing fouling rates and ensuring a constant UV output, regardless of the water temperature. Wavelength selection – The ability to select a specific output wavelength of an LED enables closer matching to peak absorption spectra of a target organism. All the above incremental benefits associated with UV-C LEDs should be accounted for when comparing the viability of implementing a UV-C LED based system vs. a traditional mercury lamp-based system. In some cases, these benefits create new applications for UV treatment technology and can even reduce the overall cost of implementation through the reduction of additional associated components.

MARKET DEVELOPMENT – UV-C LED DEVICES

The first UV-C LED devices were developed primarily in Japan, Korea and the US, as extensions to LED devices emitting in the blue and then UV-A and UV-B wavelength region. As seen in Figure 2, the increasing output power of commercially available UV-C LEDs

Table 1 – Comparison of LED UV to Conventional UV Technologies

Attribute	Conventional Mercury Lamp	UV-C LED	Implications
Mercury Content	20 — 200 mg	None	Safe disposal No special handling
Warm up Time	2 — 15 minutes	Instantaneous	Intermittent-flow friendly
Operating Surface Temp	Polychromatic ~ 600°C Monochromatic ~100°C	Same as process water	Zero-flow friendly Not fouling promoting
Architecture	Cylindrical tube	Point source	Versatile implementation
Wavelength	Polychromatic (200-300nm) Monochromatic (254nm)	Monochromatic "Selectable" (250—300nm)	Targeted performance
Voltage Current	110 — 240V AC 0.5 — 2.0 Amps	6 — 30V DC 0.02 Amps	Battery/solar realistic option
Replacement Interval	1 — 2 Years	6 Months — 5 Years	Application dependent

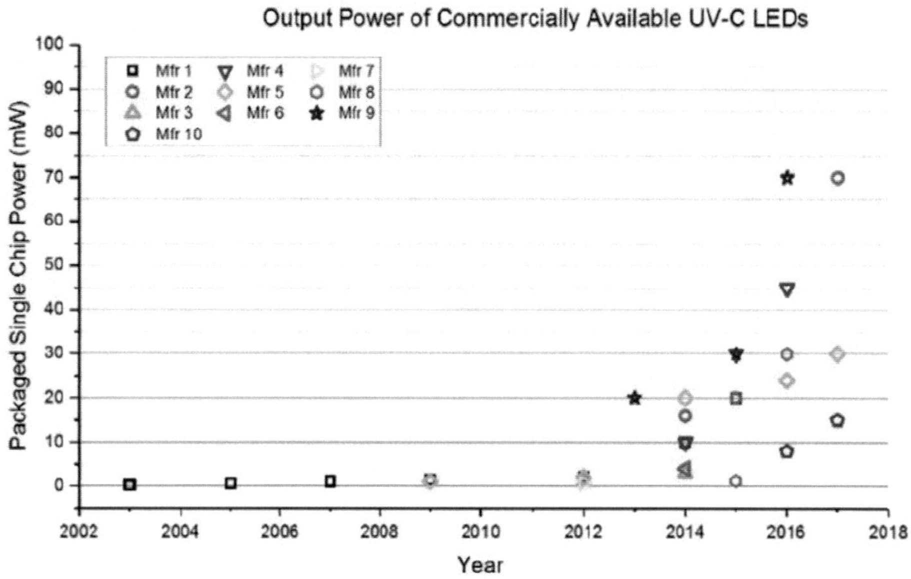

Source: AquiSense Technologies

Figure 2. Measurement data of commercially available UV-C LED devices.

has accelerated since 2012. This data is based on AquiSense internal test data of only commercially available devices and excludes "hero devices" reported in the literature and should not be considered exhaustive.

Early UV-C LED devices were primarily packaged in relatively expensive and thermally inefficient TO-type housings (Figure 3a). This mostly limited their application to analytical instrumentation as a replacement for laser, deuterium, and mercury-vapor light sources.

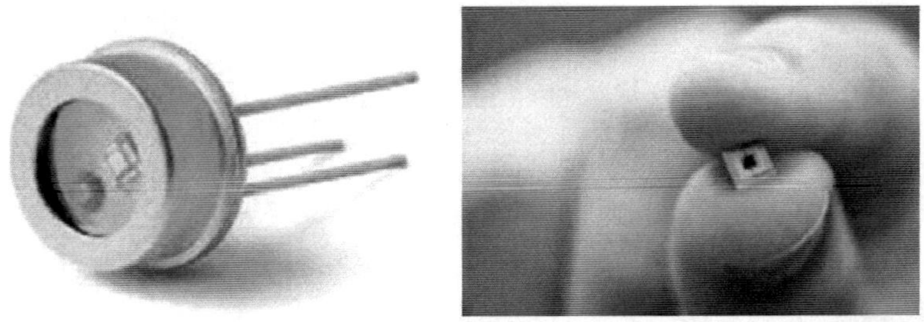

Source: AquiSense Technologies

Figures 3a and 3b. From left to right: (a) Early UV-C LED device in TO packaging. (b) Typical UV-C LED device in SMD packaging.

However, as increased investments were made in manufacturing capacities, the first Surface Mount Devices (SMDs) (see Figure 3b) became available around 2013. This enabled lower costs and higher production volumes and were more suitable for integration into disinfection products.

At least 12 corporations manufacture UV-C LED devices, although not all currently offer commercial products. Current best-in-class devices deliver UV-C optical output values in the 50 to 100 mW range, with lifetime to 70% of initial output (L70) of around 10,000 hours. It should be noted that there is a high degree of variability between the testing methods and, hence, specification reporting data of each device manufacturers. For example, UV-C optical output values are reported at a wide range of forward operating currents from 50 mA to 600 mA. This data does not always correspond directly to reported lifetime data and inherently makes datasheet like-for-like comparisons problematic. The IUVA have undertaken a project to standardize device output measurement methods, reported separately within this publication. Regardless, for a UV product designer, there should be no substitute for completing independent testing ideally using identical conditions to those within the intended product.

Figure 4 illustrates historical UV-C LED device pricing on a per mW optical output basis. This trend is related to increased manufacturing capacity, improved manufacturing yields, refinements in packaging engineering designs, and increased manufacturer competition. In

2018, the lowest market price has fallen below US $1/mW for premium, higher powered devices and below US $0.5/mW for lower powered, lower lifetime devices[1].

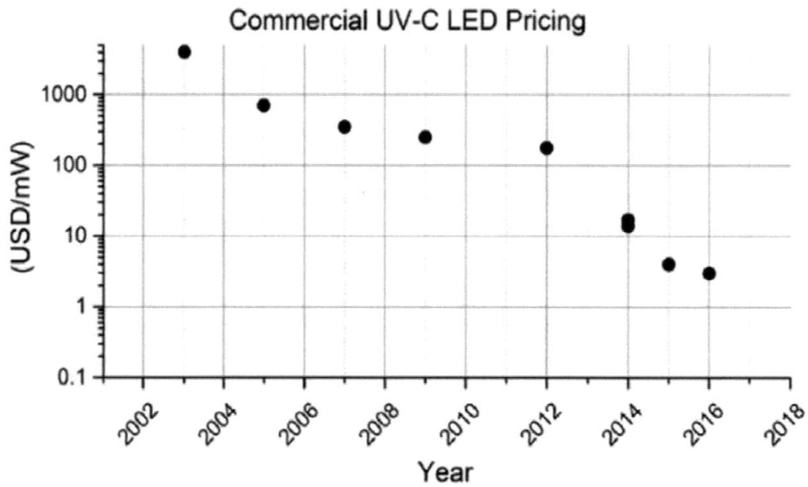

Source: AquiSense Technologies

Figure 4. Commercial UV-C LED pricing.

MARKET DEVELOPMENT – UV-C LED SYSTEMS

Early stage product development of UV-C LED-based UV disinfection systems has been challenging, due to the historical high cost of UV-C LEDs, coupled with low optical power output. Entirely new thinking and design approaches are required to fully utilize this new light source. However, despite this, some product and market development has occurred for water, surface and air disinfection products, particularly with start-up companies.

Commercially available systems are typically targeting flow rates below 20 lpm and specific applications that are challenging for gas discharge lamps. For example, the extremely small footprint potential of UV-C LED-based systems has enabled them to be incorporated directly in faucets and appliances. UV-LEDs' mercury-free status has led to these systems being incorporated into medical devices and aerospace applications where conventional UV systems were often discounted due to potential breakage and contamination risk.

Table 2 highlights example applications within key market segments where UV-C LED-based water disinfection products are currently being evaluated by OEM manufacturers globally. The technical and market drivers for this evaluation work is related to the unique characteristics of UV-C LED-based products previously discussed, but are not equal, homogeneous or even initially obvious, across each application.

Table 2 – LED UV Applications

Market segment	Example applications	UV-C LED system benefits: Top 3 important attributes valued by segment
Residential	POE Appliances Faucets	Ultra-compact footprint Plug and play (e.g. easy to retrofit) Low power draw
Commercial	Food and beverage service Water dispensers and fountains	Ultra-compact footprint Low power draw No heating of water
Health care	HAI control Dialysis Dental	Mercury-free Chemical-free Durable (e.g. vibration resistance)
Transportation	RV and boating Automotive Aviation Space	Chemical-free Durable Lightweight
Life sciences	Bio-pharma Ultrapure water	Point-of-use distribution Mercury-free Chemical-free
Defense/emergency response	Personal hydration Remote treatment	Ultra-compact footprint Lightweight Durable (e.g. vibration resistance)

Case study 1: Space exploration and habitats

Currently, 88% of the water on the International Space Station (ISS) is reused, leaving 12% lost to waste, roughly 100 ml per person per day. Monthly resupply missions are needed to bring vital resources that cannot be recycled in-orbit. This delivery service uses a 3 million horse-power rocket to catch-up and dock with the ISS as it speeds around the Earth at 17,000 mph; this is a very expensive task, costing approximately $2000 per lb of cargo.

BIOWYSE, a European Union Horizon 2020 funded project, was launched to develop and test a new type of system that uses real-time microbial monitoring (ATPmetry) and UV-C LEDs for disinfection as shown in Figure 5.

Source: AquiSenseTechnologies

Figure 5. UV-C LED disinfection module used in BIOWYSE ISS project.

The intent of the project is to develop an integrated, autonomous, chemical-free system to control and monitor biomass growth in potable water systems aboard the ISS. Two years in to the project, the consortium has developed the main components of the system, which include several advanced technologies, such as AquiSense Technologies' UV-C LED Decontamination Module. The next stage of the project will see integration followed by laboratory and field testing through to the end of 2018.

Case study 2: Biofilm control

Water cooler products for residential, commercial, and industrial applications are known to generate growth of biofilm in their static water storage components. UV-C LED lamp modules have been integrated into these products as shown in Figure 6. A review of these solutions indicates that most are using very low optical output devices in the 1 to 3 mW range. While it is not clear that the level of applied fluence is homogenous or of a high level, the commercial introduction into high-production volume products shows a demand for such technology that is likely to have a positive impact on price reduction of UV-C LED devices.

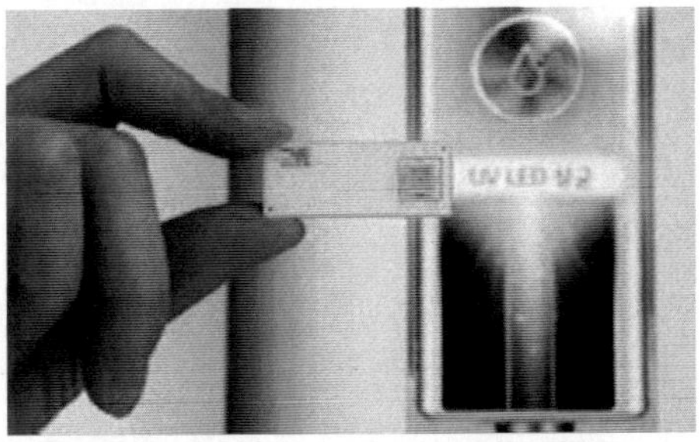

Source: LG Innotek

Figure 6.. Example of a low optical output stationary UV-C LED module.

Case study 3: Commercial steam oven

Low-temperature steam ovens are used to cook food over long periods of time, at temperature ranges of between 30°C to 300°C. This warm, moist environment has the potential to propagate the growth of harmful pathogens such as *Listeria*, *Salmonella*, *Clostridium* and *E. coli*. To mitigate this risk, Disform Food Service Technology worked together with AquiSense Technologies to integrate a disinfection module into their MyChef product as shown in Figure 7.

The key design parameters – intermittent flow, small envelope, high ambient temperature, and 4-log pathogen reduction – were used to specify the PearlAqua Micro system. Performance validation included microbiological bioassay, long-term aging, temperature and vibration tests, with volume production commencing in mid-2017.

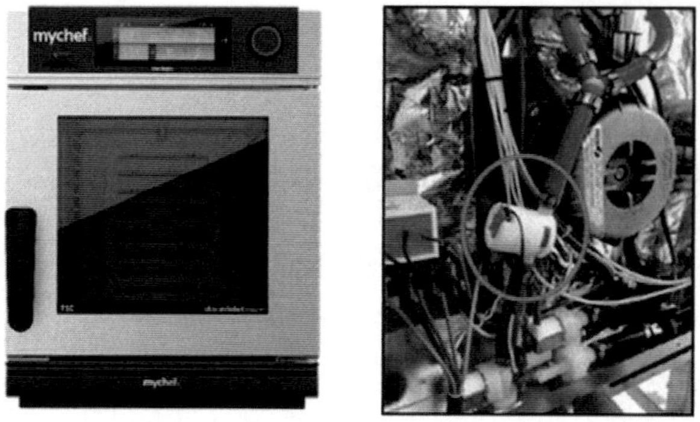

Source: Disform Food Service Technology

Figure 7. UV-C LED disinfection module used in commercial steam oven.

WASTEWATER DEVELOPMENT

AquiSense Technologies has worked with the Cincinnati Metropolitan Sewer District (CMSD) at the CMSD Mill Creek Water Treatment Plant (Figure 8) on a pilot study to investigate the use of UV disinfection technology within a wastewater treatment setting. The goals of this project were to assess the response of pathogens from this system to 254nm and 285nm UV-LED light, estimate the cost and footprint of a full-scale system, and to gather data on the effect of fouling and water temperature on the UV-LED water disinfection system.

Figure 8. Mill Creek Wastewater Treatment Plant

AquiSense provided two of its PearlAqua UV-LED water disinfection devices for this pilot. Effluent from the Mill Creek plant's secondary clarifiers was diverted to a testing area prior to entering the chlorination treatment stage as shown in Figure 8.

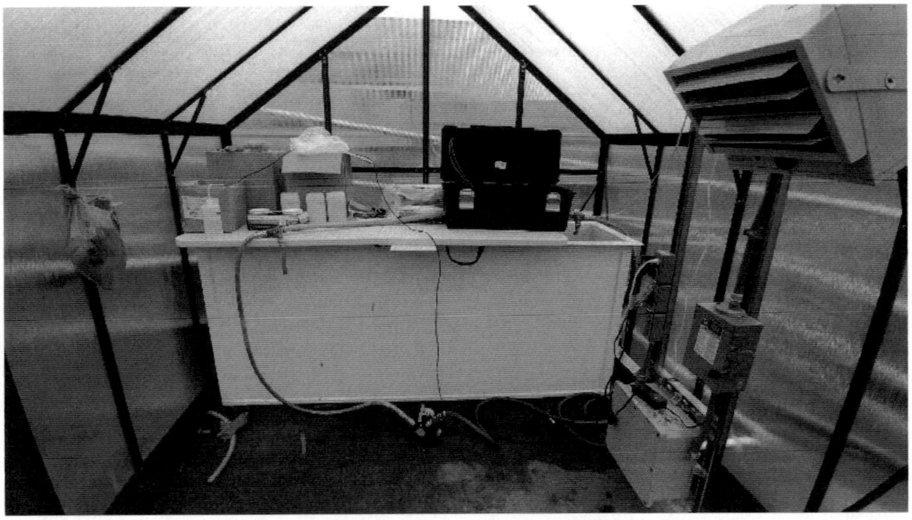

Figure 9 – UV Pilot Plant

Approximately 5 gpm was run through the two PearlAqua units in series with sampling points before and after the UV-LED systems as shown in Figure 9. To assess the response of the system's pathogens to UV light, AquiSense also supplied a PearlBeam bench-scale UV-LED collimating beam device. With the PearlBeam, a UV Dose response curve was generated in a laboratory setting as shown in Figure 10. It was found that the Mill Creek system has a response similar to that of E. Coli as published in widely available literature (Beck et al. 2016) (Figure 11).

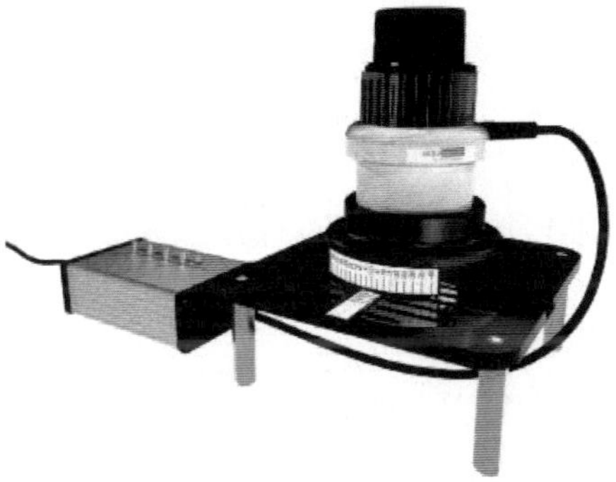

Figure 10. Collimated Beam Device

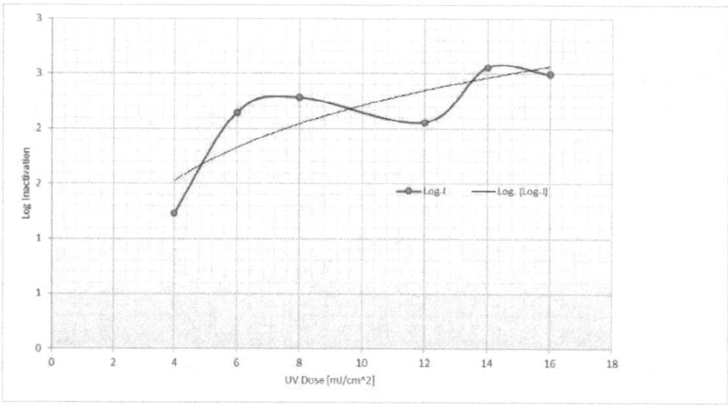

Figure 11. Mill Creek WWTP Collimated Beam Results

The disinfection performance of the PearlAqua systems was assessed with microbiological testing of samples for E.Coli and Fecal Coliforms. In all, it was found that approximately 0.5-1.0 log reduction was achieved. It is strongly believed that the reactor size and UV-Transmittance were major factors in system performance. In future studies, the reactor design will be optimized to incorporate a smaller path length for light travel. From AquiSense experience, it is believed that this will help compensate for the wastewater's low UV-Transmittance average value of 65%.

UV-C LED DEVICES AND SYSTEMS: FUTURE STATE

The future of UV-C LED devices and systems is a hotly debated topic amongst the UV community. The current focus and discussion is on the relatively low efficiencies and optical outputs of the individual UV-C LED devices reported in the literature, compared to the largest available mercury-vapor UV lamps. However, it is more helpful to consider the product performance of commercial UV-C LED-based products in the sub-20 lpm range, to similar capacity conventional mercury-vapor lamp products. Already, market experience has demonstrated that users are demanding disinfection solutions with the unique benefits offered by UV-C LED sources and are willing to integrate these new solutions. In some cases, the cost is lower, and in other cases, users are prepared to pay a premium to do so. In most cases, however, UV-C LED solutions are creating new applications for UV technology, rather than displacing existing gas discharge lamps. But, how do scientists assess the opportunity in the next two decades? How will this technology progress?

A good starting point is an examination of Haitz's Law, developed by Dr. Roland Haitz, who has been characterized as the "Godfather of LEDs." It states that every decade, the optical output of an LED will increase by a factor of 20, while the price per optical output will fall by a factor of 10.

A review of the progression of Output Powers of Commercially Available UV-C LEDs in Figure 12 and Commercial UV-C LED Pricing in Figure 12 suggests that UV-C LED devices development is following this predictable path. Based on this trajectory, the implication for 2026 is 1.4 Watt per LED at 0.1 $/mW. This position can be further validated by reviewing research literature of developmental UV-C LEDs, see figure 13.

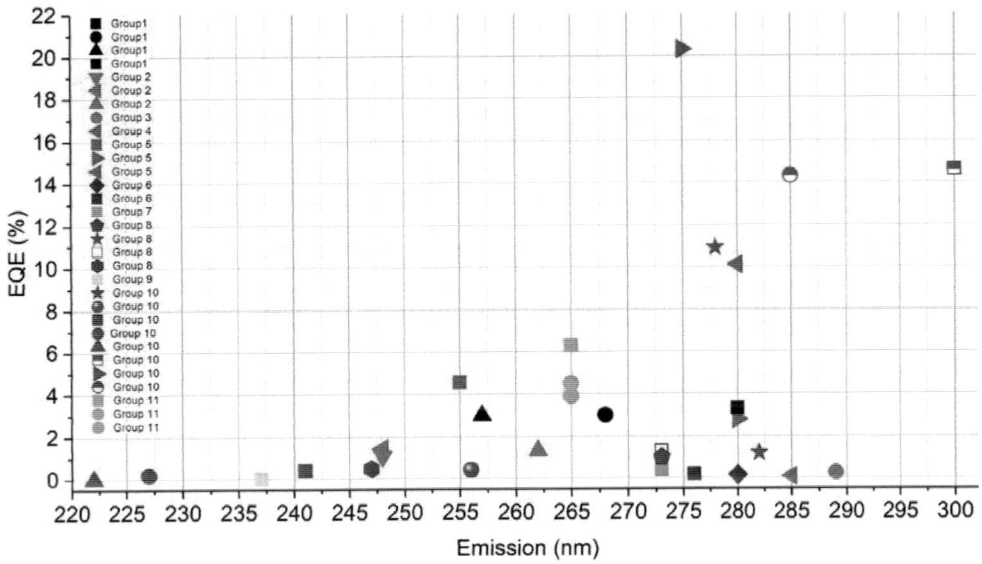

Source: AquiSense Technologies

Figure 12. Reported external quantum efficiencies for UV-C LEDs.

Internal Quantum Efficiencies (IQE) values of UV-C LEDs have long been shown to be comparable to visible LED devices, above 75%. Thus, research focus is centered on alternative light extraction techniques to increase the External Quantum Efficiency (EQE) yields of operational devices. The future for UV-C LED systems lies at the intersection of multiple factors. Optical output levels of commercial devices have now been reported in the 100 mW range, and, inevitably, LED device efficiency will increase from current levels. Costs will continue to decrease as supply increases. In parallel, system engineering for UV-C LED-based systems is evolving as system designers are beginning to realize that entirely new thinking and design approaches are required to fully utilize this new light source. The result is that larger systems will be developed as shown in Figure 13.

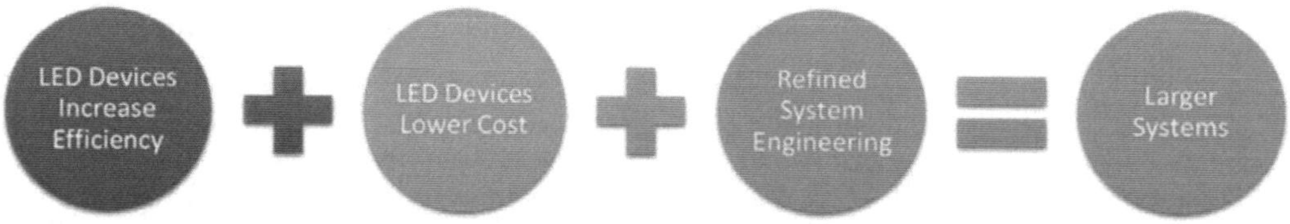

Figure 13. UV-C LEDs' path to commercially available larger systems

WILL THE MINAMATA CONVENTION DRIVE UV-C LED GROWTH?

The Minamata Convention on Mercury was initiated by the United Nations Environmental Programme (UNEP) to protect human health and the environment from anthropogenic emissions and releases of mercury. The UNEP has set the goal for mercury to be phased out of production by the year 2020. In 2013 they asked organizations and governments to discourage the use of mercury starting immediately. As of August 2017, the convention has been ratified and has officially entered into force.

Although the Minamata Convention does not specifically prohibit the manufacture and sale of UV mercury-vapor lamps, it could provide a generally positive influence for the more widespread adoption of alternative technologies. For example, if UV-C LED systems can achieve disinfection performance, capital cost and operating cost values close to those of conventional technologies, then the following user groups may respond as follows:

1. Corporations that use UV systems in products or manufacturing processes (e.g. white goods, beverage, microelectronics, life sciences, etc) may implement best practices by selecting an LED option by influence of the Minamata Convention, not by enforcement. This may especially factor into decisions on new product development where 20+ year product lifecycles are considered.
2. Original Equipment Manufacturers that currently employ gas-discharge lamps may see a conflict with the use of mercury-based products to their own environmental policies and, based on Minamata regulations, might start a transition to develop new products using mercury-free light sources again having consideration to long product lifecycles.
3. Municipalities may also follow suit in adopting UV-C LEDs, although it is more likely that they will require more time to adopt the new technology. Once a generalized mercury regulation takes place, states and municipalities will begin to critically review their processes for any potential violations and improvements.
4. There will always be a percentage of people that will look for "eco-friendly" product options. As the lamps without banning the sale or manufacturing of these lamps. Because of this, end users may choose LEDs over mercury lamps, even if they are more expensive, just as can be seen in the visible light market with CFL and LEDs.
5. Regulators will likely have a slow transition from mercury to UV-C LED lamps. Regulators will always look for viable alternatives to mercury, as it is their mandate to steer technology to the most holistically sound solution.

CONCLUSION

The introduction of UV-C LEDs will have a profound impact on the water, air, and surface disinfection markets. They may not completely replace traditional UV lamps for disinfection applications but have already created new disinfection applications as a result of the numerous incremental advantages over traditional mercury lamps. While the optical output and efficiencies of UV-C LEDs remains relatively low, it is critically important to examine the full value proposition of a UV-C LED disinfection solution by fully monetizing the unique benefits of such a solution. As the price per mW lowers, more applications and markets will become available. While the inevitable replacement of mercury vapor lamps eventually coming, UV disinfection technology has an exciting future, as it is working its way into markets and applications not previously possible.

REFERENCES

LEDinside. "Photonics West 2018 LIVE-UV LED Curing and Water Sterilization Markets start to Take Off." LEDinside, 02/02/2018, https://www.ledinside.com/showreport/2018/2/photonics_west_2018_live_uv_led_curing_and_water_sterilization_markets_start_to_take_off

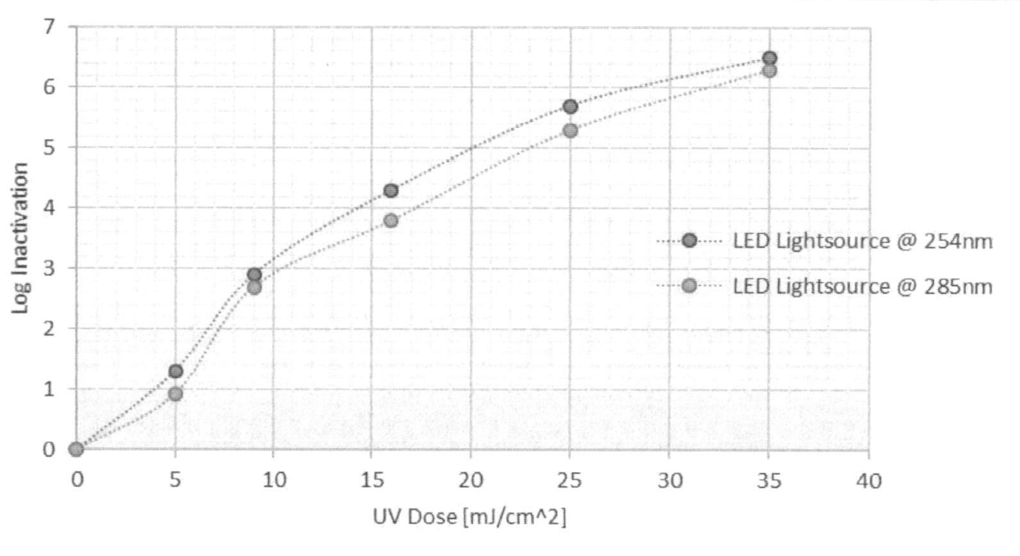

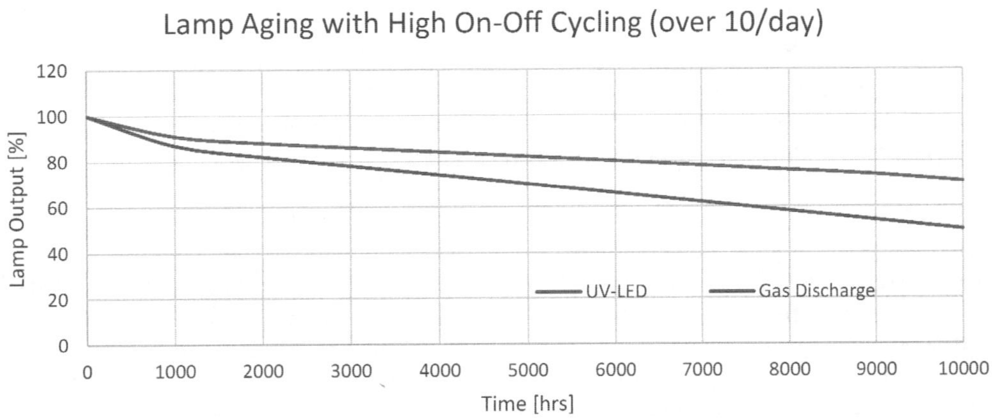

Figure 1 – History of UV Technology

Table 1 – Comparison of LED UV to Conventional UV system

Re-Birth of Ozone Based IPR Treatment Trains and the Resulting Integrated Controls for an Efficient Process

Steve Green – Xylem, Kevin Flis – Xylem, Charles Bott – Hampton Roads Sanitation District, Chris Wilson – Hampton Roads Sanitation District, Germano Salazar-Benites – Hampton Roads Sanitation District, Lauren Zuravnsky – Hampton Roads Sanitation District, Matt Poe – Hampton Roads Sanitation District

ABSTRACT

Resurfacing from the WateReuse Research Foundation (WRRF) 11-02 study that evaluated alternatives to the traditional "Full Advanced Treatment" (that typically refers to MF, RO, and UV AOP) based Potable Reuse treatment trains; ozone based alternatives have begun to emerge. This is due to the flexibility that an ozone based Potable Reuse treatment train can offer in up to 50% lower capital and operating costs and removing the concentrated brine stream.

A recent example of the flexibility that an ozone based Potable Reuse treatment train can offer is found at the Hampton Roads Sanitation District (HRSD) Sustainable Water Initiative for Tomorrow (SWIFT) Demonstration facility. The Potable Reuse treatment train arranged at the SWIFT Demonstration facility includes coagulation, flocculation, sedimentation, ozonation, followed by biologically activated filtration (BAF), granular activated carbon adsorption, and finally UV Disinfection.

The particularly unique advantage for the SWIFT Demonstration Facility is that the ozone and BAF systems are paired with different control mechanisms that allows for optimization in real time. By tracking the influent and effluent quality from the ozone contactor, the plant distributed control system (DCS) provides a feed-forward/feedback control system that adjusts the ozone dose by sending signal to the Main Ozone Control Panel. Given that the maximum design ozone dose for the SWIFT Demonstration Facility is 20 mg/L, the ability to trim the ozone dose in real time has the capability to generate real cost savings for HRSD.

Furthermore, this demonstration facility can be used as a platform to further refine the design criteria full-scale SWIFT implementation.

KEYWORDS

Indirect Potable Reuse (IPR), Sustainable Water Initiative for Tomorrow (SWIFT), Purified Water, Distributed Control System (DCS), Ozone, Biologically Activated Filtration (BAF), Total Organic Carbon (TOC), Nitrite, Ultraviolet Absorbance (UVA), ultraviolet transmittance (UVT), Virus, Cryptosporidium, Contaminants of Emerging Concern (CEC)

INTRODUCTION

Trussell Technologies, Inc. led an effort on the WateReuse Research Foundation (WRRF) 11-02 research study titled *Equivalency of Advanced Treatment Trains for Potable Reuse*. During this study six (6) different treatment trains were evaluated (see Table 1) to meet both chemical and microbiological criteria that would satisfy the needs of a potable treatment train to meet the public health standards agreed to by a National Water Research Institute (NWRI) Independent Advisory Panel (IAP).

Table 1: List of Treatment Trains Tested for WRRF 11-02. This Table is an excerpt from Table ES.3 from the Final Report of *Equivalency of Advanced Treatment Trains for Potable Reuse*

No.	Unit Processes
1	CAS[1]–UF–O_3–BAC–UV
2	CAS[1]–MF–RO–UV/H_2O_2–Cl_2
3a	CAS[1]–O_3–BAC–UF–UV
3b	CAS[1]–UF[2]–O_3–BAC–UV
4	CAS[1]–O_3–MF–RO–UV/H_2O_2
5	CAS[1]–MF–O_3–BAC–UV
6	CAS[1]–O_3–BAC–MF–UV

Ozone based treatment trains were selected in unit processes 1, 3a, 3b, 4, 5, and 6 due to lower operating and capital costs, better water recovery, and the absence of concentrated brine stream that would need to be discharged. These treatment trains were found to meet the standards set forth by the NWRI IAP for both chemical and microbiological criteria during testing performed in 2013 and 2014.

This was an important step as to that point potable reuse facilities traditionally (not always) would employ a full advanced treatment (FAT) train consisting of microfiltration (MF), reverse osmosis (RO), and UV/H_2O_2 advanced oxidation process (AOP). This FAT was very high in operational and capital costs and generated a brine stream from the RO process that would need to be discharged generally to an ocean outfall. These limitations made it challenging for other utilities to pursue this FAT approach. Fortunately, with the findings from the WRRF 11-02 study and approaches from other utilities there were now avenues to pursue alternative potable reuse trains.

One such utility that was interested in evaluating the variety of treatment trains now available was Hampton Roads Sanitation District (HRSD). A forward thinking utility, HRSD was seeking a solution to solve several challenges. They were seeing depletion of their groundwater resources, which in turn was increasing the risk of saltwater intrusion into the groundwater. Being located in coastal Eastern Virginia they are very sensitive to the conservation and restoration of the Chesapeake Bay. Finally, the Hampton Roads area was identified as the second most susceptible geography to sea level rise that was being compounded by land subsidence. Managed aquifer recharge was found to be the most effective way to solve these challenges and it would be best to accomplish this through an indirect potable reuse treatment train. Recharging the aquifer would create a barrier to prevent saltwater intrusion, limit discharges to the

Chesapeake Bay contributing to its restoration, and stop land subsidence and potentially even begin increasing the land elevations to where they once stood.

To accomplish this objective, HRSD created the Sustainable Water Initiative for Tomorrow (SWIFT) program. This began with a pilot study conducted at the York River wastewater treatment plant (WWTP). Two (2) side-by-side potable reuse treatment trains were tested. The first was coagulation-flocculation-sedimentnation-O_3-BAF-GAC-UV. The second was similar to the traditional FAT consisting of UF-RO-UVAOP. Ultimately both treatment trains were able to demonstrate contaminants of emerging concern (CEC) reduction, <4 mg/L Total Organic Carbon (TOC), and met all primary Maximum Contaminant Limits (MCLs). Due to capital cost savings potential, operating cost savings potential, and the elimination of a brine stream the coagulation-flocculation-sedimentation-O_3-BAF-GAC-UV treatment train was selected for the research center for the SWIFT program. The research center is a fully operational 1 MGD demonstration facility that provides purified water for the aquifer recharge.

The SWIFT research center is further investigating performance on 1:15 scale (rather than on a 1:2,000 scale that was performed on the pilot study). This research center will also investigate operational controls for various different treatment objectives; specifically on the Ozone-Biofiltration portion of the process.

METHODOLOGY

As mentioned prior the treatment train that was designed for the HRSD SWIFT research center demonstration facility was coagulation-flocculation-sedimentation-O_3-BAF-GAC-UV (see Figure 1 below). This preliminary design was developed by Jacobs (formerly CH2M). In an effort to condense the schedule and transition work from the pilot study to the demonstration facility, a Design-Build procurement method was selected. The Design-Build team that partnered with HRSD and Jacobs was Crowder Construction and Hazen & Sawyer.

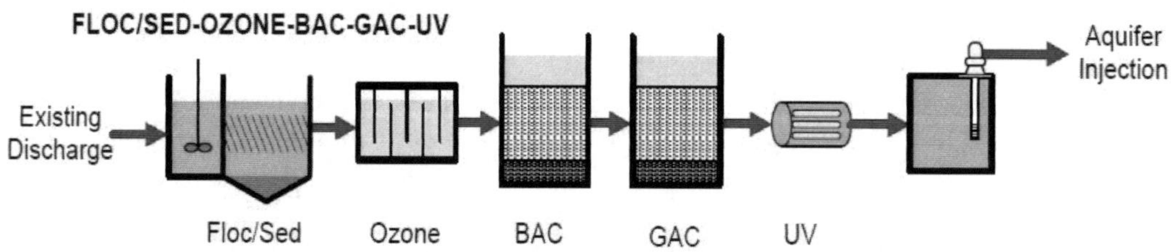

Figure 1: HRSD SWIFT Research Center Process Flow Diagram. This figure is an excerpt from *East is East and West is West – Will the Twain Meet?* presentation from Troy Walker of Hazen & Sawyer at the 2018 CA Annual WateReuse Conference and is courtesy of Jacobs (formerly CH2M) and HRSD.

During the preliminary design and Design-Build process many items regarding the implementation of the O_3-BAF process were identified. First and foremost the Ozone system would be designed for a maximum dose of 20 mg/L at 1.22 MGD and a transfer efficiency of

95%. Ultimately, the maximum Ozone production point selected for the SWIFT research center was for 200 lbs/day per Ozone Generator.

The design for the Ozone process includes a duty/standby SMOevo 610 Ozone Generator and Power Supply Unit pair as manufactured by Xylem, a duty Ozone Destruct skid, and a duty Ozone sidestream injection skid. The Ozone process utilizes a liquid oxygen (LOX) feed gas source to generate the ozone. The Ozone Generators utilize cooling water from a closed building loop cooling water source.

The BAF design is such that the unit is split into four (4) individual filter cells. The total flow rate of 1.22 MGD resulted in a surface loading rate of 4 gpm/ft^2 with one filter cell out of service. The media used for filtration was 1 foot depth of gravel, 1 foot depth of sand, and 5 feet of depth for granular activated carbon (GAC). The media was supported by the XA lateral underdrain block as manufactured by Xylem. At a nominal water depth in the BAF of 15.5 feet, the overall empty bed contact time (EBCT) was about 9 minutes.

The focus of the SWIFT Research Center was to arrange the O_3-BAF process with multiple different operating methods.

RESULTS

The focus of the study was to network the individual unit processes together to operate as a complementary process that works as a whole. There were multiple different modes of control for the process that were evaluated and listed below in Table 2. All control modes for the entire process are controlled and operated from the distributed control system (DCS) for the entire facility.

Table 2: Different Control Methods Evaluated at the HRSD SWIFT Research Center for O_3-BAF Control

No.	Control Mode
1	Applied Ozone Dose
2	Operator Entered Ozone Residual
3	UV-254 Reduction Control
4	Feed Forward Applied Ozone Dose with Ozone Residual Trim
5	Ozone Residual Setpoint with Nitrite Feed Forward Trim

The DCS was arranged to operate the preceding control modes in the following fashion.

1. Applied Ozone Dose: This is a feed forward control method that allows the operator to select the Applied Ozone Dose at the DCS. This Applied Ozone Dose along with the process flow rate will calculate the required Ozone Output in pounds per day (ppd) to the Master Ozone Control Panel (MOCP).
2. Operator Entered Ozone Residual: The operator shall select a specific analyzer in the Ozone Contactor, select an Ozone residual setpoint (mg/L). Along with these inputs, the DCS will also input Ozone Concentration to the MOCP for control of the Ozone Output to match these setpoints.

3. UV-254 Reduction Control: The operator shall select UV-254 reduction setpoint (%). This setpoint is measured in real-time with a UV-254 measurement upstream and downstream of the Ozone Contactor. The DCS shall determine if the UV-254 reduction setpoint is satisfied and as required modify the ozone output signal to the MOCP as well as provide the MOCP the Ozone Concentration.
4. Feed Forward Applied Ozone Dose with Residual Trim: With this methodology the operator is selecting a O_3:TOC ratio, a target ozone residual at a selected analyzer location, a TOC:UV-254 ratio, and a NO_2 concentration (mg/L). An algorithm is used to determine the Ozone Dose based on the individual inputs listed above. This Ozone Dose and the Process Flow Rate are used to calculate the Ozone Output (ppd) at the DCS. This Ozone Output is then communicated to the MOCP for final Ozone Dose control.
5. Ozone Residual Setpoint, an Ozone Contact Time from NO_2 sensor to Ozone Flash Reactor, an Ozone Contact Time from Ozone Flash Reactor to Ozone Residual Analyzer. The DCS will calculate the delta in NO_2 from the time the Ozone is dosed at the Flash Reactor to when the Ozone is measured at the Residual Ozone Analyzer. This change is then used to calculate the NO_2 demand on the Ozone Feed and can trim the outputted Ozone Residual Setpoint to the MOCP.

DISCUSSION

Having several different operating modes generates a lot of operating and more importantly, flexibility. What is especially unique about the HRSD SWIFT Research Center is their ability to operate in these different modes and collect different data sets. From these wide reaching data sets, the information can then be used by other utilities and industry professionals as to which operating mode(s) will best suit their projects and their individual project goals.

In particular, each data set will be evaluated for TOC reduction performance, UV-254 reduction performance, and pathogen reduction. Specifically, on pathogen reduction the Virus Log Reduction Value (LRV) will be measured. Additionally, if possible, the performance of CEC reduction across the O_3-BAF process will be measured. These performance metrics will be tracked against Ozone Dose, O_3:TOC ratio, and Empty Bed Contact Time (EBCT) to determine which control method shows strongest in each department. From there this data can then be used as a roadmap for other utilities and industry professionals (as well as HRSD on their forthcoming IPR projects) to determine the appropriate operational strategy for their specific project.

CONCLUSIONS

The first conclusion is that operating the O_3-BAF system as a single unit process streamlines the control process, offers more predictable effluent quality, and generates opportunities to save power and operating costs. This combined solution is the path forward for advanced water treatment facilities implementing potable reuse. The alternative of operating the processes independently inherently disrupts the streamlined process, eliminates opportunities for optimization, and hence requires added operational redundancy which in turn increases operating costs.

The second conclusion is that when possible adding multiple control methodologies adds flexibility for a facility to match their performance targets. This is especially important when a facility has multiple performance targets or potential future added performance targets, so that they may toggle through different operating modes as these performance targets change.

REFERENCES

Stanford, Ben; Antolovich, Abbey; 2017; Advanced Oxidation Process, Unique Hydrogeology Allow for Innovative Reuse Strategy; Journal AWWA: February 2017; Pages 59 - 63

Trusell, Rhodes R; Salveson, Andrew; Snyder, Shane; Trusell, R. Shane; Gerrity, Dan; 2016; Final Report: Equivalency of Advanced Treatment Trains for Potable Reuse; Water Environment & Reuse Foundation and IWA Publishing

Walker, Troy; 2018; East is East and West is West – Will the Twain Meet?; 2018 CA Annual WateReuse Conference

Wilson, Chris; 2018; Startup and Commissioning of a 1 MGD Advanced Treatment Demonstration Facility for Managed Aquifer Recharge; 2018 Water Research Foundation Reuse Conference

PERACETIC ACID DISINFECTION

Author: Chris Jepson, PE
Affiliation: Van Cleef Engineering

The Berkeley Heights, NJ Wastewater Treatment Plant (3.1 MGD) was faced with more stringent effluent limits for chlorine produced oxidants (CPOs) and the potential to add other disinfection byproducts at stringent levels to the upcoming NJPDES renewal for the facility. Berkeley Heights had used sodium hypochlorite for disinfection followed by bisulfite for dechlorination. Ultraviolet (UV) disinfection is currently the accepted method of wastewater disinfection to avoid issues with CPOs and disinfection byproducts caused by disinfection with chlorine. Due to the high cost of reconfiguring the existing chlorine disinfection system and the UV purchase, Berkeley Heights decided to investigate other means of disinfection. Peracetic Acid (PAA) and its possible use as the sole wastewater disinfectant, as opposed to a supplemental disinfectant, was thought to have potential. Peracetic Acid is a combination of two simple ingredients – vinegar and hydrogen peroxide. After more research and investigation, several sources of good information on the subject were obtained, which made us believe PAA could be used as a disinfectant in municipal wastewater applications. Berkeley Heights conducted a trial study in April 2016 which showed very promising results. This was discussed with NJDEP permitting and several meetings were initiated, with the location of sampling points and study protocol ascertained. This was made part of the formal Quality Assurance Project Plan (QAPP) and submitted to NJDEP for review and approval. The study protocol was approved, and a formal study took place from January to May 2017. PAA was found to disinfect the wastewater effluent to provide compliant levels of both fecal coliform and E. coli. Flow pacing the addition of PAA was required to keep disinfection at an acceptable level. Several dosing locations were tried during the trial. It was discovered that PAA cannot be added at the head of the sand filtration system, due to issues with PAA reacting with the carbonates coating the sand in the tertiary sand filtration system from chemical addition for the nitrification system utilized in the overall treatment processes at the facility. Dosing at the end of the sand filters produced satisfactory results. These results are continuing and ongoing, therefore, this location is our permanent dosing location. NJDEP has accepted the results and are making NJPDES permit modifications to enact that. There will be no more CPOs on the new permit. This is the first time PAA has been successfully used as the sole wastewater disinfectant in New Jersey. Some sources have mentioned that industrial wastewaters can have an impact on PAA effectiveness. Berkeley Heights has a minor industrial contribution, and has not seen any industrial interference.

There will be a PAA residual requirement on the new permit. PAA residual is taken the same way as chlorine residuals (DPD pillows). The only difference is that there is no waiting period before reading the color and recording the results. PAA residual readings are around 0.4 mg/l. The dosing rate is about 0.8 mg/l. Berkeley Heights is using about 9 gallons per day of PAA. PAA is provided by several suppliers and we utilize the lowest price. PAA comes in 275 gallon totes for supply. We are working with suppliers to receive larger bulk shipments to lower the cost. The delivery system is relatively the same as chlorine delivery systems – a small peristaltic pump and tubing (tubing needs to be replaced every year, or stainless steel can be used as tubing). About 20 minutes of detention time is required for effective results. An existing chlorine contact chamber works very well. The operations and maintenance budgets are kept in check versus UV disinfection. The electrical costs and maintenance on UV systems – and the fact that UV systems don't produce a completely disinfected effluent – makes PAA disinfection more than a viable alternative to UV.

Berkeley Heights has evaluated the cost effectiveness of PAA versus chlorine (and dechlorination) and has shown that PAA is more cost effective. It also readily decomposes in the environment and exhibits very little toxicity. The NJPDES required effluent toxicity reports have come back with no toxicity exhibited. It is FDA approved and has been used in the United States food industry for decades. This could be an important and cost effective alternative to UV or chlorine disinfection, and it creates a beneficial impact on operation and maintenance requirements and cost.

Any wastewater facility facing stringent limits on CPOs in their NJPDES permit should consider disinfection trials with PAA. If the results are encouraging, the next step is to contact the local regulatory agency and start the process of an alternative disinfection study, and the requirements for conducting it. Berkeley Heights has found PAA to be an effective wastewater disinfectant that exhibits very little toxicity and that is cost effective versus other disinfection methods.

Chris Jepson, P.E.
Van Cleef Engineering
CJepson@vcea.org
(O): 609-689-1100

Are TOC and COD Limits Appropriate for Regulating Potable Reuse? Analyzing Organics through the Domestic Water Cycle: Results from WE&RF Project-15-04

Larry Schimmoller[1*], Jim Lozier[1,] James Rosenblum[1]

[1]CH2MHill, 9191 S. Jamaica St., Englewood, CO 80112
*Email: Larry.Schimmoller@jacobs.com

ABSTRACT

Secondary effluent contains a vast number of organic chemicals used in everyday life, such as pharmaceuticals and personal care products, that are not regulated and do not have extensive monitoring data. Complete characterization of these myriad organic compounds, and their associated health relevance, is not practical. Consequently, various regulatory approaches incorporating surrogate bulk carbon measurements, such as total organic carbon (TOC) and chemical oxygen demand (COD), have been applied to potable reuse projects in an attempt to control potentially hazardous organic chemicals. These regulatory approaches range from a TOC limit of 0.5 mg/L in California for direct groundwater injection to COD limits of 10 mg/L and 18 mg/L in Virginia and Georgia, respectively for surface water augmentation. In some cases, these limits are significantly lower than the conventional water supplies in the same region. In addition, regulating to an extremely low TOC or COD limit for potable reuse can produce significant unintended consequences, without necessarily improving the protection of public health. For example, recent research showed that mandating a TOC limit of 0.5 mg/L, which has driven the adoption of reverse osmosis (RO) technology, increases the net present value costs for a 20-mgd potable reuse plant at an inland location by more than 200% because of the high cost of concentrate disposal. Consequently, as potable reuse continues to grow, practical, yet safe, guidance regarding organics removal is desperately needed by those considering its application. The Water Research Foundation Project 15-04 (Reuse-15-04) was funded to begin development of this guidance.

KEYWORDS: direct potable reuse, organics limit, total organic carbon (TOC)

REGULATING ORGANICS IN POTABLE REUSE PROJECTS

The benefits of potable reuse are broad, however there are still concerns over the reuse of wastewaters that may contain constituents that possess human and environmental health concerns; such as harmful biological agents or anthropogenic chemicals (Scruggs et al., 2017). Biological agents include pathogens (e.g. virus, protozoan, bacteria) to anti-microbial resistant genes, which can pose significant direct (i.e. salmonellosis) health hazards if not removed in a reuse treatment train regardless of treatment technique. Chemical constituents may include pharmaceuticals, personal care products, disinfection by-products to heavy metals. Biological organisms and their target removals (i.e. log removal requirements) follow a similar consensus among reuse plants to date. However, the level of agreement is vastly different for organic chemical limits and the specific constituents of concern. This is demonstrated by the Environmental Protection Agency (EPA) Water Reuse Guidance document identifying a TOC target of 2 mg/L for indirect potable reuse, California having set a TOC limit of 0.5 mg/L for subsurface direct injection indirect potable reuse projects, Texas not having a TOC limit, and

Florida setting TOC and TOX limits of 3.0 mg/L and 0.2 mg/L, respectively for groundwater recharge (EPA Reuse 2012 and 2017). Table 1 summarizes organics limits for a variety of potable reuse projects throughout the United States. This illustrates the range for TOC limits and the lack of consensus towards what is considered "safe" for TOC levels in potable reuse water. This is compounded by the nature of organic matter present in water and wastewater that contain various fractions that may be harmful or benign.

Table 1 – Organics Limits for Full-Scale Potable Reuse Projects

Location/Project	Organics Limit
California – 100% groundwater injection	TOC < 0.5 mg/L
Hampton Roads Sanitation District – Sustainable Water Initiative for Tomorrow (groundwater injection)	TOC < 4 mg/L
Gwinnett County; Georgia	COD < 18 mg/L
Texas (El Paso; Big Spring)	None
Florida	TOC < 3mg/L; TOX < 0.2 mg/L
EPA Guidelines (2012)	TOC < 2mg/L; TOX < 0.2 mg/L

APPROACH

The focus of The Water Research Foundation's project Reuse-15-04 was to investigate potential organics limits for controlling direct potable reuse projects. As part of this project, six utilities from around the world that practice potable reuse participated in a comprehensive sampling program to track the fate of organics through the domestic water cycle. Numerous advanced organic analytical tests were conducted including regulated and unregulated DBPs (e.g., haloaectonitriles), excitation emission matrix (EEM) testing, size exclusion chromatography with on-line carbon detection (SEC-OCD), trace organic compound (TOrC) quantification, in vitro bioassays, and non-targeted analysis using heat maps and cluster analysis. For each utility, samples were collected for analysis at multiple locations within the domestic water cycle (see Figure 1), including the drinking water source, drinking water, wastewater effluent, and at multiple points in their advanced potable reuse treatment process (e.g., RO, ozone, GAC). Samples were analyzed by Dr. Shane Snyder at the University of Arizona and Dr. Bill Mitch at Stanford University. Analytical data from these numerous tests is being used to develop a scientifically defensible framework to properly describe and quantify the exposure to organics contaminants for potable reuse, rather than using bulk parameters such as TOC and COD that may be inadequately precise.

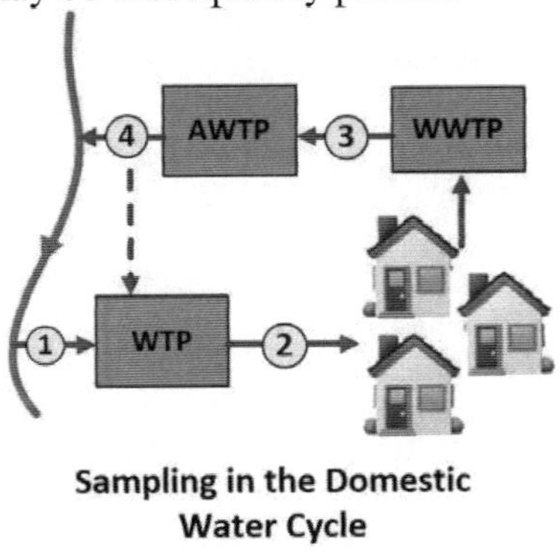

Figure 1 – Sampling Locations

RESULTS

Figures 2 shows the dissolved organic carbon (DOC) results for the various samples. The different utilities are denoted by a letter (A through F) and the sampling events are denoted by a number (1 through 4). Significant variation exists in the source water DOC concentrations between utilities, indicating large differences in watersheds. Potable reuse water DOC concentrations in some locations is lower than the associated drinking water DOC; in other locations the drinking water DOC concentration is less than the potable reuse water DOC.

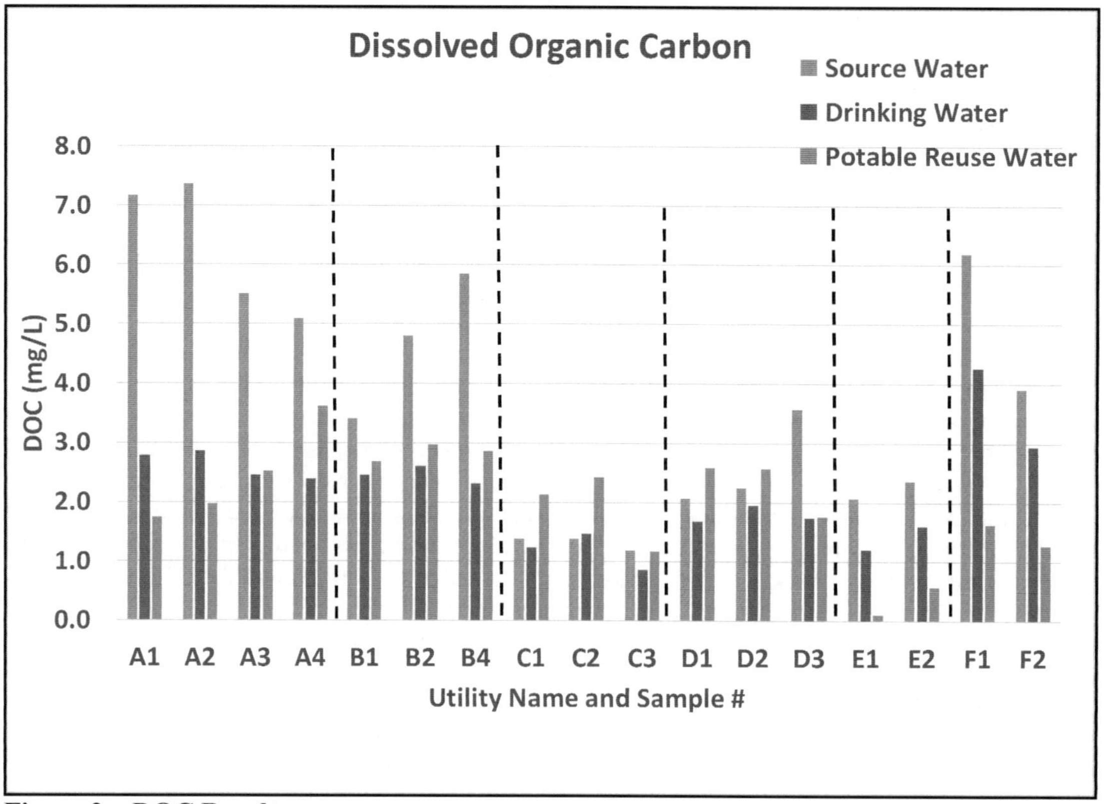

Figure 2 – DOC Results

Figure 3 shows the disinfection byproducts (DBPs) formed at the various locations under uniform formation conditions to allow for direct comparison between drinking water and potable reuse water. The results include both regulated and unregulated disinfection byproducts (see Table 2). Similar to the DOC results, in some locations the total DBPs were lower in the potable reuse water and in some locations the drinking water DBPs were lower.

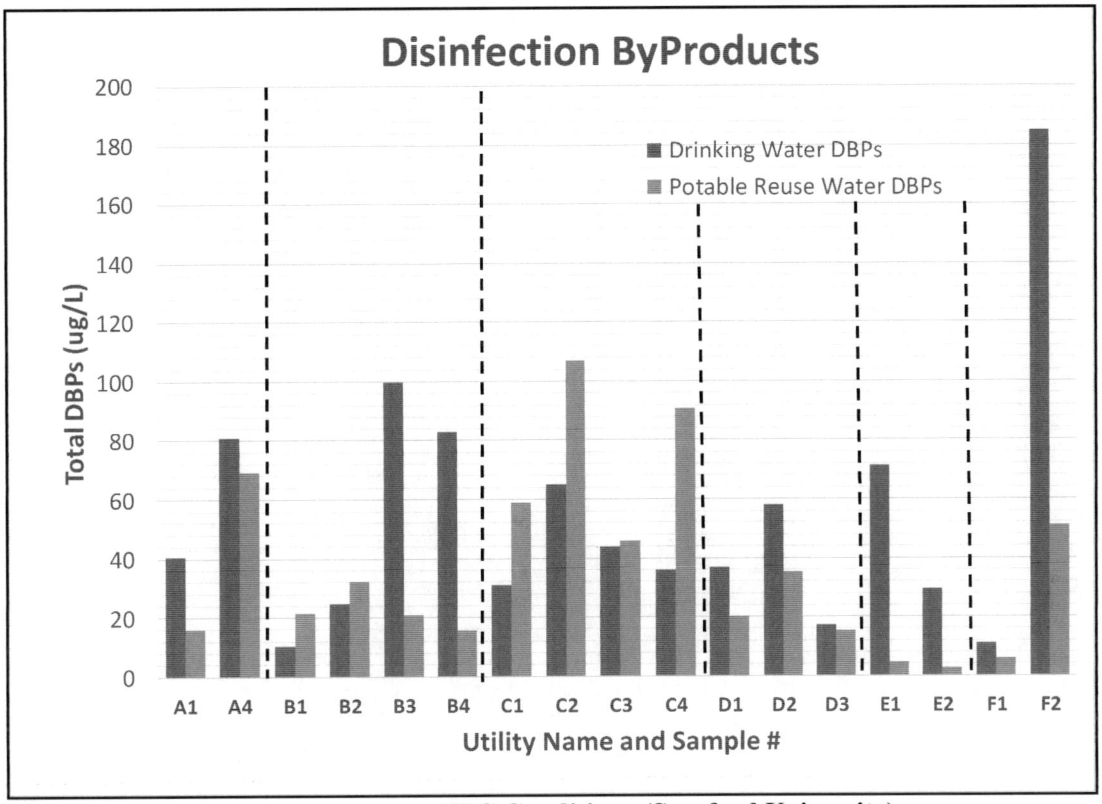

Figure 3 – Disinfection Byproducts, UFC Conditions (Stanford University)

Table 2 – Regulated and Unregulated DBPs Measured in this Project

DBP Class	DBP	Acronym
Trihalomethanes (THMs)	Chloroform	TCM
	Bromodichloromethane	BDCM
	Dibromochloromethane	DBCM
	Bromoform	TBM
Haloacetic acids (HAAs)	Chloroacetic acid	CAA
	Bromoacetic acid	BAA
	Dichloroacetic acid	DCAA
	Trichloroacetic acid	TCAA
	Bromochloroacetic acid	BCAA
	Dibromoacetic acid	DBAA
	Bromodichloroacetic acid	BDCAA
	Dibromochloroacetic acid	DBCAA
	Tribromoacetic acid	TBAA
	Iodoacetic acid	IAA
Iodinated THMs (I-THMS)	Dichloroiodomethane	DCIM
	Bromochloroiodomethane	BCIM
	Dibromoiodomethane	DBIM
	Chlorodiiodomethane	CDIM
	Bromodiiodomethane	BDIM
	Iodoform	TIM

DBP Class	DBP	Acronym
Halonitromethanes	Chloropicrin	TCNM
Haloketones	1,1-dichloropropanone	1,1-DCP
	1,1,1-trichloropropanone	1,1,1-TCP
Haloacetaldehydes (HALs)	Trichloroacetaldehyde	TCAL
	Bromodichloroacetaldehyde	BDCAL
	Dibromochloroacetaldehyde	DBCAL
	Tribromoacetaldehyde	TBAL
Haloacetonitriles (HANs)	Dichloroacetonitrile	DCAN
	Bromochloroacetonitrile	BCAN
	Dibromoacetonitrile	DBAN
	Trichloroacetonitrile	TCAN
Haloacetamides (HAMs)	Dichloroacetamide	DCAM
	Bromochloroacetamide	BCAM
	Dibromoacetamide	DBAM
	Trichloroacetamide	TCAM
Nitrosamines (NAs)	N-nitrosodimethylamine	NDMA
	N-nitrosomorpholine	NMOR

Figure 4 show the total mass at the various sampling locations for trace organic compounds (TOrCs). 43 TOrCs were measured in each collected sample and the detection limit for most compounds was less than 10 ng/L. The advanced treatment processes employed at the potable reuse facilities removed more than 90% of the TOrCs at all utilities, except Utility B which removed approximately 40%. The most prevalent TOrCs remaining in the potable reuse water was sucralose (70% - 90% of total mass) and iopamidol (5% - 20%), which was very similar to the TOrCs remaining in the drinking water.

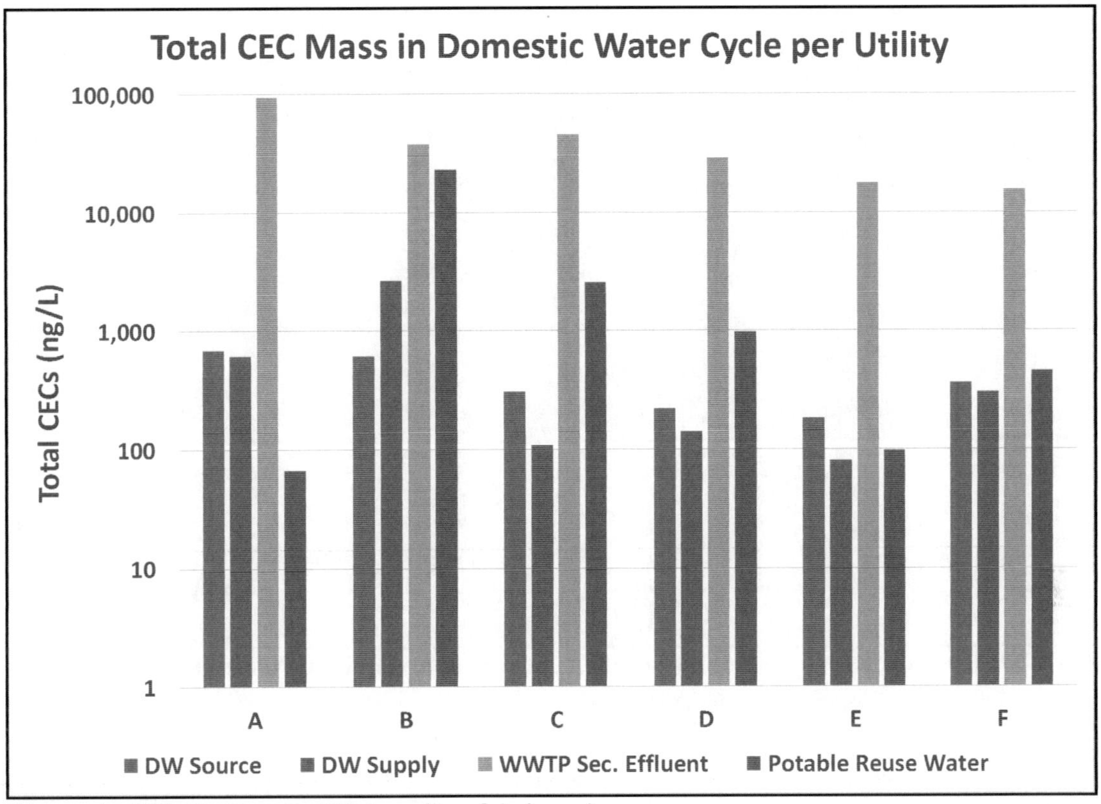

Figure 4 – TOrCs Results (University of Arizona)

Six bioassays were conducted on each collected sample: Estrogen Receptor (ER), Glucocorticoid Receptor (GR), Aryl hydrocarbon Receptor (AhR), p53, and HepG2. No positive bioassay hits were measured in any potable reuse water. Advanced treatment at each utility was very effective as shown in Figure 5.

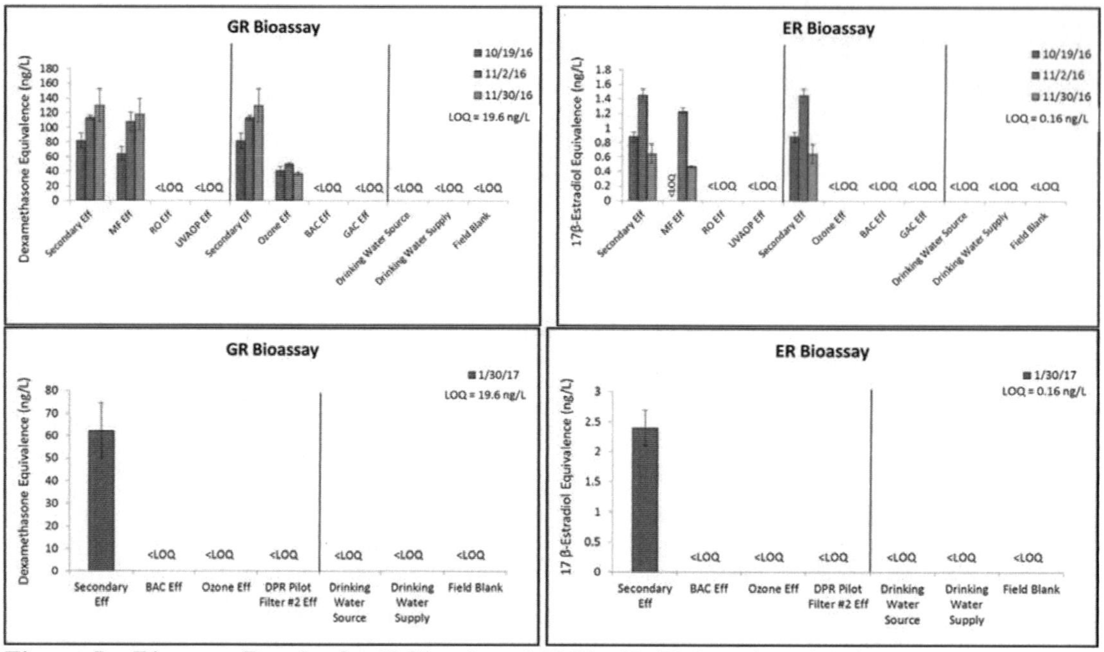

Figure 5 – Bioassay Results for Utility A and Utility B (University of Arizona)

REFERNCES

Scruggs, C. E., & Thomson, B. M. (2017). Opportunities and challenges for direct potable water reuse in arid inland communities. *Journal of Water Resources Planning and Management*, *143*(10), 04017064.

U.S. Environmental Protection Agency. (2012). Guidelines for Water Reuse.

U.S. Environmental Protection Agency. (2017). Potable Reuse Compendium.

Using Large Data Sets and Statistical Tools to Define "Failure" and Demonstrate the Safety of DPR

Austa Parker
Carollo Engineers

Yan Qu (Trussell Technologies), David Hokanson (Trussell Technologies), Jeff Soller (Soller Environmental), Eric Dickenson (Southern Nevada Water Authority), Bob Angelotti (Upper Occoquan Service Authority), Julian Inoue (Carollo Engineers), Shane Trussell (Trussell Technologies), Andrew Salveson (Carollo Engineers)

Abstract

Extensive research has been conducted showing the ability of advanced treatment processes, including ozone, biologically active filtration (BAF), microfiltration (MF), reverse osmosis (RO), UV advanced oxidation (UV AOP) and chlorine disinfection, to treat secondary effluent to the highest potable water standards for direct potable reuse (DPR). Each treatment process has a range of performance, meaning that, over a short interval of time, the performance of that process is potentially well below the anticipated average or setpoint value. From a holistic treatment train perspective, if a low level of performance for one key process occurs simultaneously with lowered performance of an additional key process, this may pose an increased risk to potable water quality and thereby public health.

In order to assess the risk caused by simultaneous reductions in treatment performance, a research study funded by the Water Environment and Reuse Foundation (WE&RF 14-16), utilized Quantitative Microbial Risk Assessment (QMRA) to quantify the acute risk of pathogen infection though DPR treatment processes *and* to show which processes require increased attention to monitoring and response.

Failure and risk evaluation included (1) gathering data and creating probability distribution functions (PDF), (2) tracking operational and monitoring failures, (3) performing a fault tree analysis, and (4) running QMRA to simulate a range of failure events across a one year period of DPR treatment.

Online monitoring data was provided by 12 water recycling and conventional water treatment facilities, all utilizing one or multiple advanced treatment processes: ozone, BAF, MF, RO, UV AOP, and chlorine disinfection. Four treatment trains were studied for their ability to remove by pathogens and microorganisms, shown as Figure 1.

Monitoring data was analyzed by setting a baseline performance for the process, and identifying when a monitoring parameter fell outside of the confidence interval for a process, causing an "off-spec" event. The cause of each off-spec event was identified, and the *frequency and duration* of each event was recorded and used in the QMRA scenario.

Fault tree analyses were conducted on each treatment process, resulting in a list of the most likely and highest impact failure events to occur in each process. This analysis provided the *impact* on pathogen removal performance of each off-spec event.

The QMRA analysis simulated treatment in 15 minute intervals for 1 year, with the failure events for each process included in the 1 year of treatment. The risk output for each DPR treatment train includes the annual risk of infection by select pathogens, showing which pathogen(s) drives the highest risk for each DPR treatment train in a 1 year period (Figure 2).

Preliminary results indicate that for short duration off-spec events, the risk posed by pathogens can increase over a lifetime of exposure. All work and results are expected to be complete by May 2018.

This work represents a critical step in outlining how to safely implement direct potable reuse treatment trains. With a comprehensive understanding of process failure modes and resulting water quality implications, the industry can better design, operate, and maintain treatment plants with increased confidence for public health protection.

A Framework for Regulating Raw Water Augmentation – Results from a Pathogen Benchmarking Study

Dave MacNevin, PhD, PE; Matthew Azarian, EI; Andrea Netcher, PhD, PE – Tetra Tech, Inc.

ABSTRACT

California, Texas, and Florida, have been leaders in innovative reuse applications through groundwater replenishment reuse projects (GRRPs), direct potable reuse (DPR) projects, and unrestricted public access reuse for irrigation. Concerns over water scarcity in each state are driving increased interest in development of regulations for raw water augmentation projects.

The occurrence and concentration of pathogens, including enteric viruses (especially norovirus), *Cryptosporidium,* and *Giardia,* are among the most important data sets needed to help guide the development of regulations. Although past studies[i-ii] have evaluated pathogens in Florida reclaimed water from the perspective of unrestricted public access reuse for irrigation; a few Florida pilot studies have evaluated pathogen data from the potable reuse perspective[iii]. Therefore, developing regulatory criteria for raw water augmentation will require a broader and more comprehensive survey that uses current pathogen data.

This paper presents the results from a new pathogen benchmarking study that investigates the occurrence of protozoa in Florida reclaimed waters from the Florida Department of Environmental Protection's (FDEP) large pathogen database (2002-2017). The database is one of the largest datasets comprising more than 6,500 pathogen data points. Florida's reuse rules require certain domestic wastewater treatment plants to monitor for the protozoan pathogens, Cryptosporidium and Giardia and report the results to FDEP.

Key results of this study include the following.

- Updated Cryptosporidium concentration distributions in Florida reclaimed waters
- Updated Giardia concentration distributions in Florida reclaimed waters
- Evaluation of impact of secondary treatment type (conventional activated sludge, contact stabilization, extended aeration, biological nutrient removal, pure oxygen/fine bubble activated sludge, sequential batch reactor, modified activated sludge, oxidation carousel basins, A/O biological treatment process).
- Evaluation of impact of tertiary filtration type (i.e., deep bed single media, shallow bed automatic backwash, fabric filter, deep bed multimedia, upflow (e.g. Dynasand), diatomaceous earth filter, slow rate sand filter, deep-bed/auto backwash, and disk filter)
- Recommended pathogen LRVs using a uniform LRV requirement and site-specific requirement, starting at reclaimed water
- Data gaps in the Florida pathogen dataset for evaluating raw water augmentation applications include the following:

- Limited information on the pathogen concentrations in raw wastewater
- Lack of data on enterovirus (especially norovirus) concentrations

KEYWORDS

Cryptosporidium, Giardia, protozoa, reclaimed water, direct potable reuse, indirect potable reuse

INTRODUCTION

Protozoa have been a major concern for drinking water in the past 25 to 30 years. Two major public health outbreaks in the United States caused by protozoa focused public attention on protozoa in drinking water, prompting the United States Environmental Protection Agency (EPA) to enact regulations pertaining to surface water treatment.

This introduction briefly discusses the following items relating to protozoa:

- Two (2) major public health outbreaks (1987, 1993) in drinking water attributed to protozoa, and the regulatory response.
- An influential 2004 dataset of pathogens (including protozoa) from six (6) United States wastewater treatment plants[iv]
- Florida's decision in the early 2000's to require all reuse systems to sample for protozoa in reclaimed water.

Over the past 18 years, nearly 300 water reclamation facilities in Florida have continued routinely collecting reclaimed water protozoa data. As of 2017, the data set included 1,828 observations each for total and viable *Cryptosporidium* and total and viable *Giardia*. This paper examines this data set, evaluating the occurrence of protozoa across reclaimed waters from numerous facilities. Protozoa concentrations are compared against drinking water goal concentrations to observe what level of protozoa log reduction may be required to utilize reclaimed waters as a drinking water source of supply.

Significance of Protozoa to Water Treatment

In January 1987, a Georgia surface water treatment plant (WTP) treating river water experienced a major public health outbreak, attributed to *Giardia lamblia,* that was coincident with a raw sewer overflow occurring 6 hours upstream of the WTP intake. Nearly 20% of the service population (about 13,000 out of 65,000) became ill. After the outbreak, the utility performed $700,000 (2018 dollars) in WTP upgrades to provide additional treatment. In response, EPA initiated the 1989 Surface Water Treatment Rule which applied to all public water systems (PWSs) using surface water sources or ground water sources under the direct influence of surface water (GWUDI). The Rule required most water systems to filter and disinfect water, while also establishing maximum contaminant level goals (MCLGs) for viruses, bacteria, and *Giardia*. It also included treatment technique (TT) requirements for filtered and unfiltered systems to protect against adverse health effects of exposure to pathogens.

In early 1993 (March-April), a large surface WTP in Wisconsin taking water from Lake Michigan experienced a major *Cryptosporidium* outbreak. Seasonal runoff increased turbidity loads to the WTP. Half of the service population became ill (about 400,000 out of 825,000), and more than 100 deaths were attributed to the event. The WTP was shut down and received $170M (2018 dollars) in upgrades. In response the EPA took several steps to expand the SWTR, including the Interim Enhanced Surface Water Treatment Rule (IESWTR) (1998) which regulated *Cryptosporidium*, the Filter Backwash Recycling Rule (FBRR) (2001), the Long Term 1 Enhanced Surface Water Treatment Rule (LT1ESWTR) (2002), and the Long Term 2 Enhanced Surface Water Treatment Rule (LT2ESWTR) (2006).

Protozoa in Wastewater and Reclaimed Water

Another outcome of the major protozoa outbreaks was an increased focus on protozoa and other pathogens in wastewater and reclaimed water. In 2004, a landmark study[iv] was published describing the results of an intensive sampling program for pathogens at six (6) United States water reclamation facilities (WRFs) in Florida, California, and Arizona. Samples were taken at the following locations: wastewater influent, secondary effluent, filter effluent, and disinfection effluent. Pathogens sampled included enteric viruses, *Giardia*, and *Cryptosporidium* (total and infective). Indicator bacteria and alternative indicators included total coliforms, fecal coliforms, enterococci, coliphage 15597, and coliphage 700891. Five (5) to seven (7) samples were collected for each site at each WRF within a 10-month sampling period.

Across the six (6) WRFs a total of 33 protozoa samples (for both *Giardia* and *Crypto*) were collected in the raw wastewater influent, while a total of 32 protozoa samples (for both *Giardia* and *Crypto*) were collected in the secondary, disinfected, filtered wastewater (reclaimed water). Figure 1 shows the observed distributions of total protozoan density for raw wastewater and reclaimed water on a log-normal probability plot. For both locations, and both protozoa, a linear fit is apparent, indicating that the variation in protozoa density of the population of interest (6 WWTPs) can be described according to a log-normal probability distribution. The method used to generate this plot will be described shortly. Similar log-normal trends are seen for other data sets also. The *Crypto* data indicate that the WRFs sampled provided nearly 2.36 LRV for *Crypto* and 3.26 LRV for *Giardia* between the raw wastewater point and the reclaimed water sample point.

The log reduction value requirement to reduce protozoa densities to drinking water goals (DWG) was based upon a *Cryptosporidium* drinking water goal of $<3\text{x}10^{-5}$ oocysts/L[v] and a *Giardia* drinking water goal of $<6.8\text{x}10^{-6}$ cysts/L[vi]. Both goals are associated with an estimated 10^{-4} risk level for acute gastroenteritis. A protozoan concentration can be converted to a point LRV value by the following equation $LRV=\log_{10}(Observed/DWG)$.

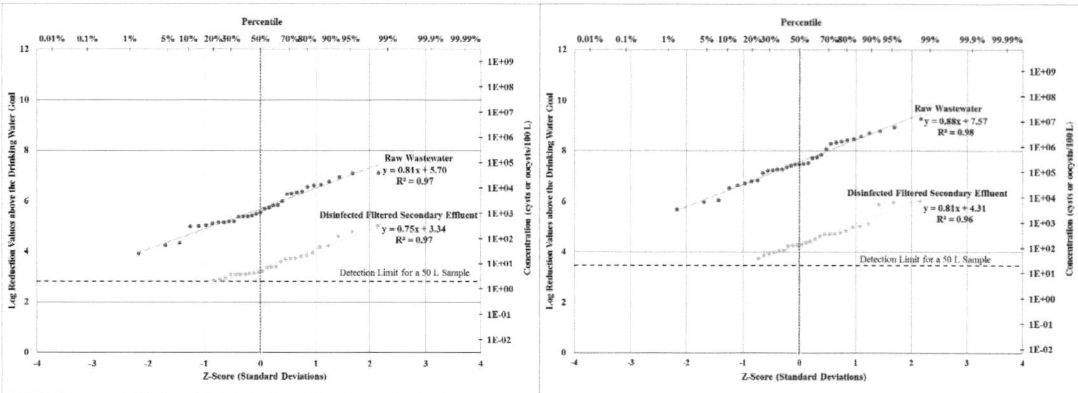

Figure 1 Observed Distributions of Total Protozoa (*Crypto*, Left) (*Giardia*, Right) from Six (6) WWTPs by Rose et al. 2004

This data set was one of the sources considered for the 12-10-10 regulations from the California State Water Resources Control Board's groundwater replenishment rule, which required 12-log virus reduction, 10-log *Crypto* reduction, and 10-log *Giardia* reduction between the headworks of the wastewater plant and the point of water recovery for potable purposes. It is evident from Figure 1 that a uniform LRV goal of 10 would be sufficient for both *Crypto* and *Giardia*, although it appears more conservative for *Crypto*.

As interest in direct potable reuse (DPR) has increased in recent years, so has the need for additional reliable pathogen data. In 2016, California's DPR Expert Panel, commissioned by the National Water Research Institute (NWRI) recommended[vii] that the State Water Board (SWB) gather additional pathogen data, specifically calling for DPR projects to (Recommendation #2) utilize probabilistic quantitative microbial risk analysis (QMRA) with inputs (including pathogen concentrations) derived from the literature. The report also recommended (Recommendation #3) including monitoring requirements for protozoa and viruses in raw wastewater feeding a DPR system. The SWB has initiated this data collection effort across several facilities in the state of California.

In the State of Florida, since the early 2000's, reuse facilities have been required by the Florida Department of Environmental Protection (FDEP) to collect protozoa data in the treated reclaimed water after a rule revision in 1999 (F.A.C. 62-610) required all reuse treatment systems >1 MGD to collect protozoa samples on a 2-year sampling interval. The recommended detection limit was one (1) (oo)cyst per 100L. Early (2003, 2004) during the sampling of this data, papers[i,viii] were written summarizing various trends in the concentration of protozoa based upon various characteristics of the water reclamation facility (WRF). At the time, 156 observations each for *Giardia* and *Crypto* were available in the data set. However, since that time, no additional papers were found that discussed the data.

In late 2017, the authors of this paper contacted FDEP to inquire about the status of the protozoa database, looking to use the data to assist PWSs in planning for the potential pathogen LRV requirements that might apply under a future Florida DPR regulation. More details about the updated dataset and review methodology are provided in the following section.

METHODOLOGY

Database Characteristics

FDEP has continued to maintain the protozoa pathogen database as a Microsoft Access database, which as of 2017 had increased to nearly 1,828 observations each of total and viable *Crypto* and *Giardia*, from 292 facilities, with some facilities having data as far back as 18 years. The database did not include any pathogen data from the raw wastewater but only from the filtered and disinfected reclaimed water. All data analyses were performed with Microsoft Excel.

The database included several fields within each record, including, but not limited to, the following:

- Facility Site ID
- Nitrified? (Nitrifying, Non-Nitrifying, Mixed)
- Type of Secondary Treatment Provided (Conventional Activated Sludge, Biological Nutrient Removal, Contact Stabilization, and Extended Aeration)
- Type of Filtration Provided (Deep bed multimedia, deep bed single media, fabric filter, membranes, shallow bed automatic backwash, slow rate sand filter, and upflow (including Dynasand))
- Type of Disinfection Provided (Chlorine gas, sodium hypochlorite, UV disinfection)
- Pathogens (*Cryptosporidium (*total and viable), *Giardia* (total and viable)
- Water quality (Turbidity, total suspended solids, total chlorine residual)
- Date
- Amount filtered (for pathogens)

Data Processing

For this paper, the data analysis focused on total *Crypto* and *Giardia*. Viable protozoa data (determined by DAPI staining) was found to be a fraction of the total, and was not considered further. All concentrations were normalized against the DWG concentrations (as described in the introduction) to calculate an instantaneous point log reduction value (LRV) requirement for every observation.

For *Crypto* and *Giardia* respectively, all protozoan densities for were sorted in ascending order, and the percentile for each data point was estimated using the Weibull percentile formula where $p_i=n_i/(n_t+1)$, where p_i is the percentile of the ith point. n_i is the number of the ith point. n_t is the total number of data points. After the percentile was calculated for each point, a corresponding z-score was calculated for every point, according to the assumption that the protozoan density distribution is log-normal. The Z-score for the ith point, Z_i was calculated using the Excel function NORM.S.INV, which returns the inverse of the standard normal cumulative distribution, which is graphed in Figure 2. For example, the Z-score for a percentile of 0.83 (83%) is approximately 1.0, indicating that if the data set is normally distributed, the point at the 83rd percentile would be approximately 1 standard deviation greater than the mean. The fit of the distribution can be assessed by plotting the logarithm of each value versus its z-score. If the plot falls on a straight line, then the data is lognormally distributed.

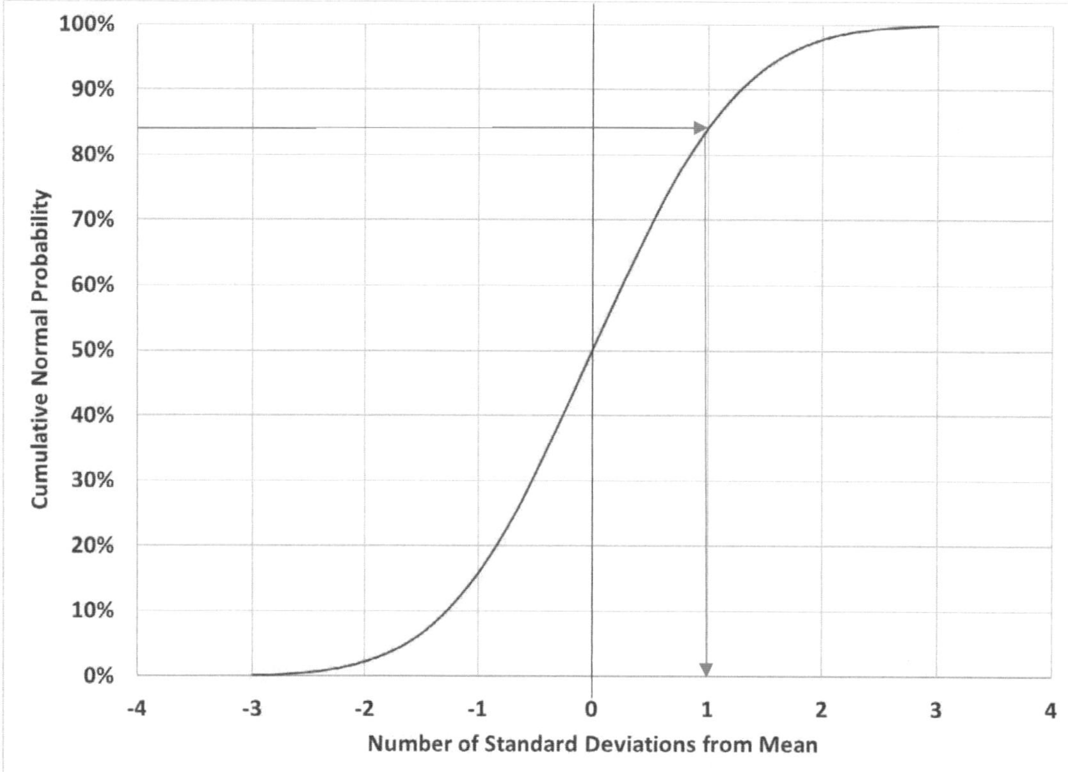

Figure 2 Standard Normal Cumulative Distribution

Cumulative distributions were generated for the entire data set from all WRFs, looking at reclaimed water *Crypto* and *Giardia* distinctly. This analysis also examined the effect of various treatment factors including nitrification, secondary treatment type, filtration type, and disinfection type, in order to discern which factors may have the most significant influence on observed protozoa levels, and to observe the general variability of pathogen concentrations. Note, these sub analyses of treatment type excluded facilities with changing treatment over the 18-year period of record. Finally, the reclaimed water protozoa data were compared to other reclaimed water data sets and surface water data sets.

RESULTS

All WRFs

Figure 3 shows the overall distribution of *Giardia* and *Crypto* over all 292 reclaimed water facilities, comprising 1,828 total observations for each protozoan. The left-hand portion of each distribution is relatively level, corresponding to non-detect readings at various detection limits. The right-hand portion appears roughly linear and corresponds to detections of protozoa.

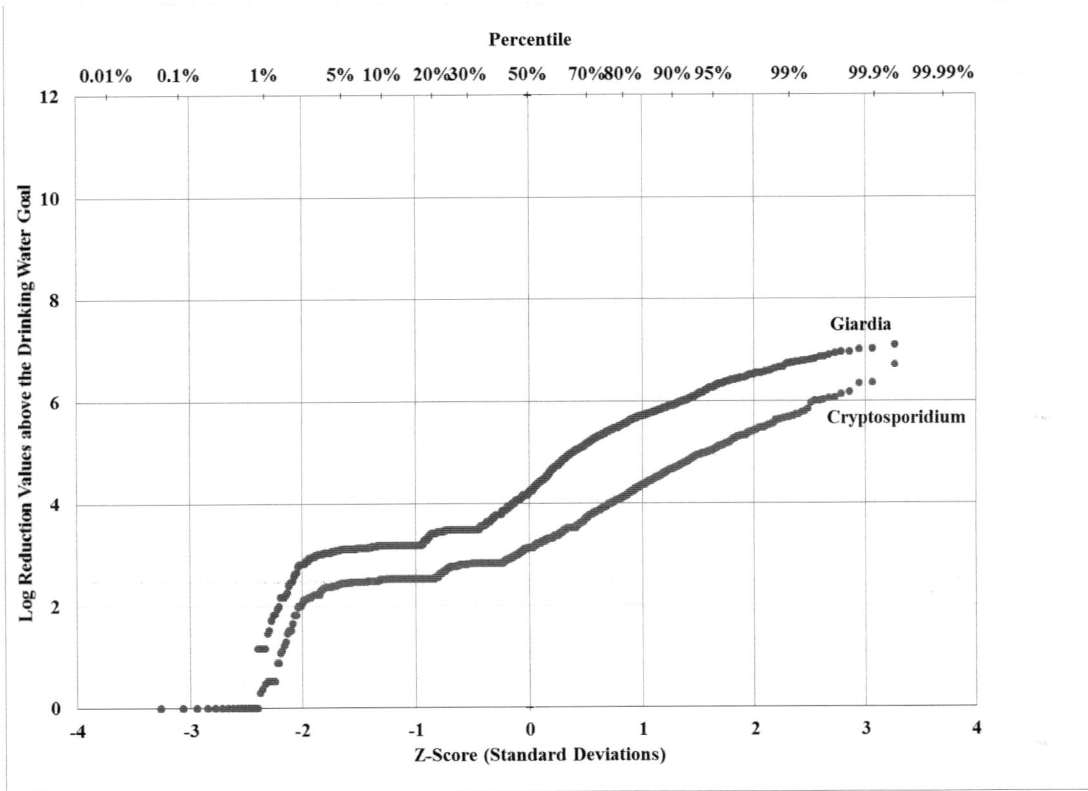

Figure 3 Observed Distribution of Total Protozoa in Florida Reclaimed Waters (N=1,828)

Figure 3 shows the overall distribution of *Giardia* and *Crypto* over all 292 reclaimed water facilities, comprising 1,828 total observations for each protozoan. The left-hand portion of each distribution is relatively level, corresponding to non-detect readings at various detection limits. The right-hand portion appears roughly linear and corresponds to detections of protozoa. Most facilities used a 100 L filtration sample however, there were also many facilities that used 50L filtered sample. Therefore, a detection limit of 2 (oo)cysts per 100L was assumed for purposes of masking the data to mark the cutoff point between detects and non-detects.

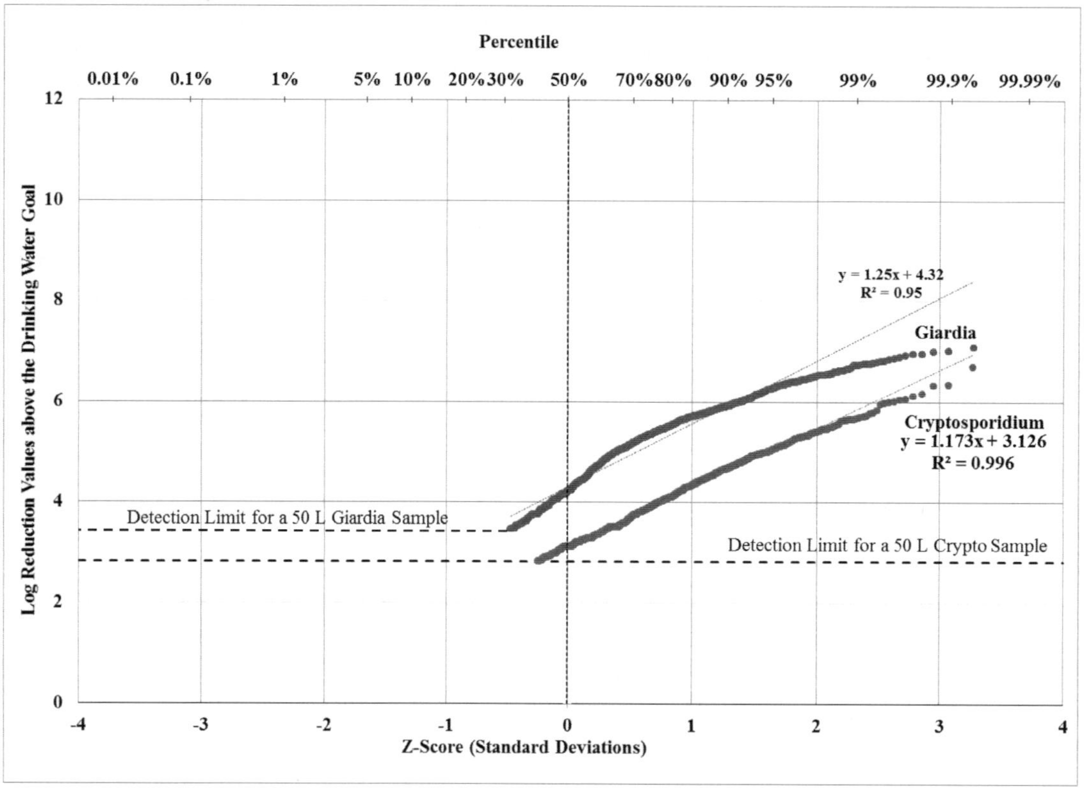

Figure 4 Checking for Fit of a Log-Normal Distribution to Total Protozoa Detections in Florida Reclaimed Waters (N=1,828)

The detection data were used to check the log-normality of the data (Figure 4). The Cryptosporidium densities are well described by a log normal distribution (R^2=99.6%). While *Giardia* also shows a high R^2 value (95%), the *Giardia* densities appear less well described at higher densities, potentially indicating an issue with recovery and enumeration of higher densities of *Giardia.* In general, the *Giardia* levels were typically higher, in terms of required LRV, than the *Crypto* levels.

Effects of Treatment

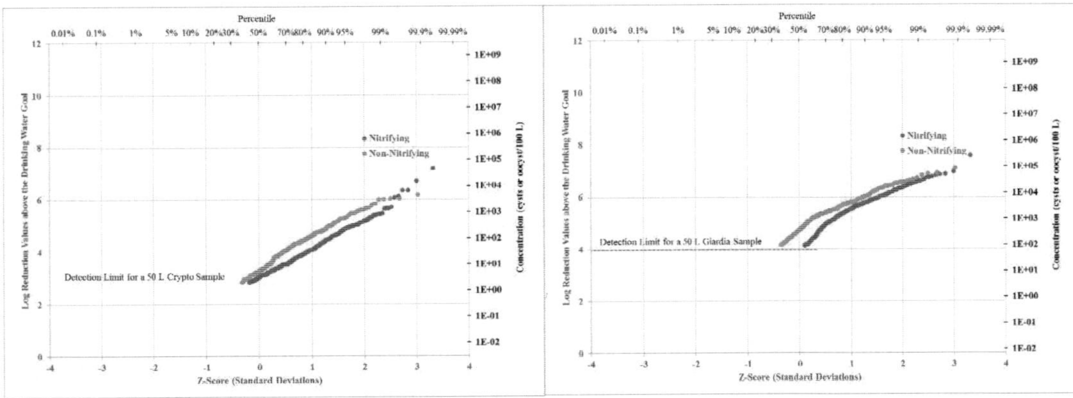

Figure 5 Protozoa Densities for Nitrifying & Non-Nitrifying Florida Reclaimed Waters (Crypto, Left) (Giardia, Right)

The data set was split between nitrifying facilities and non-nitrifying facilities to observe any potential relationship between nitrification and protozoa levels (Figure 5). Previous investigators had reported that facilities that nitrified tended to have lower *Giardia* levels[viii]. Facilities practicing nitrification typically had lower levels of both *Crypto* and *Giardia*; however, this effect was small, typically being less than 0.5 LRV in magnitude.

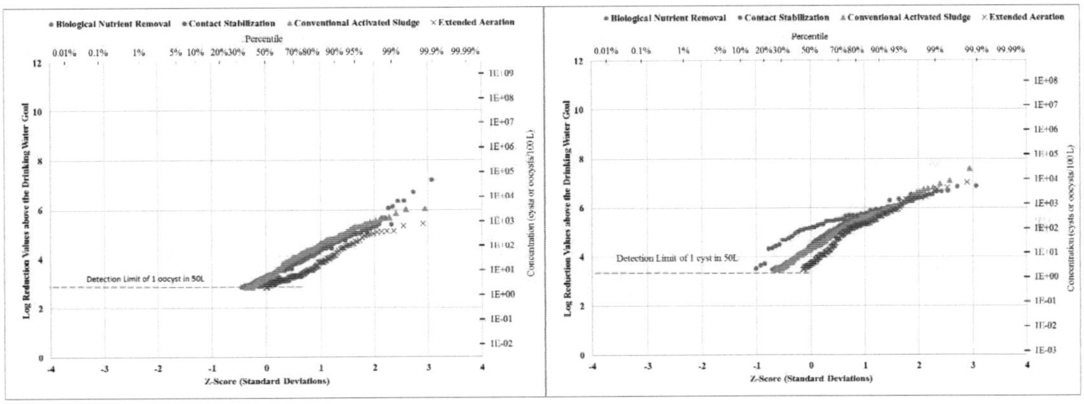

Figure 6 Protozoa Densities by Secondary Treatment Type in Florida Reclaimed Waters (Crypto, Left) (Giardia, Right)

The data set was again divided on the basis of different types of secondary treatment (Figure 6). The observed effect of treatment varied by protozoan, with the lowest densities observed for extended aeration for both protozoa. However, the overall difference by treatment type was minor, within a band of about 1 LRV for *Crypto* and 2 LRV for lower levels of *Giardia,* converging at higher levels.

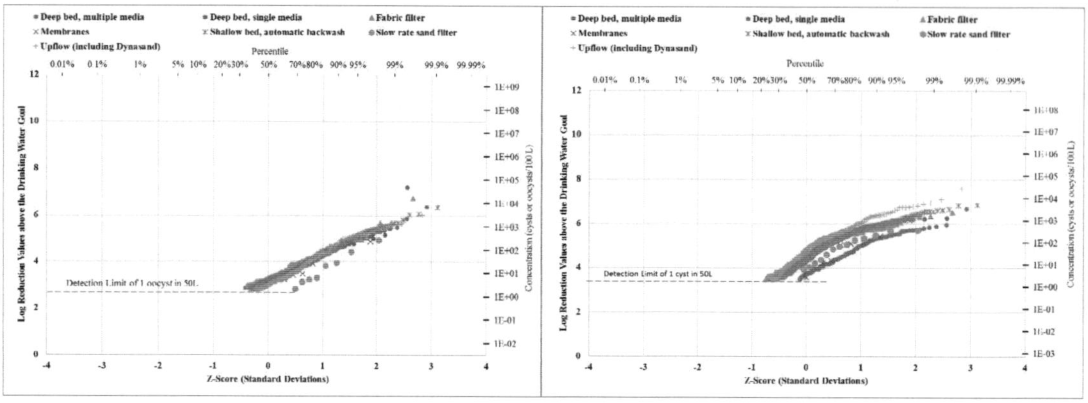

Figure 7 Protozoa Densities by Filter Type in Florida Reclaimed Waters (Crypto, Left) (Giardia, Right)

The data set was analyzed looking at impact of filter type. Filter type appeared to have had a greater impact on *Giardia* densities than *Crypto*, as shown in Figure 7. While both protozoa indicate lower levels with slow rate sand filters, *Giardia* had lower levels observed from facilities with deep bed multimedia filters. Nevertheless, the width of this differential for *Giardia* by filter type is only about 1.0 LRV.

The impact of disinfection was also checked, comparing UV disinfection against chlorine gas and sodium hypochlorite (not shown). No significant difference in protozoa levels was observed according to disinfection type.

Individual Facilities

The analysis was extended beyond the aggregate analysis to look at the top 20 individual facilities (with the most observations) out of the nearly 300 facilities in the database. This top 7% of sites in the dataset comprised nearly 20% of all protozoa observations in the database. *Crypto* densities for the 20 selected individual facilities are shown in Figure 8. *Giardia* densities for the 20 selected individual facilities are shown in Figure 9

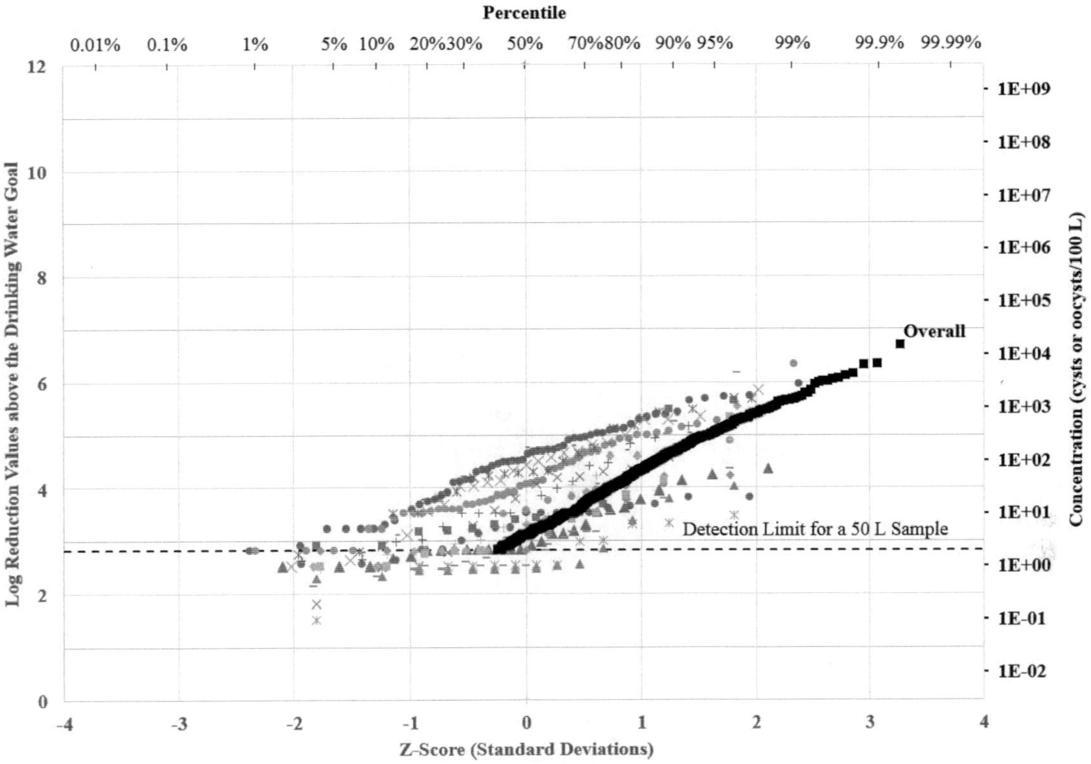

Figure 8 *Crypto* Densities Among Top 20 Individual Facilities with Most Observations

In general, the maximum *Crypto* density was about 1x10^4 oocysts/100L (Figure 8); however, most individual sites were well below this value. The spread in *Crypto* by individual facility was about ±2 LRVs. Therefore, a site-specific approach based on sampling may be justified and reduce treatment costs.

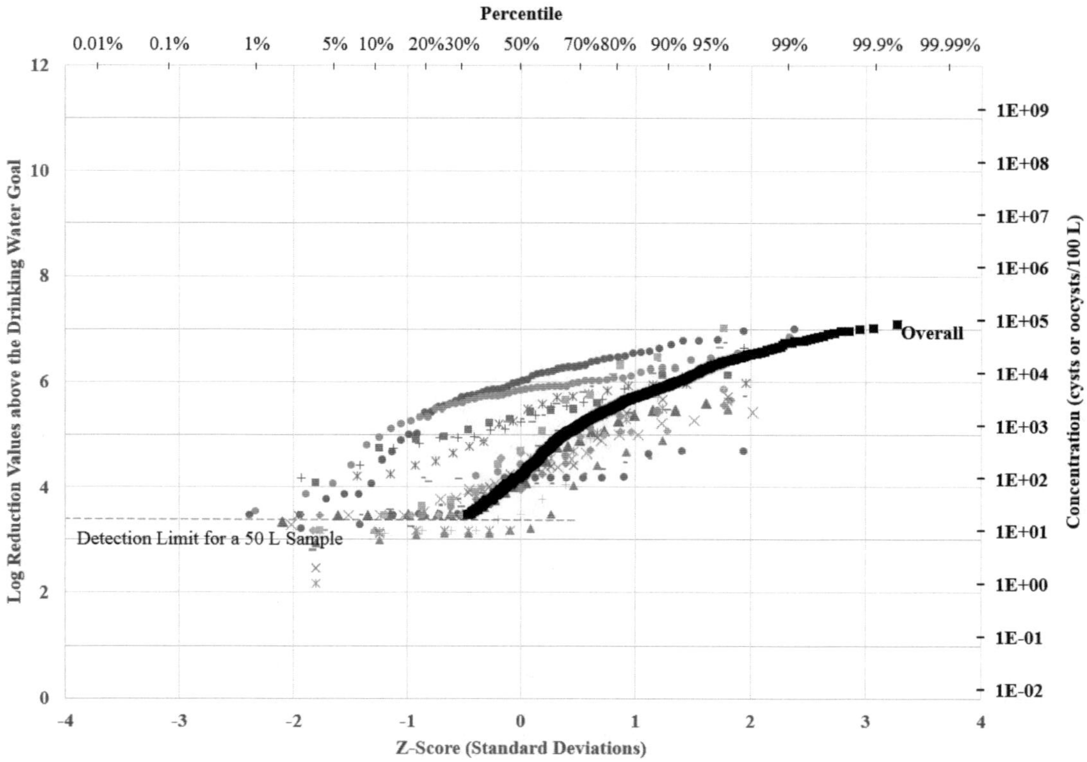

Figure 9 *Giardia* Densities Among Top 20 Individual Facilities with Most Observations

In general, *Giardia* concentrations were higher than *Crypto.* with the maximum *Giardia* density being nearly an order of magnitude higher near $1x10^5$ oocysts/100L (Figure 9); however, most individual sites were well below this value. Notably, the curvilinear trend was observed for multiple individual facilities. The spread in *Giardia* by individual facility was up to about ±3 LRVs. Therefore, a site-specific approach based on sampling may be justified and reduce treatment costs.

DISCUSSION

Comparing Potential Uniform or Site-Specific LRV Requirements for Protozoa

While Florida has not yet established the laws and rules governing treatment of pathogens for direct potable reuse (DPR), the state has amassed a valuable dataset that could be used to test different regulatory approaches. The following discussion attempts to make some observations regarding where the starting points may be for regulating protozoa.

No data were available for enteric viruses or noroviruses, so no discussion of viruses is included. Additional sampling for noroviruses in reclaimed water is recommended to provide the state with a basis for rulemaking. It also may be helpful to collect protozoa and virus samples in the raw wastewater to reduce the number of non-detects and better gauge the actual pathogen loads challenging water reclamation facilities (WRFs).

Having reviewed this data by individual facility, the following section discusses what LRV treatment requirements might be justifiable to protect public health by achieving a better than 10^{-4} annual level of risk of illness from protozoa. Since the dataset is based upon reclaimed water pathogen data, it follows the Texas approach of calculating LRV between reclaimed water and the finished drinking water. This is not to say that Florida should use this approach (instead of the California approach which begins at the raw wastewater). It just reflects the fact that the available data is for the reclaimed water. Note, the actual LRV requirement that a treatment train is designed for may also vary based upon the regulators' desired amount of redundant LRVs.

If a uniform LRV criterion is imposed for *Crypto*, 6-7 LRV from the reclaimed water to the finished drinking water may be sufficient to meet the drinking water goal density for *Crypto*. However, a uniform criterion may be somewhat conservative for several facilities since the *Crypto* density varies between sites by about ±2 LRV.

Similarly, if a uniform LRV criterion is imposed for *Giardia*, 7-8 LRV from the reclaimed water to the finished drinking water may be sufficient to meet the drinking water goal density for *Giardia*. However, a uniform criterion may be somewhat conservative for several facilities since the *Giardia* density varies between sites by about ±3 LRV.

Differences in total protozoa counts by nitrification, secondary treatment type, and filtration type, were all relatively small, usually less than one order of magnitude. These differences are almost insignificant when compared to the capabilities of advanced water treatment technology barriers, which can provide multiple log-reductions of pathogens.

Comparison of Reclaimed Water Protozoa Counts to Surface Water Protozoa Counts

To put the protozoa results in context, it is helpful to compare the densities against values taken from untreated surface water. Figure 10 shows the density of *Crypto* observed in raw surface water samples collected from 50 WTPs under LT2 sampling as adapted from a prior study[ix]. In general, the data appear to indicate that at least 5 LRV of *Crypto* would be required to confidently treat all locations, and that there is a variation of about ±2 LRV in *Crypto* densities from location to location.

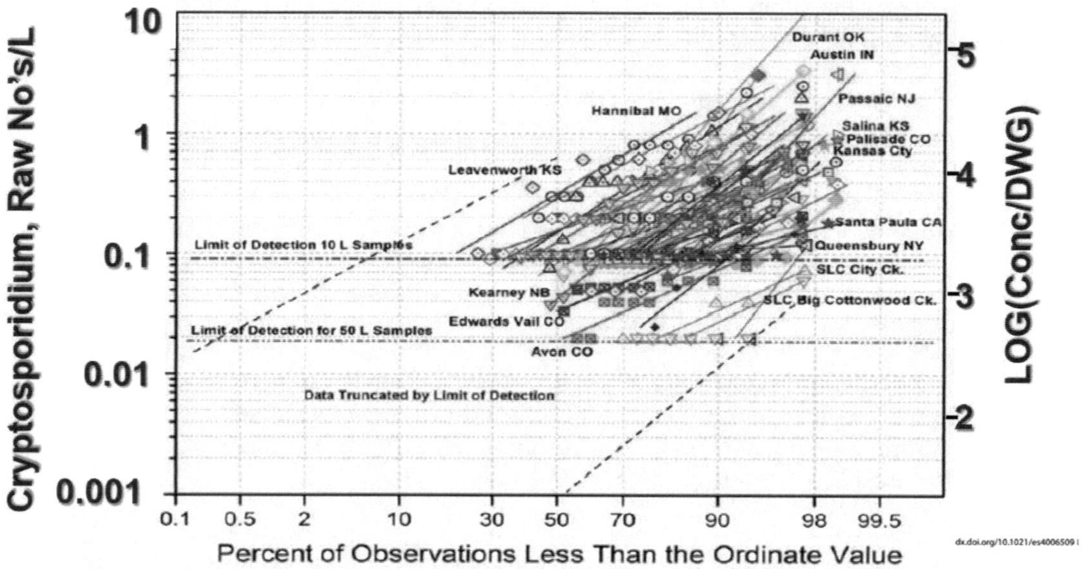

Figure 10 Representative LT2 Raw *Crypto* Data from 50 Water Treatment Plants (Image Adapted from Ongerth 2013[ix])

Table 1 EPA *Crypto* Bin Classifications for Filtered Water PWSs

Bin	"Mean" Concentration	"Calculated LRV" in this Study	LT2 Additional (& Total) *Crypto* Treatment Requirement at Filtration Plants
1	<0.075 oocysts/L	<3.4	None (3- log total)
2	<1 oocyst/L	3.4-4.5	1-log (4-log total)
3	<3 oocysts/L	4.5-5.0	2-log (5-log total)
4	>3 oocysts/L	>5.0	2.5-log (5.5-log total)

The previous result is in line with EPA's *Crypto* Bin classifications for filtered water PWSs (Table 1). Looking at the reclaimed water *Crypto* data again (Figure 12) in comparison with Table 1, it appears that many individual facilities would appear to receive adequate treatment from the LT2 requirement alone while only about 1.5-2.0 additional LRVs of treatment appear necessary to encompass the facilities with the highest *Crypto* densities, between $1x10^3$-$1x10^4$ oocysts per 100 L (Figure 11).

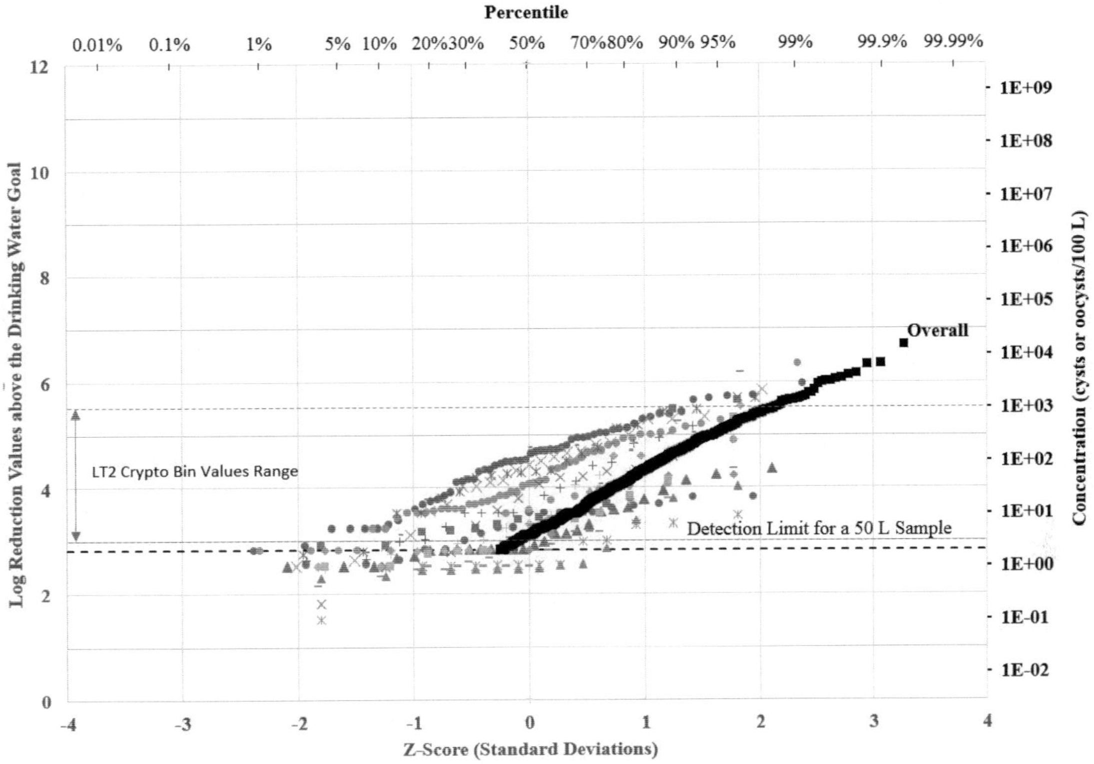

Figure 11 Comparison of *Crypto* Densities to Surface Water LT2 *Crypto* Bin Values

Taking this data and comparing (Figure 12) with other reuse water data sets and surface water data sets[x], it is apparent that *Crypto* densities in reclaimed waters are somewhat comparable to surface water protozoa densities. A similar correspondence was apparent for *Giardia* (Figure 13). This is reassuring because it indicates that, while individual water quality can vary from location to location, in aggregate, reclaimed water is of comparable microbial quality, for protozoa, to surface water supplies. It is also important to note that for the reclaimed waters, a large fraction for the protozoa are considered non-viable since chlorine or UV disinfection does not reduce the number of total protozoa (oo)cysts detected.

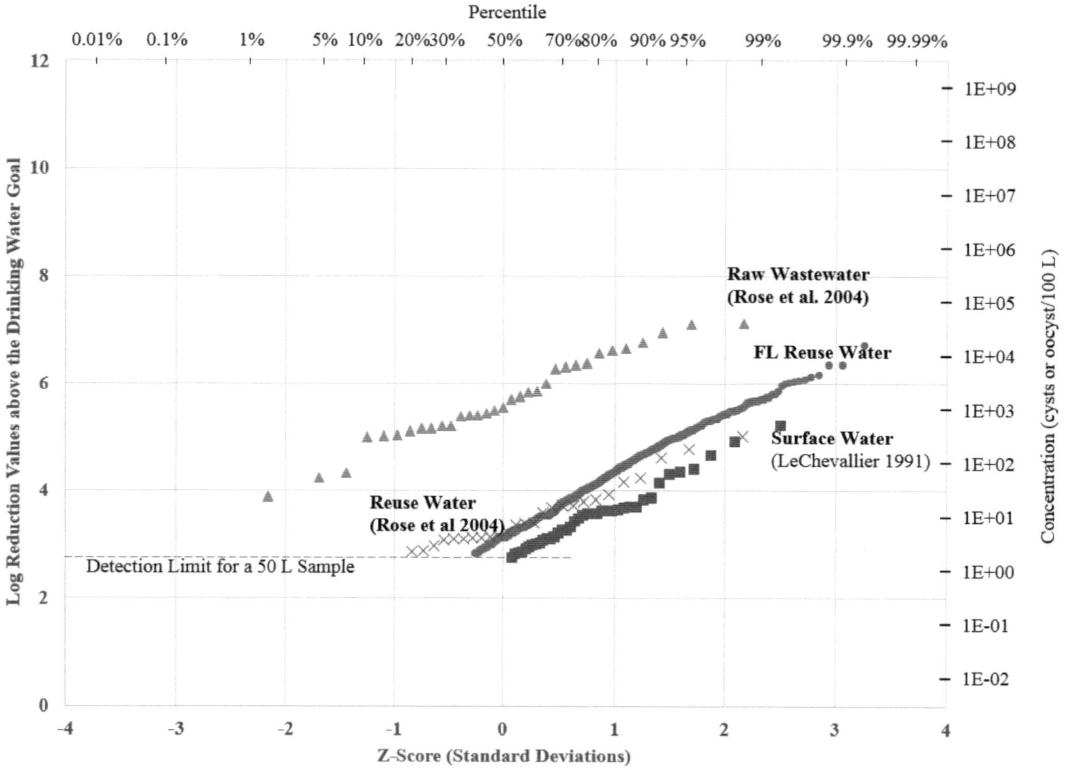

Figure 12 Comparison of Reclaimed Water *Crypto* Densities to Surface Water *Crypto* Densities Shows Similarity

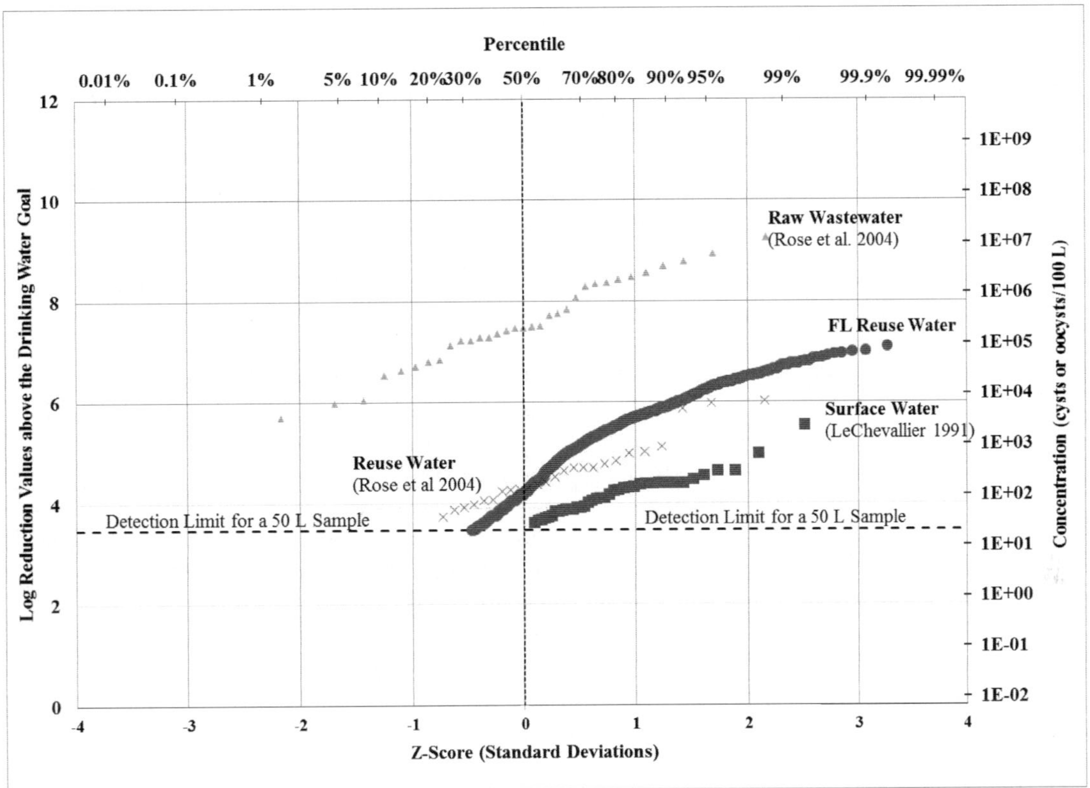

Figure 13 Comparison of Reclaimed Water *Giardia* Densities to Surface Water *Giardia* Densities Shows Similarity

CONCLUSIONS

Concerns over the health impacts of elevated levels of protozoa in drinking water have shaped regulations in the United States. An increasing interest in using reclaimed water as a source of potable supply through direct potable reuse (DPR) has increased the importance of collecting large bodies of reliable pathogen data in raw wastewater and reclaimed water for better assessing the potential occurrence and risk from pathogens.

- This paper reviewed Florida's reclaimed water protozoan database for *Cryptosporidium* and *Giardia*, which is among the largest data sets of its kind in existence. Comprising 1,828 individual observations of each protozoa, sourced from 292 water reclamation facilities, with some facilities providing data from as far back as 1999 (18 years of coverage).

- *Crypto* densities were very well described by a log-normal distribution for the entire data set, and at the individual facility level. *Giardia* densities appeared log-normal at lower concentrations but appeared to taper off at higher concentrations, potentially due to a sampling or method issue. This tapering off effect was observed for the entire data set, and at the individual facility level.

- If a uniform LRV requirement for *Crypto* were enacted, the required log reduction value (LRV) between the reclaimed water and finished drinking water would be at least between 6-7 LRV to meet the drinking water goal density for all the facilities analyzed in the data set. However, protozoa levels vary significantly from facility to facility. An analysis of 20 individual facilities compared with the entire data set showed ±2 LRV in differences between individual facilities. Therefore, a site-specific approach based on sampling may be justified and reduce treatment costs.

- If a uniform LRV requirement for *Giardia* were enacted, the required log reduction value (LRV) between the reclaimed water and finished drinking water would be at least between 7-8 LRV to meet the drinking water goal density for all the facilities analyzed in the data set. However, protozoa levels vary significantly from facility to facility. An analysis of 20 individual facilities compared with the entire data set showed ±3 LRV in differences between individual facilities. Therefore, a site-specific approach based on sampling may be justified and reduce treatment costs.

- Minor differences in protozoa density were observed based on an aggregated analysis of nitrification status, secondary treatment type, and filtration type across all facilities. No significant difference in protozoa density was observed between facilities based on disinfection type (UV disinfection, sodium hypochlorite, and chlorine gas disinfection). However, the aforementioned differences typically amounted to less than ±1 LRV of each protozoa, which is relatively insignificant compared to the multiple log reductions that can be achieved with advanced treatment technologies.

- In engineered treatment trains, current advanced treatment technologies are more than able to handle the levels of protozoa observed in reclaimed water, reducing densities to levels better than drinking water goals.

- Pathogens do not necessarily need to be regulated starting at reclaimed water; however, this approach is easier to analyze for Florida reuse systems, since much more protozoa data is currently available for reclaimed water than raw wastewater.

- In aggregate, protozoa densities in reclaimed water are comparable to protozoa densities in surface water supplies.

- Additional enteric virus (especially norovirus) data is needed to guide the development of treatment requirements for viruses.

REFERENCES

---

[i] York, D.W., L. Walker-Coleman, and P. Menendez. 2002. "Pathogens in Reclaimed Water: The Florida Experience." Proceedings of Water Sources 2002, Las Vegas, NV, AWWA and WEF. 2002.

[ii] Slifko, T. and Gilliam, A. 2004 "Occurrence, Removal, and Risk of Pathogens in Reclaimed Water." AWWA Annual Conference and Exposition.

[iii] MacNevin, D. 2016. "Where Wastewater Ends and Water Treatment Begins: Evaluating the Viability of Potable Reuse" WateReuse Research Conference. Denver.

[iv] Rose, J.B., et al. 2004. Reduction of pathogens, indicator bacteria, and alternative indicators by wastewater treatment and reclamation processes. WERF. Final Report. IWA Publishing.

[v] Regli et al. 1991. "Modeling the Risk from Giardia and Viruses in Drinking Water." Journal AWWA, Nov. 1991. 76-84.

[vi] Haas, C. et al. 1999. "Quantitative Microbial Risk Assessment." John Wiley & Sons.

[vii] NWRI 2016. "Memorandum: Expert Panel Draft Key Research Recommendations Related to the Development of Uniform Water Recycling Criteria for Direct Potable Reuse - June 30, 2016" http://www.waterboards.ca.gov/drinking_water/certlic/drinkingwater/documents/recycled_water/ep_research_rec.pdf

[viii] York, D. et al. 2003. "Monitoring for Protozoan Pathogens in Reclaimed Water: Florida's Requirements and Experience." https://floridadep.gov/sites/default/files/protozoan.pdf

[ix] Ongerth, J. 2013" LT2 Cryptosporidium Data: What do They Tell Us about Cryptosporidium in Surface Water in the United States?" ES&T. dx.doi.org/10.1021/es4006509

[x] LeChevallier, M. et al. 1991. Occurrence of Giardia and Cryptosporidium spp. in surface water supplies. *Applied and Environmental Microbiology*, *57*(9), 2610-2616.

Stormwater Reuse: Aquifer Storage and Recovery Evaluation Using Municipal Residential Stormwater in the Pacific Northwest

Jason Melady, RG[1a], Andrew Davidson, PE[1b], Jadene Stensland, PE[2], David Winship PE[3]

[1]GSI Water Solutions, Inc., Portland, Oregon

[2]Clean Water Services, Hillsboro, Oregon

[3]City of Beaverton, Oregon

[1a]Email: jmelady@gsiws.com

[1b]Email: adavidson@gsiws.com

[2]Email: stenslandJ@CleanWaterServices.org

[3]Email: dwinship@beavertonoregon.gov

ABSTRACT

In the Pacific Northwest, wintertime urban stormwater represents an ample, cool, clean source of water. In most locations, stormwater is discharged to surface water at rates that are far higher than the natural (undeveloped) hydrologic regime, which can be detrimental to stream channels, water quality, and aquatic life. The technical and regulatory feasibility of using urban residential stormwater for artificial storage and recovery (ASR) in a deep basalt aquifer in Beaverton, Oregon, has been characterized and evaluated. The project is a collaboration between the City of Beaverton and Clean Water Services (CWS), a regional stormwater and wastewater management utility, and proposes to collect, treat, and store stormwater during the winter for reuse in the summer for non-potable purposes, such as irrigation and streamflow enhancement. This assessment—which included a review of existing regional stormwater quality datasets and site-specific water quality and quantity data collected in 2016-2017, and an evaluation of water treatment options, hydrogeologic characteristics, and regulatory viability—indicated that the project appears feasible.

The following is a summary of key feasibility information:

- The ability to use stormwater as a source for ASR is primarily dependent on compliance with Oregon underground injection control rules, which generally require that drinking water quality standards (one-half of the primary drinking water standard for most constituents) be met prior to injection. Results of the feasibility assessment confirm that a regulatory model exists for the proposed application.
- Of the approximately 170 constituents analyzed over the course of four stormwater sampling events in 2016-2017, only eight constituents, including turbidity and microbiological components [coliform (total and fecal), Escherichia coli (E. coli), and enteric viruses] exceeded regulatory criteria. Approximately 75 additional regulated analytes were detected, but at concentrations below relevant regulatory criteria [i.e. one-half of the maximum contaminant level (MCL) or below one times the secondary maximum contaminant level (SMCL)].
- Artificially recharged stormwater is anticipated to remain within 300 meters (1,000 feet) of the ASR well even under conservative storage scenarios and should not affect other groundwater users.

- Following treatment, stormwater quality is anticipated to be similar in character to treated surface water used successfully on many ASR projects hosted in the Columbia River Basalt Group (CRBG) aquifer. As such, mixing of treated stormwater with native groundwater is not anticipated to result in adverse chemical reactions.

The project would use ASR technology to store stormwater belowground in the aquifer and recover it during the summer months. This concept—of collecting, treating, storing, and repurposing stormwater—is an innovative application of ASR and has yet to be implemented in Oregon.

Keywords:

Stormwater reuse, aquifer storage and recovery, ASR, managed aquifer recharge, MAR, Columbia River Basalt Group, CRBG

INTRODUCTION

Like many cities in the western U.S., the Portland, Oregon, metropolitan area is experiencing a population boom and rapid residential expansion. In the suburb of Beaverton, Oregon, a major residential development is planned in the South Cooper Mountain area, near the steep and elevated areas of Cooper Mountain, to address this growth. Approximately 3,400 housing units and associated infrastructure are slated for construction over the next 20 years, with potential for development of another 4,000 housing units beyond the current planning horizon.

Effective stormwater management in this area will be essential to protecting surface water supplies throughout the Tualatin River Basin. In much of western Oregon, wintertime urban stormwater is discharged to surface water at rates that are far higher than the natural (undeveloped) hydrologic regime, which can be detrimental to stream channels and aquatic life. Stormwater runoff from impervious urban areas also reduces natural groundwater recharge and cool base flow, which in turn can impair summertime streamflows. Stormwater management is particularly challenging in the area of the planned South Cooper Mountain development because of the steep topography, which causes excessive stormwater flow velocity during wintertime storm events that is damaging to local streams.

Meanwhile, in the summer, streams in western Oregon are impaired from high water temperatures and low flows. Segments of the Tualatin River are covered under the Oregon Department of Environmental Quality (DEQ) 303(d) list of impaired or threatened waters and have established Total Maximum Daily Loads (TMDL) for a number of water quality parameters, including temperature. Landscape irrigation in this state-designated Critical Groundwater Area (CGWA) compounds the strain on groundwater and surface water supplies in the basin and diverts precious drinking water resources.

Although stormwater reuse is a relatively common practice across the southwestern U.S. and ASR has been implemented by a number of Oregon municipalities for drinking water storage, the use of treated stormwater as a source for ASR storage during the winter and use as nonpotable supply in the summer has not been assessed, and is the focus of this evaluation. Use of stormwater as a source for ASR would have the combined benefits of (1) reducing reliance on drinking water for landscape irrigation by creating a sustainable and non-potable supply (2) enhancing groundwater supply with a source that does not require obtaining new water rights; (3) providing a means for managing urban stormwater runoff, which can negatively affect surface water hydrology and quality; and (4) allowing the water to be retrieved at times of the year when demands are greater to supplement and benefit instream flows with cooler groundwater. Based on the feasibility assessment presented in this report, use of stormwater as a source for ASR appears to be feasible.

The stormwater ASR feasibility assessment was conducted at the Sterling Park Stormwater Quality Facility (Sterling Park Site), which is owned and operated by the City of Beaverton, Oregon (City). The site was selected for this project because it has the basic elements that are needed for a detailed feasibility assessment: (1) a deep well not currently in use (and, therefore, was available as a possible location for stormwater recharge), (2) a nearby small-diameter monitoring well that could be used for data collection during a pilot study, and (3) an onsite stormwater quality treatment facility that receives runoff from residential neighborhoods, with existing outfalls that could easily be routed to the recharge well if pilot testing is pursued.

Nearby Summer Creek, the discharge location for stormwater from the Sterling Park Site, is anticipated to be used as a discharge point of stored water to help mitigate summertime stream temperatures. Additionally, the City intends to use the deep well at the Sterling Park Site in the near future as source for "purple pipe" irrigation in nearby developing residential neighborhoods of South Cooper Mountain, and is interested in the possibility of using the reclaimed stormwater instead of native groundwater for this purpose. Since the site is in the Bull Mountain-Cooper Mountain CGWA, offsetting groundwater usage would prove especially beneficial.

PROJECT DESCRIPTION

The project evaluates the potential use of residential municipal stormwater as a source for ASR. Stormwater collected from a residential neighborhood in Beaverton, Oregon, that currently discharges into a stormwater water quality facility is being considered by the City and CWS as a source of water for planned pilot testing of stormwater ASR. Stormwater would be recharged into an existing and currently unused ASR test well (known as ASR 3) owned by the City and located adjacent to the stormwater water quality facility, which is referred to as the Sterling Park Water Quality Basin. The project site is shown in Figure 1.

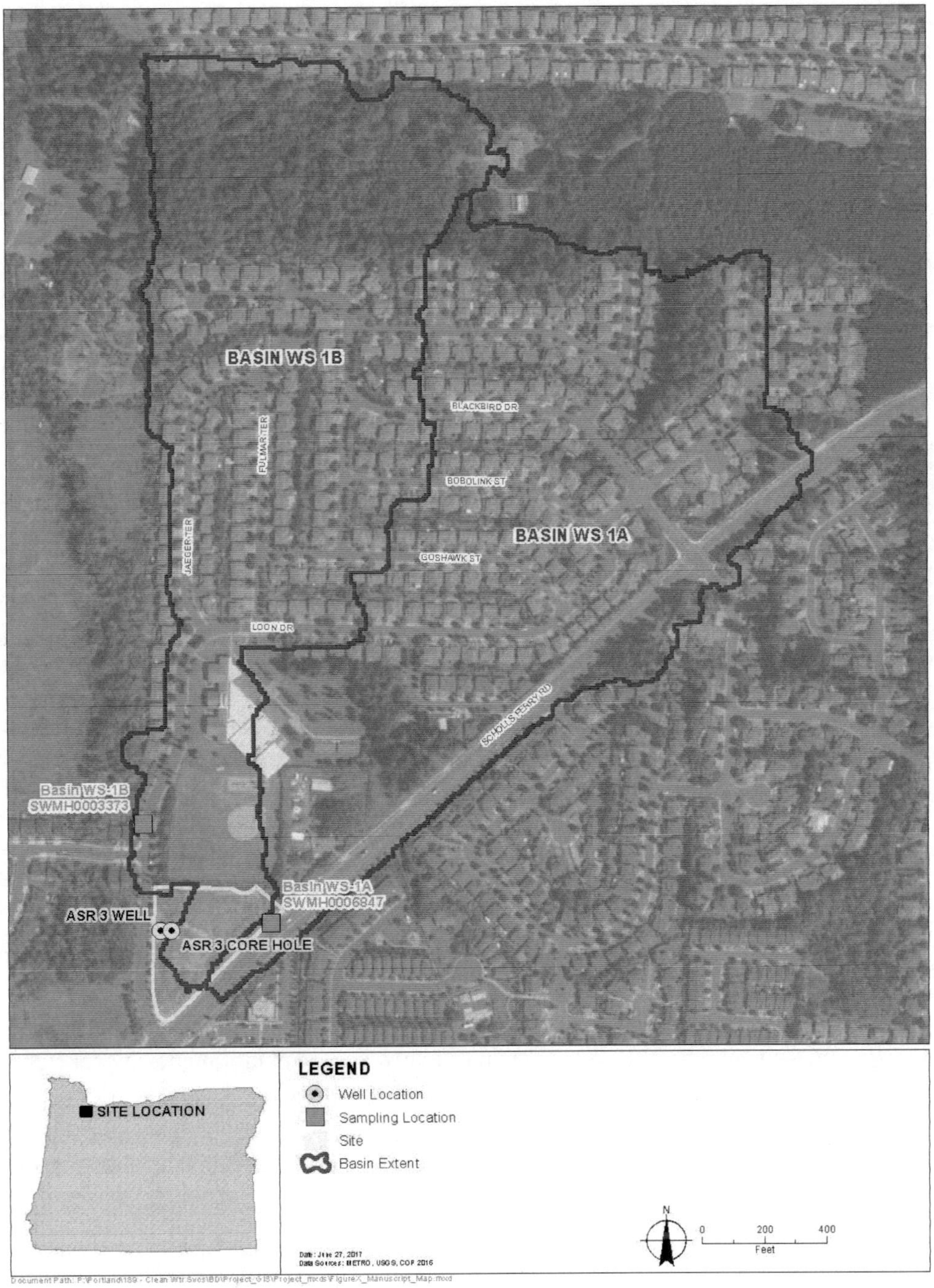

Figure 1. Project Location Map

ASR 3 extends to a total depth of approximately 305 m (1,000 feet) below ground surface (bgs). The well is 20.3 cm (8 inches) in diameter from the surface to 137 m (450 feet) bgs and 15.2 cm (6 inches) in diameter for its remaining depth. ASR 3 is hosted in the CRBG aquifer, which is host to several successful municipal ASR projects, including one operated by the City. The CRBG has been shown to be capable of storing millions of liters of recharged water and recovering the same stored water. Also located at the project site is a 302 m (992 feet) deep, 5.1 cm (2 inch) diameter well (known as ASR 3 Corehole), initially drilled to evaluate the CRBG section in this area; the well has a screen interval between approximately 247 and 302 m (809 and 992 feet) bgs. While ASR 3 Corehole is too small to be used for recharge of meaningful quantities of water, it could be used for water quality or water level monitoring purposes during ASR operation at this site. The locations of these wells are shown in Figure 1.

The aquifer in which ASR 3 is completed and that would be used for stormwater storage is considered a drinking water aquifer. Stormwater recharge into this aquifer would be designed to protect the highest beneficial uses of the aquifer, which is drinking water. Stored stormwater would be recovered from ASR 3 for non-potable uses, including irrigation and streamflow mitigation in nearby Summer Creek.

The conceptual plan is to capture stormwater between approximately November and April at the Sterling Park Site for recharge into ASR 3. Stormwater in the area is derived primarily from residential roads, sidewalks, driveways, and roofs. As discussed in subsequent sections, evaluation of representative stormwater quality data indicates that this residential stormwater runoff will meet all applicable water quality criteria with minimal treatment. Collection and analysis of site-specific stormwater quality data are being completed to confirm representative stormwater quality data. During potential future pilot testing, stormwater quality and flow at the project site would be monitored in accordance with applicable DEQ and Oregon Water Resources Department (OWRD) permit requirements to ensure protection of the CRBG aquifer for its highest beneficial use: drinking water.

Some of the potential benefits of this project are listed below. By extension, if the project is successful, these types of benefits could be realized at other locations in the region. Benefits include:

- Reducing reliance on highly treated municipal drinking water for landscape irrigation, thereby reducing demand on surface water and native groundwater.
- Providing direct recharge to the local basalt aquifer (CRBG), thereby enhancing groundwater supply with a source that does not require obtaining new water rights.
- More closely mimicking the natural hydrologic cycle by reducing unnaturally large runoff volumes from impervious surfaces to surface water during periods of high flow, and mitigating the negative impacts to streams from rapid changes to streamflow (e.g. elevated solids concentrations and bank erosion).
- Maintaining summer flows and reducing stream temperature in nearby surface water bodies, such as in Summer Creek, near the Sterling Park Site, through the discharge of recovered stormwater in the summer months.
- By developing the area around ASR 3, infiltration of stormwater may preclude the need to install or increase the capacity of piped stormwater infrastructure in this area.

REGULATORY FEASIBILITY

Identifying applicable regulations and standards that need to be met was an important step in evaluating the feasibility of stormwater ASR in general and for this proposed project specifically. Preliminary discussions with state regulators were completed to assess potential regulatory concerns related to using stormwater as a source of water for an ASR project. The proposed project would fall under the following general sets of regulations, as discussed below: ASR, Underground Injection Control (UIC), and municipal separate stormwater system permitting. If the stored stormwater is withdrawn and discharged to Summer Creek to supplement streamflow or to mitigate stream temperatures, this discharge would be done in accordance with all applicable water quality criteria.

ASR Regulatory Requirements

ASR in Oregon is administered by OWRD in consultation with DEQ and Oregon Health Authority (OHA). OWRD's rules governing ASR are described in Oregon Administrative Rules (OAR) 690-350. General requirements include authorization of recharge source water, typically through a water right; recharge source water quality requirements; and hydrogeologic assessments necessary to evaluate the viability of a proposed project and the potential for injury to other groundwater users.

Authorization of recharge source water for this proposed project would be different from most ASR projects because OWRD does not require a water right for use of stormwater, which is defined as "precipitation collected from an artificial impervious surface" under Oregon Revised Statute (ORS) 537.141 (h). Water quality requirements for source water for ASR projects are based on drinking water quality standards that reference U.S. Environmental Protection Agency (EPA) maximum contaminant levels (MCLs) and secondary MCLs (SMCLs). Specifically, allowable concentrations are limited to one-half the MCL for most constituents, with the exception of analytes with an SMCL and disinfection by-products (DBP), which allow recharge water with concentrations up to the SMCL (OAR 690-350-0020(5)(i)). Microbiological constituents in recharge source water also are regulated on the basis of drinking water requirements of 4-log inactivation.

Federal and state rules prohibit the construction, operation, maintenance, conversion, plugging, or abandonment of any type of injection system or activity that would allow the direct or indirect movement of contaminated fluids into groundwater if the presence of the contamination may cause a violation of the Safe Drinking Water Act (SDWA) MCLs. In addition, the injection system must comply with DEQ's Groundwater Quality Protection Rules in OAR 340-040, which require that discharges meet existing background water quality at a compliance point that DEQ chooses (usually the property boundary).

UIC Regulatory Requirements

The proposed recharge system would also fall under DEQ's UIC rules. DEQ regulates the UIC Program under OAR Chapter 340, Division 44; these rules regulate all groundwater as a potential source of drinking water. According to the UIC rules, the burden of proof is on the owner/operator of the system, not DEQ, to prove that an injection activity does not have the potential to cause a violation of the primary drinking water standards or adversely impact groundwater quality, human health, or the environment. Before operating the proposed system,

DEQ would require the owner/operators of the injection system to register the injection system and gain written DEQ approval to operate by either of the following:

- Qualifying as a rule-authorized UIC.
- Receiving a Water Pollution Control Facility (WPCF) permit (i.e., UIC permit).

Under current regulations, the proposed system would not qualify as a rule-authorized UIC, given that the proposed system is deeper than 30.5 meters (100 feet) and discharges to groundwater. Therefore, DEQ is expected to require a WPCF permit to manage the use of stormwater for aquifer recharge.

In summary, the existing regulatory framework provides a structure and process for permitting and operating a stormwater ASR system.

STORMWATER QUALITY

A key component of this feasibility evaluation was to assess whether the general quality of stormwater that would be recharged into ASR 3 is suitable for injection. Specifically, because the beneficial use of the target CRBG aquifer is for drinking water, any stormwater considered for ASR injection must be of a quality that would protect the native groundwater as a drinking water source. Site-specific stormwater quality data were collected and reviewed along with stormwater quality data collected from similar regional municipal and residential area studies. Data were compared to screening values to identify stormwater chemicals/analytes of interest (COIs) that may be present at concentrations that could adversely affect the native groundwater and thus require treatment or removal before ASR injection. Consistent with the regulatory requirements discussed above, stormwater COIs were screened against one-half their respective MCL, with the exception of analytes with an SMCL, which were screened at the full SMCL value.

STORMWATER SAMPLING AND ANALYSIS

Four stormwater sampling events were conducted in 2017 within drainage basins WS 1A and WS 1B. Samples were collected from two stormwater manholes just upstream of outfalls discharging to the Sterling Park site. Locations of the two sampling points within basins WS 1A and WS 1B are depicted in Figure 1. Minimum storm event criteria[1] were targeted, but samples were collected from a range of seasonal and storm conditions, including "first flush[2]" conditions, to determine the range of COI concentrations that may be present in stormwater discharges to the Sterling Park Water Quality Basin throughout the rainy season. Time series grab samples were collected at both locations during two storm events (Events 1 and 3) to assess COI concentrations and concentration trends for the course of the storm events, and single grab samples were collected from both locations during the other two events (Events 2 and 4). Samples were collected in accordance with a sampling and analysis plan and analyzed for either (1) a comprehensive "full" suite of COIs including all chemicals for which an MCL, SMCL, or treatment technology (TT) is available as well as additional emerging contaminants (e.g., new pesticides), and a variety of viral, bacterial, and protozoan pathogens; or (2) a shorter list of indicator analytes deemed to be representative of common COIs that could be encountered in the

---

[1] OAR 340-044-0018 storm event criteria that were targeted included (1) antecedent dry period of at least 72 hours with less than 0.1 inch rain, (2) minimum predicted rainfall volume greater than 0.1 inch per event, and (3) expected duration of storm event of at least 3 hours.

[2] "First flush" is defined in OAR 340-044-0018 and DEQ's *Guidance for Evaluating Stormwater Pathways at Upland Sites* (DEQ, 2009, Updated 2010) to mean within the first 30 minutes of stormwater discharge.

Sterling Park area or similar residential land use areas. Table 1 provides a summary of the stormwater sampling events and associated analyses.

Table 1. Summary of Stormwater Sampling and Analyses

Sampling Event	Date	Time Series	Stormwater Drainage Basin	
			WS-1A (Scholl's)	WS-1B (Loon)
1	2/15/2017	Sample 1*	♦	♦
		Sample 2	•	•
		Sample 3	•	•
2	6/8/2017	Sample 1	•	•
3	10/11/2017	Sample 1*	•	♦
		Sample 2	•	•
		Sample 3	•	•
4	11/8/2017	Sample 1	•	•

Notes:

* Targeted first flush

♦ Indicates sample was analyzed for comprehensive list of COIs identified in the SAP (CWS and GSI, November 2016).

• Indicates sample was analyzed for list of indicator COIs identified in the SAP.

COI = chemical of interest

SAP = sampling and analysis plan

STORMWATER DATA SCREENING

In general, many of the COIs that were analyzed as part of the sampling events were not detected above method reporting limits (MRLs), and concentrations of COIs that were detected in stormwater samples were generally below screening criteria limits of one-half the MCL or one times the SMCL.

Of the approximately 170 COIs analyzed in the comprehensive or "full" suite, eight COIs exceeded their respective screening level value in one or more sample:

- Turbidity
- Apparent color
- Total aluminum
- Total iron
- Fecal coliform
- Total coliform
- E. Coli
- Culturable enteric viruses

Three of these analytes (apparent color, total aluminum, and total iron) do not have MCLs, but instead have non-mandatory SMCLs, which are developed for aesthetic considerations such as taste, color, and odor. The remaining COIs, including turbidity and the microbiological parameters (coliform bacteria and enteric viruses), are ubiquitous in residential stormwater runoff. Concentrations of many COIs are expected to increase with increasing concentrations of suspended solids and particulates entrained in the stormwater runoff, and indeed many of the constituents that exceeded screening criteria have a positive correlation with increasing suspended solids concentrations. Accordingly, it is expected that many of these COI concentrations could be greatly reduced with an effective solids removal process. However, it should be noted that coliform data appear to have less association with suspended solids concentrations than with increasing temperatures.

Microbiological constituents are an important consideration in drinking water quality standards, but typically are of less interest in stormwater management because of the end use of routing stormwater to existing source waters[3]. Preliminary discussions with regulators and recent work completed in support of various municipal UIC programs indicates that microbiological constituents present in stormwater attenuate after injection and may not require treatment before recharge. However, this evaluation included analysis of potential impacts to the aquifer from microbiological constituents resulting from stormwater ASR. Potential survival of pathogens, particularly viruses, is a concern, particularly for a period of time long enough to impact the beneficial use of groundwater as a drinking water resource (e.g., potential impact to a hypothetical nearby private or municipal well by the stored water). For these reasons, disinfection or other effective antimicrobial treatment processes before ASR injection have been considered.

It is worth noting that while concentrations of zinc never exceeded the SMCL of 5,000 micrograms per liter (μg/L), zinc and other metals may need to be further reduced to meet requirements of CWS's basin-wide individual permit for discharge into the Tualatin Basin. Because stormwater would be used to augment and cool adjacent streams during summer months, it would be subject to these requirements and a metals removal processes (e.g., via adsorptive media) should be considered.

REVIEW OF MUNICIPAL URBAN STORMWATER DATA

In addition to site-specific stormwater data, municipal urban stormwater data in Oregon were reviewed as a further evaluation of anticipated water quality in stormwater that would be used for ASR. Urban stormwater data were obtained from two sources:

- Kennedy/Jenks (2009), a compilation of stormwater data from 15 public agencies in Oregon
- DEQ (2015), a municipal UIC database that includes the Cities of Portland, Gresham, Redmond, Bend, and Keizer

The above-referenced studies focus on approximately 40 permit-defined COIs. Of these 40 permit-defined COIs, 10 were detected at concentrations greater than their respective MCLs in one or more sample. Of those 10, five COIs were detected at concentrations that exceeded MCLs in approximately 1 percent or more of the samples:

- benzo(a)pyrene (BaP)
- total chromium

[3] The current NPDES benchmark for E. coli at active landfills and sewage treatment plants is 406 counts/100ml.

- di(2-ethylhexyl)phthalate (DEHP)
- total lead
- pentachlorophenol (PCP)

Concentrations of PCP and total lead were observed to exceed their respective MCL values most frequently, with 10 to 15 percent of the samples exceeding MCLs. The other constituents exceeded the MCLs in less than 1 percent of the samples included in the studies.

EVALUATION OF PORTLAND METROPOLITAN AREA RESIDENTIAL STORMWATER DATA

Additionally, 9 years of the City's UIC Program stormwater data were reviewed. Data were collected from locations with single-family residential land use similar to that in the Sterling Park residential development area. UIC stormwater samples from these areas were collected at the point of discharge into individual UIC facilities in public rights-of-way, and reported to DEQ in compliance with the City's municipal WPCF permit (i.e., UIC permit). The City's UIC permit required that any analyte detected at any concentration be reported (i.e., ancillary pollutants). Of the 100+ permit-required and ancillary COIs analyzed during the 9-year period, the following constituents were detected at concentrations greater than one-half the MCL in at least one sample:

- Antimony (total)
- Arsenic (total)
- Benzene
- Benzo(a)pyrene
- Cadmium (total)
- Chromium (total)
- DEHP
- Lead (total)
- PCP

This list of COIs exceeding screening values is similar to COIs observed in the two municipal stormwater studies. Five of the nine COIs listed above (antimony, arsenic, benzene, cadmium, and chromium) were detected in less than 1 percent of the samples. The four remaining COIs—benzo(a)pyrene, DEHP, total lead, and PCP—were detected at concentrations greater than one-half of the MCL in more than 1 percent of the samples. However, geometric mean concentrations for these four COIs during the 9-year period did not exceed one-half of their respective MCL.

STORMWATER QUALITY EVALUATION SUMMARY: SUITABILITY FOR RECHARGE

The concentrations of stormwater COIs observed in site-specific stormwater samples and those evaluated from applicable local and regional stormwater datasets indicate that (1) local municipal residential stormwater generally meets screening criteria, and (2) recharge of this water into the aquifer would not be expected to adversely affect the beneficial use of the aquifer if coupled with minimal treatment. The COIs exceeding screening criteria in site-specific stormwater samples and the reviewed regional stormwater data are ubiquitous in the environment (City of Portland Bureau of Environmental Services, 2008; DEQ, 2015). A large fraction of the COI concentrations likely is associated with stormwater particulates that could be removed via an effective suspended solids removal process, and the remaining dissolved fractions could be removed through treatment steps, such as activated carbon adsorption. Concentrations of particulates and COIs that do migrate to the aquifer would have limited mobility in the subsurface and would be expected to be further reduced through the processes of mechanical filtration, degradation, dispersion, and adsorption.

Microbiological data (bacteria, protozoa, and viruses) were not analyzed in the Portland UIC data. However, these constituents are known to be ubiquitous in the environment and were observed in site-specific samples. Disinfection or other anti-microbial treatment processes (e.g., ultraviolet [UV] irradiation) would need to be employed before ASR injection to meet primary drinking water standard treatment technology requirements and safeguard the aquifer's beneficial use.

TEMPERATURE

In addition to chemical and biological COIs, water temperature is an important physical water quality property and of interest for the project because the recovered stormwater may be used for streamflow mitigation during warmer months. Maintaining cool stream temperatures in the summer is important for supporting aquatic ecosystems. Cool stormwater, stored in the aquifer, can be pumped from the subsurface, and discharged to mix with surface water, thereby potentially reducing the overall thermal load to a stream.

Temperature data collected during the site-specific stormwater sampling and from Portland's UIC data were evaluated to estimate typical temperatures of the stormwater proposed for recharge. Site-specific temperature data ranged from 6.96 to 9.09 degrees Celsius (°C). During the 9-year period of UIC data reviewed for this evaluation, stormwater temperatures ranged from 6.4 (in January) to 14.4°C (in October) (see Table 2).

Table 2. Geometric Mean Stormwater Temperature in UICs (residential land use, Portland, Oregon)

Month	Number of Samples	Temperature (°C)
October	75	14.4
November	90	9.6
December	62	7.6
January	79	6.4
February	83	6.9
March	110	8.4
April	63	10.5
May	15	13.4

Notes:

Blue highlight = likely months of stormwater recharge

°C = degrees Celsius

UIC = underground injection control

The site-specific and UIC temperature data indicate that the temperature of banked stormwater used to recharge ASR 3—particularly runoff generated during cooler months—could be used to cool streamflow (i.e., streamflow mitigation) in nearby Summer Creek. Summer Creek streamflow temperature data are not available, but temperature data from Fanno Creek downstream from its confluence with Summer Creek indicates an average summertime (June-September) streamflow temperature of approximately 19°C (USGS, 2015). Observation from other regional ASR projects indicate that stored water quality characteristics, including temperature, are generally maintained during ASR storage. Given the temperature differential between wintertime stormwater and summertime streamflow temperatures, a significant thermal benefit is anticipated from the project.

STORMWATER TREATMENT PRIOR TO RECHARGE

Based on a review of the preliminary data collected from the Sterling Park Water Quality Basin and a review of local municipal residential stormwater data, the majority of COIs present in stormwater discharging to Sterling Park are likely to be at concentrations below ASR regulatory criteria. However, several groups of constituents, including metals, turbidity and nuisance parameters (e.g., odor, color), pathogens, pesticides, PAHs, and phthalates, may be present in stormwater discharges above the regulatory requirement of less than one-half the MCL or one times the SMCL during some storm events. A substantial fraction of these constituents likely is associated with suspended stormwater particulates that could be reduced through an effective solids removal process before recharge. For the dissolved fraction of COIs, adsorptive materials, such as biochar or activated carbon, may be required to reduce COI concentrations to less than regulatory levels. Table 3 presents COIs detected at levels exceeding the screening criteria in the site-specific, local, and regional stormwater quality studies along with several potential treatment options that were researched and could be employed for COI removal.

Table 3. Stormwater COIs Potentially Requiring Treatment at Sterling Park Water Quality Facility

Data Source			Treatment Options
Sterling Park Site-Specific Stormwater Data (2017)	**Oregon Municipal Stormwater Studies (Kennedy/Jenks, 2009; DEQ Municipal Database, 2015)**	**DEQ Municipal Database (2015)**	
Solids, Nuisance, Odor			
Turbidity, Apparent Color, Odor	Not analyzed	Not analyzed	Turbidity may be removed with flocculants/coagulants, filtration, sedimentation, adsorption, or some combination of all of these processes. Water quality issues associated with color and odor may be greatly reduced after suspended solids removal processes.
Metals			
Zinc (total), Aluminum (total), Iron (total)	Antimony, Arsenic (total), Cadmium (total), Chromium (total), Lead (total), Zinc (total)	Antimony, Arsenic (total), Cadmium (total), Chromium (total), Lead (total)	Particulate portion of the total metals can be reduced through filtration processes (e.g., sand filter). Concentrations of dissolved metals may be removed through activated carbon/biochar adsorption, ion exchange processes, ultra or membrane filtration, precipitation, electrodialysis, or distillation.
PAHs, Phthalates, and Other Organics			
No COIs above screening levels	Benzo(a)pyrene, DEHP	Benzo(a)pyrene, DEHP, Benzene	Activated carbon is particularly effective in removing concentrations of dissolved organics such as PAHs, phthalates, and benzene.
Nitrate-Nitrogen			
No COIs above screening levels	Nitrate-Nitrogen	No COIs exceeding the screening level	Likely would require ion exchange units, reverse osmosis, or distillation. Some nitrate-nitrogen may be removed through filtration and adsorptive treatment steps or biological processes
Pesticides			
No COIs above screening levels	PCP	PCP	Activated carbon and other adsorptive media, such as organoclay, are often effective in removing pesticides such as PCP.
Potential Pathogens			
Fecal Coliform, Total Coliform, E. Coli, Culturable Enteric Viruses	Not Analyzed	Not Analyzed	Various effective means of disinfection are readily available including chlorine, chloramines, ozonation, ultraviolet irradiation, etc.

Notes:

COI = chemical of interest

DEHP = di(2-ethylhexyl)phthalate

DEQ = Oregon Department of Environmental Quality

PAH = polycyclic aromatic hydrocarbon

PCP = pentachlorophenol

PRELIMINARY TREATMENT SCHEME AND LAYOUT

Based on a review of various treatment processes that may be needed to eliminate concentrations of COIs potentially exceeding regulatory criteria, a conceptual preliminary treatment train was generated for a gravity-fed system that incorporates the key components of effective treatment for the observed flows and water quality: settling, filtration, adsorption/treatment, and disinfection.

The preliminary treatment train was used to explore treatment options including construction of a custom stormwater treatment system for the Sterling Park Water Quality Facility and installing pre-fabricated stormwater treatment systems. Key criteria used to evaluate these options included capital costs, operation and maintenance (O&M) costs, expected lifetime of the system, flow capacity, reliability of the system, and access to installation and operation support. In addition to constructing a site-specific treatment system, pre-fabricated stormwater treatment systems from Contech Engineered Solutions, Crystal Stream Technologies, Lakeside Equipment Corporation, and StormwaterRX were evaluated.

Although more detailed treatment design options will be assessed and fine-tuned during a pilot treatment system, a StormwaterRX treatment system was found to be cost-effective and particularly applicable based upon the options that were researched in this study. The future pilot test program will allow for a more cost-effective evaluation of full-scale implementation options.

For feasibility costing and treatment purposes, the following treatment components were assumed:

- An Aquip® Model 400SBE filtration system, enhanced stormwater filtration system designed for a treatment flow rate of 1,514 liters per minute (lpm) (400 gpm) with an aboveground system footprint of 4 x 15 meters (13 x 48 feet). Includes pre-treatment chamber and integrated oil skimmer, and down-flow filtration chamber with layered inert and adsorptive biofiltration media for particulates, organics, and dissolved pollutants.
- A standard treatment pump and bypass system for a 400SBE system including 1.5 horsepower (208 volt/1Ph/15A) pumps operating at 1,514 lpm (400 gpm) with less than 8.5 meters (28 feet) of total dynamic head, float switches, and check valves.
- A Purus® Bacteria Model 400-volt UV disinfection and polishing system designed to disinfect flows up to 1,514 lpm (400 gpm) with an aboveground foot print of 0.5 x 3.7 meters (1.5 x 12 feet). Includes 32 UV lamps with a total system continuous power requirement of 2,800 watts.

PROJECT IMPLEMENTATION PLAN

To begin development of an ASR system with use of treated stormwater at the Sterling Park Site are provided below several proposed actions are being targeted in the near term.

Preliminary Design Pilot Testing

Before full-scale construction, initial pilot testing is anticipated to confirm the efficacy of the proposed treatment technique and to confirm water quality elements during aquifer storage. The following is a general description of anticipated tasks and schedule during this phase of implementation:

- Stormwater treatment design pilot testing:

- The stormwater treatment pilot system would consist of a scaled treatment system (anticipated to be approximately 5 to 10 percent of full-scale design capacity) with filtration and UV disinfection capabilities.
- Water quality analysis of pre- and post-treated stormwater at the Sterling Park Site would be completed to confirm the level of treatment and to develop full-scale design parameters.

- Recharge pilot testing:
  - A scaled recharge cycle at ASR 3 would be completed with treated stormwater anticipated to be approximately 24 hours of recharge, a 24-hour storage period, and a 24-hour recovery period.
  - Water quality analysis of native groundwater at ASR 3, treated stormwater, stored water, and recovered water would be completed to evaluate water quality changes during artificial recharge.
  - Aquifer parameters would be evaluated during recharge testing to confirm the ability to recharge at projected rates and volumes.

Full-Scale Treatment System Design, Construction, and Implementation

Based on information obtained during initial pilot testing, the next steps are anticipated to be: develop project regulatory permitting documents, complete design, and construct full-scale treatment and conveyance infrastructure from the existing stormwater system to the ASR 3 wellhead. The following is a general description of anticipated tasks and schedule during this phase of implementation:

- Regulatory permitting: ASR limited license and UIC permit.
- Stormwater treatment and conveyance system design and construction.
- Initiate ASR testing including detailed evaluation of treated stormwater water quality at full scale, baseline native groundwater quality, and recovered water quality. Groundwater level monitoring and assessment of aquifer response to recharge and recovery.

Operation and Maintenance

Full-scale system O&M is anticipated to consist of water quality analysis, groundwater-level monitoring, treatment system and filter media maintenance, electrical costs, and annual regulatory reporting.

GROUNDWATER MIXING

Mixing stormwater runoff with native groundwater has the potential to result in precipitation of constituents that could result in clogging of the well. Although no direct stormwater quality data are available to assess mixing compatibility with the native groundwater, based on a review of representative data presented above and experience with other ASR projects in the area, it is believed that no adverse reactions would occur to cause clogging. This is predicated on the understanding that stormwater chemistry is similar, except for turbidly, to source water used to recharge current ASR wells hosted in the CRBG aquifer. The latter has not resulted to date (in more than 15 years of ASR activity) in clogging of ASR wells. There is a good possibility that recharging an ASR well with local stormwater, assuming it is stripped of turbidity, will not result in chemical precipitation of constituents and clogging. As part of the pilot test assessment of this potential ASR recharge project, local water quality data currently are being collected and mixing of this water with native groundwater in the target aquifer will be completed to assess compatibility between the waters and the potential for clogging.

STORMWATER FLOW ESTIMATE

This section provides a summary of stormwater flow information obtained at the Sterling Park Site to evaluate stormwater flow rates, variability, and annual volume. Understanding these elements is necessary to (1) size various conveyance and treatment options for the proposed project and (2) confirm that a sufficient volume of stormwater is available on an annual basis to make ASR viable to offset anticipated non-potable groundwater use.

Stormwater flow from the two contributing stormwater drainage basins (Basin WS 1B and Basin WS 1A; see Figure 1) was monitored continuously from October 2016 to November 2017 with Hach® flow loggers (model F1901) equipped with a FLO-DAR, model 4000, radar/ultrasonic sensor. The flow loggers were monitored monthly by CWS personnel and were inspected at the start of each stormwater sampling event to ensure the logger was functioning properly. In addition to stormwater flow, precipitation data during this same period was reviewed from the nearest City of Portland HYDRA Rainfall Network[4] at the Sylvania Portland Community College (PCC) rain gauge (approximately 11 km [7 miles] from the Sterling Park Site). Based on the precipitation data, a total of approximately 100 discrete storm events[5] were observed during the monitoring period. Total flow volumes measured during this period were approximately 144 million (M) liters (38 million gallons [MG]) at the Basin WS 1A monitoring location and approximately 72 M liters (19 MG) at the Basin WS 1B monitoring location, with a combined volume between the two basins of approximately 216 M liters (57 MG).

Stormwater flow varied at the two stormwater basins between several 10 lpm (1-3 gpm) and more than 3,785 lpm (1,000 gpm) during peak periods of precipitation. Based on the flow rates observed:

- Approximately 50 percent of the total volume from Basin WS 1A discharged at rates less than 662 lpm (175 gpm), which represents a volume of 72 M liters (19 MG).
- Approximately 50 percent of the total volume from Basin WS 1B discharged at rates less than 322 (85 gpm), which represents a volume of 36 M liters (9.5 MG).
- Approximately 50 percent of the combined total volume from both basins discharged at a combined rate of approximately 984 lpm (260 gpm), which represents a volume of 108 M liters (28.5 MG). The highest flow rates observed (> 7,571 lpm or >2,000 gpm) were relatively uncommon and accounted for less than 10 percent of the combined flow and a total duration of less than 2 percent of the period of observed stormwater flow.

Stormwater treatment anticipated for the proposed project typically is sized with a design flow rate, with larger and more expensive treatment required at higher flow rates. Without any pre-treatment system storage, a treatment system capable of accepting flow rates up to 757 lpm (200 gpm) could have treated 98 M liters (26 MG) of stormwater over the flow monitoring period, and a system capable of accepting flow rates up to 1,514 lpm (400 gpm) could have treated up to 172 M liters (33.5 MG) of stormwater from October 2016 to November 2017.

---

[4] The HYDRA Rainfall Network is a collection of 39 gauges operated and maintained by the City of Portland's Bureau of Environmental Services.

[5] Discrete storm events are defined by rainfall lasting at least 1 hour and separated by at least 12 hours with no precipitation.

Approximately 162.6 cm (64 inches) of rain were observed over the flow monitoring period, which is atypically wetter than average stormwater years. The average annual rainfall in Beaverton is 101.6 cm (40 inches) per year. Despite the unusually wet monitoring period, the discharge volumes from both basins indicate that a large portion of the City's anticipated non-potable groundwater use could be offset with captured, treated, and injected stormwater on an annual basis. For context, the City is anticipated to use approximately 197 M liters (52 MG) of groundwater [based on a 1,514 lpm (400 gpm) design pumping rate for a 90-day irrigation season] to meet non-potable demands on an annual basis, as well as potentially provide streamflow enhancement along Summer Creek. Capturing and treating stormwater with flow rates less than 1,514 lpm (400 gpm) has the potential to offset as much as half of the anticipated groundwater usage on an annual basis. Additional volumes of stormwater could be captured and treated if stormwater detention or storage is built into the conveyance system prior to treatment. Excess stormwater volume due to flow rates exceeding treatment capacity rates would be stored in available detention as design treatment flow rates are metered into the treatment system. Maximizing stormwater capture and treatment will be a key aspect of the final design phase.

POTENTIAL FOR STORED WATER MIGRATION

This section evaluates the potential migration of treated stormwater during ASR storage and assesses the potential for stored water to interact with other groundwater users. To evaluate groundwater flow, a previously developed 3-dimensional numerical groundwater flow model (GSI, 2011) was modified to predict flow paths of water recharged at ASR 3. The model uses the U.S. Geological Survey (USGS) MODFLOW-2000 finite-difference groundwater modeling software (Harbaugh et al., 2000), and the Groundwater Vistas graphical user interface (ESI, 2007) is used to manage the modeling process. The model is calibrated to historical data from regional ASR programs. MODPATH particle tracking software with forward particle tracking to determine the advective transport of the recharged stormwater was used to delineate zones of influence from water recharged at ASR 3 based on the following assumptions:

- Based on stormwater flow data from basins WS-1A and WS-1B in 2017, 185 M liters (49 MG) of treated stormwater were conservatively assumed as an annual recharge volume. Recharge would occur between October and May at rates up to 1,514 lpm (400 gpm).
- Two predictive model scenarios were selected to predict the most conservative conditions with respect to travel distance of the naturally attenuated surface water:
    - Scenario 1: Assumed a total recharge volume of 185 M liters (49 MG) of treated stormwater from October through May for 1 year without any recovery pumping. No recovery pumping for 1 year is conservative and represents a potential pump failure after the recharge has occurred.
    - Scenario 2: Assumed a total injection volume of 185 M liters (49 MG) of treated stormwater from October through May for 4 consecutive years (total volume of 742 M liters (196 MG) in 4 years) without any recovery pumping. No recovery pumping represents a conservative scenario and unlikely potential pump failure for multiple years following each of the annual recharge periods.
- Stormwater treatment is anticipated to include UV treatment for microbiological inactivation. In the event inactivation is incomplete during the treatment process, based on natural microbiological attenuation within the aquifer as described previously, 4-log attenuation is expected for the most conservative viral inactivation rate within 500 days of entering the groundwater system.

Based on these assumptions, Table 4 summarizes potential migration distances of stored water from ASR 3 may occur at 500 days following recharge for the two scenarios. A safety factor of 2x (1,000 days) and 3x (1,500 days) are included for comparison.

Table 4. Potential Stored Water Migration Distance from ASR 3

Travel Time	Potential Migration Distance from ASR 3	
	Scenario 1	Scenario 2
500-day	253 m (830 ft)	326 m (1,070 ft)
1,000-day (2x Safety Factor)	268 m (880 ft)	460 m (1,510 ft)
1,500-day (3x Safety Factor)	283 m (930 ft)	549 m (1,800 ft)

A review of nearby existing water well logs from OWRD databases was completed to assess the distance between the Sterling Park Site and any potential groundwater wells in the areas. Two well logs were identified, indicating two wells were constructed in 1971 and located approximately 229 meters (750 feet) south of ASR 3, based on tax lot information from OWRD. Given the conversion of nearly all parcels in this area from small farms to dense residential development since these wells were constructed, it is likely that they are no longer in use and have been abandoned. OWRD well records indicate that 25 water wells have been abandoned within the one square mile surrounding ASR 3 (Township 2 South, Range 1 West, Section 5). Specifically, two abandonment logs appear to be associated with the decommissioning of these two nearby wells based on the address, tax lot, and well construction information. No other wells were identified in a search of OWRD's water well records within the extent of potential stored water migration predicted within 1,500 days of storage.

Similarly, in the surrounding vicinity, high-density residential development is underway and most wells that previously were constructed in this area for domestic use are no longer in use and have been abandoned. Based on OWRD records, the four sections encompassing and surrounding the Sterling Park Site, which represents the area within approximately 1.6 km (1 mile) of the Site, historically has contained 72 wells (at least 2 of which appear to be mislocated) and 67 wells have been decommissioned since the early 1990s. Based on the dense residential development that has occurred and is being initiated in the vicinity, former domestic water supply wells likely are no longer in use and have been decommissioned. Beyond the area within 1 mile of the Sterling Park Site, several municipal water supply agencies use groundwater wells for ASR in the region; none of these are closer than 3.2 km (2 miles) from the Sterling Park Site, which is well beyond the area of potential influence of treated stormwater proposed for recharge at this location.

As previously stated, OWRD's Critical Groundwater Area declaration has restricted existing groundwater use and prohibited issuance of new groundwater rights and construction of wells for irrigation or domestic use on properties of less than 10 acres in this area since 1971. Ultimately, the Critical Groundwater Area limits the number of potential nearby wells that could capture recharged water from ASR 3 in the future.

CONCLUSIONS

This section provides a summary of findings related to the use of stormwater as a source water for ASR at the Sterling Park Site. Based on the evaluation of the applicable regulations and the assessment of the physical and chemical parameters, stormwater recharge at the Sterling Park Site appears feasible, with minimal treatment, to meet regulatory requirements. Specifically:

- Regulatory feasibility for use of stormwater as a source for ASR is primarily dependent on compliance with drinking water quality standards (one-half of the primary drinking water standard for most constituents).
- Analysis of continuous stormwater flow data collected between October 2016 and November 2017 indicate a volume of nearly 216 M liters (57 MG) of stormwater was discharged to the Sterling Park Site.
- Stormwater sampling conducted for this study indicates drinking water standard exceedances for a small number of constituents including turbidity, apparent color, aluminum, iron and microbiological components (coliform, fecal, E. coli, and enteric viruses). Several other regulated analytes were detected, but at concentrations below one-half of respective drinking water standards.
- Temperature data indicate that wintertime stormwater used for recharge typically ranges from 6.4 to 9.6°C. This cool water could be recovered during the summer and used to mitigate the temperature in Summer Creek and provide streamflow enhancement. Summer Creek streamflow temperature data are not available, but it discharges to Fanno Creek, which has an average summertime streamflow temperature of approximately 19°C (USGS, 2015).
- Stormwater quality treatment would be needed to meet regulatory requirements and is anticipated to consist of filtration, activated carbon/biochar adsorption, and disinfection (e.g., chlorine or UV irradiation).
- Following treatment, stormwater quality is anticipated to be similar in character to treated surface water used successfully on many ASR projects hosted in the CRBG aquifer. As such, mixing of treated stormwater with native groundwater is not anticipated to result in adverse chemical reactions.
- Artificially recharged stormwater is anticipated to remain within 305 meters (1,000 feet) of ASR 3 even under conservative storage scenarios and would not impact other groundwater users.

Next steps for implementation of the proposed project include stormwater treatment pilot testing, regulatory permitting application, and final treatment design and construction.

REFERENCES

City of Portland, Bureau of Environmental Services (BES). 2006a. *Stormwater Discharge Monitoring Plan* [consists of Sampling and Analysis Plan and Quality Assurance Project Plan]. February 2006. Version 1.

BES. 2008. City of Portland, Bureau of Environmental Services (BES), *Decision Making Framework for Groundwater Protectiveness Demonstrations, Underground Injection Control System Evaluation and Response,* June 2008.

City of Portland, Bureau of Environmental Services (BES). 2012a. *Stormwater Discharge Monitoring Plan* [consists of Sampling and Analysis Plan and Quality Assurance Project Plan]. December 2012. Version 2.

City of Portland Bureau of Environmental Services. 2014. *Annual Stormwater Discharge Monitoring Report—Year 9—October 2013-May 2014.* November 2014.

CWS and GSI. 2016. Sampling and Analysis Plan. 2016/2017 Stormwater Investigation, Sterling Park Stormwater Quality Facility, Groundwater Recharge Feasibility Evaluation, Beaverton, Oregon. Prepared for Clean Water Services (CWS) and the City of Beaverton. Prepared by CWS and GSI Water Solutions, Inc. November 2016.

DEQ. 2009. Guidance for Evaluating Stormwater Pathways at Upland Sites. Oregon Department of Environmental Quality (DEQ).

DEQ. 2015. *Public Review Draft - Permit Evaluation Report General Permit for Class V Stormwater Underground Injection Control Systems.* Oregon Department of Environmental Quality (DEQ) released on August 18, 2015, in a Public Notice for Public Hearing for proposed Underground Injection Control (UIC) Water Quality Permit: Sept. 23 Portland.

ESI. 2007. *Guide to Using Groundwater Vistas*, Version 5. Environmental Simulations, Inc. (ESI).

GSI. 2004, *10-Day Aquifer Test Results for the ASR No. 3 Pilot Well – City of Beaverton Aquifer Storage and Recovery Project*, August 14, 2004. GSI Water Solutions, Inc. (GSI).

GSI, 2011, Aquifer Storage and Recovery Limited License Application and Workplan, Joint Water Commission, Submitted to OWRD in October 2011.

Harbaugh, A.W., E.R. Banta, M.C., Hill, and M.G. McDonald. 2000. *MODFLOW-2000, the U.S. Geological Survey Modular Ground-Water Model – User Guide to Modularization Concepts and the Ground-Water Flow Process.* U.S. Geological Survey Open File Report 00-92.

Kennedy/Jenks. 2009. *Compilation and Evaluation of Existing Stormwater Quality Data from Oregon.* Prepared for the Oregon Association of Clean Water Agencies.

Keswick, B., C. Gerba, S. Secor, and I. Cech. 1982. *Survival of Enteric Viruses and Indicator Bacteria in Groundwater.* J. Environ. Sci. Health, A17. pp. 903-912.

James, R.B. 2003. *TSS-A Viable Measure of Storm Water Pollutants?* In: Proc. Stormcon 2003 Conf., San Antonio, Texas, July 2003

Muthukrishnan, S. 2005. *Treatment of Heavy Metals in Stormwater Runoff Using Wet Pond and Wetland Mesocosms.* Urban Watershed Management Branch, Water Supply and Water

Resources Division, National Risk Management Research Laboratory, Office of Research and Development, U.S. EPA, Edison, NJ 08837

Pedley, S., W. Yates, J. Schijen, J. West, G. Howard, and M. Barrett. 2006. *Pathogens: Health Relevance, Transport and Attenuation*, in, Schmoll, O., G. Howard, J. Chilton, and I. Chorus (eds.), *Protecting Groundwater for Health: Managing the Quality of Drinking-Water Sources,* World Health Organization, IWA Publishing, London, pp. 64-80.

Rice, T., C. Johnson, D. Wild, and D. Reasoner. 1992. *Survival of Escherichia coli O157:H7 in Drinking-Water Associated with a Waterborne Disease Outbreak of Hemorrhagic Colitis*. Letters in Appl. Microbiol. V. 13, pp. 38-40.

Sansalone, J. J. 2003, *The Physical and Chemical Nature of Urban Stormwater Runoff Pollutants,* in: Wet-Weather Flow in the Urban Watershed: Technology and Management, Field, R. and D. Sullivan, ed. Lewis Publishers, Boca Raton, FL, pp. 43-65

Snyder, D.T., D.S. Morgan, and T.S. McGrath. 1994. *Estimation of Groundwater Recharge from Precipitation, Runoff into Drywells, and On-site Waste Disposal Systems in the Portland Basin, Oregon and Washington.* Water Resources Investigations Report WRIR 92-4010. http://or.water.usgs.gov/pubs_dir/orrpts.html5HDR6.

USGS, 2015, *USGS 14206950 Fanno Creek at Durham, OR,* https://waterdata.usgs.gov/usa/nwis/uv?site_no=14206950.

ViralT. 1994. *Documentation and User's Guide for the Program ViralT (3.0).* U.S. Environmental Protection Agency, Office of Drinking Water, Washington, D.C.

Yates, Marylynn V., C. Gerba, and L. Kelley. 1985. *Virus Persistence in Groundwater.* Applied and Environmental Microbiology, v. 49, no. 4, pp. 778-781.

Telling the story of ONE WATER a pint at a time: What four years of brewing (& drinking) beer made from wastewater has taught us about the language of water.

Co-Authors: Mark Jockers, Government and Public Affairs Manager and Mark Poling, Business Operations Director, both of Clean Water Services, Hillsboro, Oregon

Little did we know that an innocent conversation before a Clean Water Services Citizen's Advisory Commission would launch a national and now international phenomenon in Water Reuse. One of our citizen members, Art Larrance (who also happens to be one of the founders of craft brewing in Portland) said "if you want to get people talking about water you ought to make beer."

From that conversation Pure Water brew was born. Working with the local home brewers club, the Oregon Brew Crew an international conversation was sparked about reuse by brewing beer. Started in 2014, the Oregon Brew Crew's Sustainable Water Challenge/Pure Water Brew competition has inspired utilities, water quality professionals, brewers and beer lovers throughout the U.S. to jump start the conversation about the nature of water.

When the District strategized on how to make this idea of making beer from wastewater into a reality, we realized we had three fronts where to work on to advance the conversation about the nature of water, the water cycle, and how we can make water fit for any purpose.

The first area is the technology. We realized that this must address all the concerns of our regulatory partners so a decision was made to make a very high purity water by employing ultrafiltration, followed by reverse osmosis, and then high rate oxidation with peroxide and ultraviolet light disinfection. The resulting water was non-detect for almost all parameters tested including drinking water parameters and those recommended for direct potable reuse standards. The few compounds remaining were all well below drinking water standards. Figure 1 shows the treatment train.

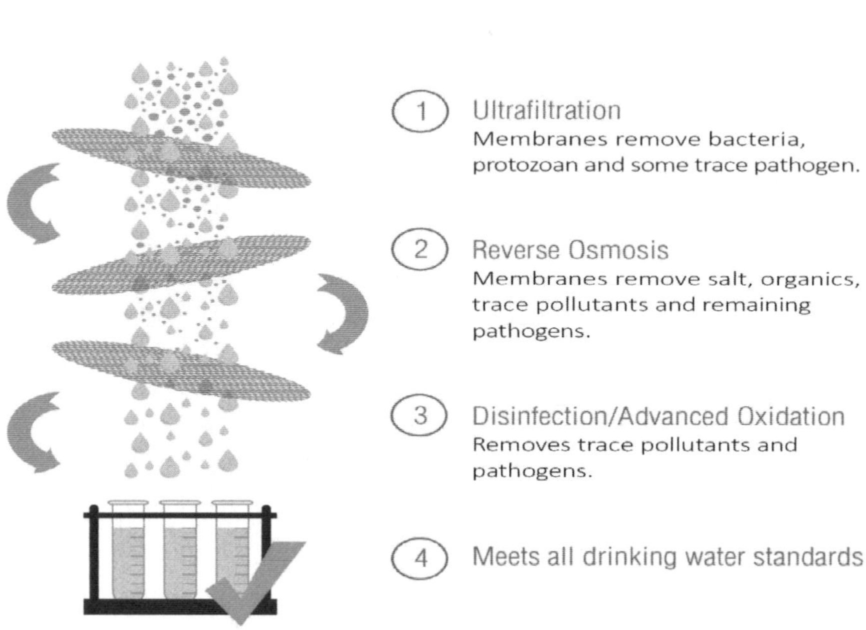

Figure 1 – Representation of the processes employed for creating the high purity water for brewing

The next area was regulatory challenges. In Oregon, the reuse rules have a prohibition against direct potable reuse. The District first sought support from the regional state regulatory office and then from the Environmental Quality Commission, the appointed board that oversees the Oregon Department of Environmental Quality. With strong support from the Commission to move this proposal forward, a public hearing was held. Fortunately, we had also worked with local and state-side NGO's to garner their support for this project. At the hearing, all public comment enthusiastically supported the idea. We received support from the EQC to move forward and then worked with the regional office of the DEQ to develop the specifics of the regulatory authority. A Reuse Plan modification was selected as the vehicle to create the necessary authority for the pilot project. A second public hearing was required where we once again had very supportive public comment. Regulators, who were initially hesitant to approve the project due to concerns about public response, agreed to move forward. The modified reuse plan at the end of the process now allows for limited production of beer from the purified water resource recovery facility effluent.

And last but not least, the court of public opinion. Little did we know that when the email notice shown below went to the standard distribution list (including media outlets) there would be the most effective press release imagine? Within an hour the phone calls started coming in to the District's Government and Public Affairs Office. Not only local and state-wide media but national and then international soon followed. While headline writers couldn't resist potty humor, the majority of stories were fair, positive and conveyed our message. To date, more than 500 stories and coverage has been generated. A study[1] done by Decision Research on the media coverage found that:

- Media coverage leading up to and following the 2015 Pure Water Brew event resulted in an estimated 500+ television, radio, print, and electronic stories between January 20 and October 20, 2015.
- In 2016, CWS contracted with Decision Research to analyze the media response to the Pure Water Brew project and assess whether CWS, in its outreach efforts, successfully utilized the recommended vocabulary from the WateReuse Research Foundation *Talking About Water* report.

The researchers observed the following. "There were two noticeable differences between the media response of the Pure Water Brew project and potable reuse programs generally. Overall, coverage of the Pure Water Brew project had the tone of a human-interest story; the purpose of the media reports were to entertain as well as inform. Coverage of the Pure Water Brew project was often fun, light, and humorous. By contrast, media stories about other potable reuse programs had a more scientific, emotionless tone, and seemed typical of a journalistic piece on an emerging technology. In short, reading or listening to coverage of the Pure Water Brew project made the competition seem fun, while coverage of other programs was merely informative."

From: DEQ Online Subscriptions [mailto:ordeq@public.govdelivery.com]
Sent: Tuesday, January 20, 2015 9:09 AM
To: jockersm@cleanwaterservices.org
Subject: Request for Comments on Recycling Water for Drinking Uses

DEQ is currently accepting comments on a proposed approval for Clean Water Services limited reuse of recycled water for use in an alcoholic beverage.

Documents that describe this proposal can be found at http://www.oregon.gov/deq/docs/022015drinkable.pdf.

Written comments due: 5 p.m., Friday, February 20, 2015.

Need more information?

Contact Avis Newell
Phone: 503-229-6018 or 800-452-4011
Fax: 503-229-6957
Email: Newell.Avis@deq.state.or.us

SUBSCRIBER SERVICES:
Manage Subscriptions | Unsubscribe All | Help

Figure 1 – the DEQ Listserv Email Notification for Drect Potable Reuse in an "alcoholic beverage".

So where are we today? The "crazy" idea of brewing beer from wastewater has spawned a nationwide Pure Water Brew movement. The thirst for Pure Water Brew quickly spread to the Milwaukee Metropolitan Sewer District, Hillsborough County, Florida and CMD Smith in Colorado before becoming a statewide caravan with Arizona Pure Water Brew partnering with 26 commercial brewers. In 2017, these early adopters and a pioneering group of utilities, consultants, professional associations, and craft brewers formed the Pure Water Brewing Alliance who will take this compelling clean water story on the road to the National Craft Brewers Alliance in 2018. And Pure Water Brew made by Craft Brewer Cascade Brewing in Portland is pouring at the Singapore International Water Week.

Beyond the beer and the bling, the Pure Water Brewing Alliance has been able to creatively tell the story about the nature, source and history of water while:

- Starting a conversation about reuse and the nature of water
- Demystifying water purification and the urban water cycle
- Showcasing innovative water technology to inform how water can be used
- It sparked a conversation and challenged the taboo of direct potable reuse

The Pure Water Brewing movement has shown that with a little creativity and a willingness to take a chance, we can start a conversation and begin overcoming the social and regulatory barriers to water reuse.

References

1 - "Media Analysis of the Pure Water Brew Project," Decision Research, Andrew Quist, C.K. Mertz, Paul Slovic, 2016

Disinfection By-Product (DBP) Formation and Control in Potable Reuse Applications

Caroline Russell, Pranjali Kumar, David Ammerman, Justin Sutherland, Andrew Salveson
(Carollo Engineers, Inc.)

ABSTRACT

Decades of research has elucidated the precursor characteristics, formation mechanisms, and control strategies for regulated disinfection by-products (DBPs) and some unregulated DBPs in conventional water treatment plants (WTPs). However at potable reuse facilities, the prevalence and type of precursors, formation mechanisms, and control strategies can all be different than at conventional WTPs. The impact of ozonation is a good example. At conventional surface WTPs, ozone degradation of chloramines-reactive n-nitrosodimethylamine (NDMA) precursors tends to result in a net reduction of finished water NDMA concentrations. In contrast, NDMA formation hypothesized to be via ozone-catalyzed nitrosation reactions results in increased NDMA concentrations at advanced water treatment plants for potable reuse applications. Data from three potable reuse studies illustrates the importance of carefully assessing DBP concentrations in purified water, factoring in proposed treatment, blending strategies, and chlorination conditions to accurately anticipate resulting DBP concentrations and any required adjustments for regulatory compliance and to meet site-specific water quality goals.

Keywords: Disinfection by-products, potable reuse, total trihalomethanes (TTHM), sum n-nitrosodimethylamine (NDMA), bromate

INTRODUCTION

Chloroform was identified as a by-product formed during disinfection of drinking water with chlorine in the mid-1970s (Rook, 1974). The discovery triggered decades of research and eventual promulgation of the Stage 1 and 2 Disinfectants and Disinfection By-product Rules (D/DBPR; USEPA 1998 and USEPA 2006; Figure 1). Collectively, the Stage 1 and 2 D/DBPRs resulted in an 80 microgram per liter (μg/L) maximum contaminant level (MCL) for total trihalomethanes (TTHM), and a 60 μg/L MCL for the sum of five haloacetic acids (HAA5), both based on locational running annual average concentrations in the distribution system. The Stage 1 D/DBPR includes a 10 μg/L MCL for bromate, and a 1 mg/L MCL for chlorite.

NDMA, a chloraminated DBP, was first detected in drinking water in 1989 (Barrett et al., 2003). USEPA included NDMA along with five other nitrosamines on the second Unregulated Contaminant Monitoring Rule. USEPA indicated in the third Six Year Review that any future revisions to the Stage 1 and 2 D/DBPR should be considered in conjunction with an evaluation of impacts to NDMA formation.

Research on regulated halogenated DBPs, i.e., TTHMs and HAA5, has illustrated the importance of disinfection pH, disinfectant type and dose, contact time, natural organic matter characteristics, and bromide to dissolved organic carbon (DOC) ratio on TTHM and HAA5 speciation and concentrations. Schreiber and Mitch (2006) and others have identified chloramination conditions that impact NDMA formation. This past research presents a treasure-trove of information that should be mined when considering DBP formation during potable reuse applications. However, other factors that are less relevant during disinfection of conventional WTP finished water also warrant consideration. Factors influencing DBP formation at potable reuse facilities include:

- Sewershed and wastewater treatment plant (WWTP) characteristics;
- Treatment processes incorporated in the potable reuse facility;
- Disinfectant type and contact time; and
- Blending location and percentage.

DOC concentrations and reactivity with disinfectants are often different in wastewater as compared to traditional source waters such as surface water and groundwater. Potable reuse treatment process selection also impacts DBP formation and control. For example, ozone can form bromate and NDMA. Biologically active filtration (BAF) removes NDMA and THM and HAA precursors. Granular activated carbon (GAC) and reverse osmosis (RO) removes THM and HAA precursors (and chloramines-reactive NDMA precursors). UV photolysis degrades NDMA; however, NDMA reformation following chloramines addition can be a concern. Additionally, potable reuse treatment can impact the concentration of DOC and bromide, impacting the ratio of DOC and bromide and the relative proportion of brominated to chlorinated DBPs that are formed.

Data from three potable reuse studies are reviewed in this paper to illustrate the different trends that may be observed during disinfection of potable reuse supplies.

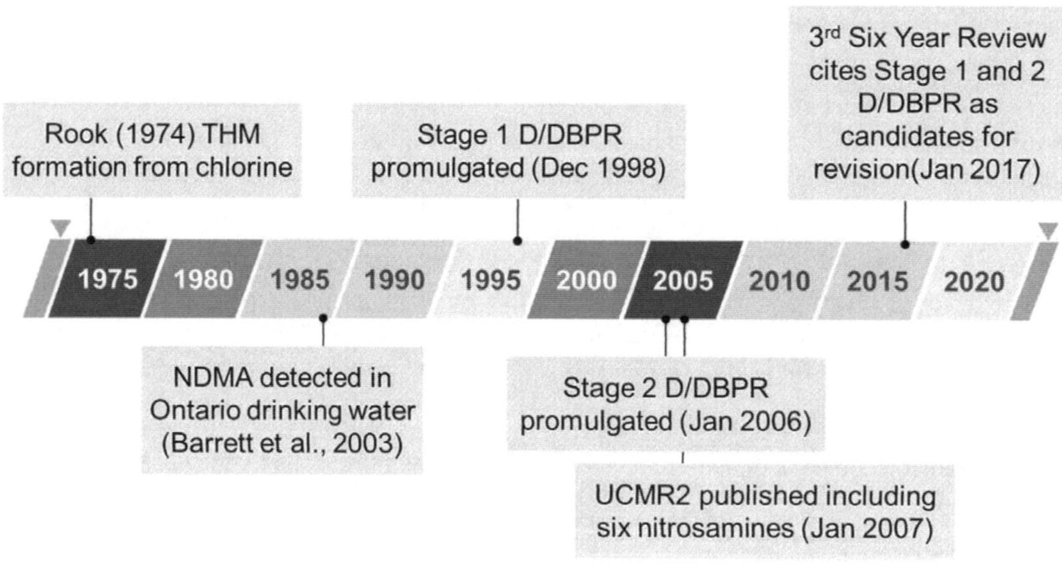

Figure 1: Milestones for DBP Detection and Regulation in Drinking Water

METHODOLOGY

Disinfection by-product and DBP precursor concentrations were measured during three different potable reuse projects with varying characteristics. Table 1 provides a description of each project along with the DBPs analyzed.

Table 1: Summary of Projects Reviewed for this Manuscript

Project	Testing Scale	Description	DBPs analyzed
Project A	Pilot-Scale	Direct-to-distribution potable reuse pilot plant Unchlorinated secondary effluent treated through membrane filtration, nanofiltration (NF) or RO membrane desalination, UV/peroxide advanced oxidation (AOP), GAC for peroxide quenching, with free chlorine disinfection. Pre-treatment with ozone was also tested.	• TTHM, HAA5, nitrosamines, bromate, halonitromethanes (HNMs), total organic halides (TOX) • Precursors analyzed under simulated distribution system (SDS) conditions
Project B	Demonstration-Scale	Demonstration of a non-RO treatment train capable of meeting all drinking water quality standards. Treatment train includes: ozonation (O_3), biologically-active filtration (BAF), ultrafiltration (UF), GAC, and UV AOP, free chlorine disinfection.	• Bromate, NDMA, TTHM, HAA5, TOX • Precursors analyzed under formation potential (FP) conditions and uniform formation conditions (UFC)
Project C	Bench-Scale, coupled with Full-Scale Sampling	Water Research Foundation Project #4536, Blending Requirements for Water from Direct Potable Reuse Treatment Facilities Evaluated impact of three blending variables: • Different proportions of recycled water to raw water, • Different qualities of purified water (i.e., different treatment process trains) • Different points of introduction / blending	• TTHM, HAA5, NDMA • Precursors analyzed under formation potential (FP) conditions

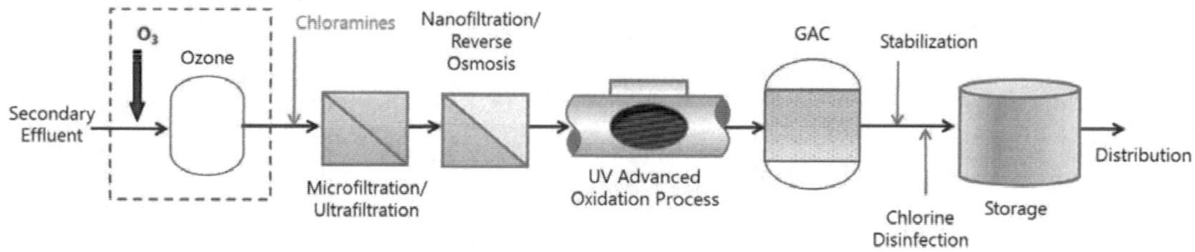

Figure 2: Process Schematic for Project A Pilot Treatment Plant

RESULTS

Key results from each project are summarized in the following paragraphs.

Project A, Direct-to-Distribution Potable Reuse Pilot Study (MF/UF-NF/RO-UVAOP-GAC-Cl2)

TTHM concentrations in the unchlorinated secondary effluent used as the source water to the potable reuse pilot treatment plant were non-detect (0.5 µg/L detection limit) and HAA5 concentrations ranged from 2 to 3.5 µg/L. NDMA concentrations in the source water for this proposed potable reuse facility range from less than 2 to 60 nanograms per liter (ng/L) despite diversion prior to chlorination, attributed to upstream industrial discharges.

Despite the use of free chlorine as the disinfectant, TTHM and HAA5 SDS concentrations in the pilot treated water were below the respective MCLs, in most case by more than a 60% safety margin, not accounting to compliance averaging. The highest TTHM SDS concentrations measured in the purified water correlated to the higher bromide concentrations and increased concentrations of the more brominated THMs (i.e., bromoform and chlorodibromomethane). TOX and HNM concentrations were also low in the purified water, reflecting low precursor concentrations following treatment through RO.

TTHM and HAA5 SDS tests were conducted at both ambient pH and at a higher pH targeted for corrosion control during full-scale treatment. TTHM concentrations were higher at pH 7.5 - 8 than at ambient pH in the UVAOP effluent, but remained well below the MCL. HAA5 concentrations were similar irrespective of SDS pH. The results highlight the importance of factoring in the impact of pH on DBP formation, particularly for waters with higher TOC than Project A purified water which had very low TOC concentrations as a result of RO treatment.

Bromate concentrations increased during use of ozone for pretreatment but were below the 10 µg/L. A 0.4 to 1 ozone to total organic carbon (TOC) ratio was targeted. Bromate concentrations decreased through RO. Ozonation increased NDMA concentrations by up to 40 ng/L.

Results from the pilot study illustrated the ability to meet Stage 1 and 2 D/DBPR requirements with free chlorine as a disinfectant and project goals for NDMA using UV photolysis.

Project B, Demonstration Scale Potable Reuse Study (O_3-BAF-UF-GAC-UVAOP-Cl2)

In Project B, ozone was applied at a higher ozone to TOC ratio than in Project A. Bromate concentrations measured when ozone was applied at a 1.3 to 1 ozone to TOC mass ratio were non-detect in biofiltration effluent and 1.3 μg/L in UVAOP effluent. Average bromide concentrations in the source water to the demonstration potable reuse plant were 56 μg/L.

NDMA concentrations increased by 28 to 43 ng/L during ozonation, but were non-detect at a 2 ng/L detection limit following biofiltration. TTHM concentrations increased by approximately 10 μg/L through the treatment train from exposure to chloramines added for membrane biofouling control.

NDMA, TTHM and HAA5 formation measured using formation potential (FP) and uniform formation condition (UFC) tests illustrated the value of including UFC or SDS tests to identify DBP concentrations expected to form under conditions representative of full-scale disinfection practices. While FP tests illustrate important trends across treatment processes, UFC or SDS tests are needed to assess whether DBP compliance or treatment goals will be compromised without any additional control strategies.

Project C, Impact of Blending on DBP Formation

Water Research Foundation (WRF) project #4536 focused on evaluating the impact of potable reuse treatment provided and blending location on blended water quality. Overarching trends from an evaluation of DBP concentrations for different source waters (including groundwater and surface water), potable reuse treatment (O3-BAF and MF-RO-UVAOP), and blending location are as follows:

- The utility using chlorine had the highest TTHM formation, compared to three other utilities included in the project that used chloramines.
- All blends for each utility were below regulated values for TTHM and HAA5. However, in some cases potable reuse treated waters increased TTHM and HAA5 concentrations when blended and in other cases decreased TTHM and HAA5 concentrations when blended.
- Treatment through the RO-based treatment train tended to decrease TTHM FP, whereas the O_3-BAF treatment train tended to increase the TTHM FP.
- NDMA was only partially removed during O_3-BAF and the RO-based treatment train, highlighting the need to evaluate NDMA concentrations when planning for a potable reuse project (and resulting blends) to make sure water quality goals are met.

CONCLUSIONS

With an increasing number of water utilities considering potable reuse to augment and diversify their water supply portfolio, a better understanding of the formation of regulated and emerging DBPs is needed. Results from the three projects presented above highlight the need to evaluate DBP concentrations under site-specific treatment and blending conditions to determine if treatment goals and regulatory compliance goals can be met. Past research highlights the tools

(FP, UFC, and SDS) and water quality variables (e.g., pH, bromide, disinfectant type) that should be considered when developing experimental plans to measure DBPs during potable reuse bench-, pilot, or demonstration-scale studies.

ACKNOWLEDGEMENTS

The authors would like to acknowledge the significant contributions from JoAnn Jackson and Mark Honadle from Altamonte Springs, and Eric Dickenson of Southern Nevada Water Authority.

REFERENCES

Barrett, S., Hwang, C., Guo, Y., Andrews, S.A., Valentine, R., 2003. Occurrence of NDMA in Drinking Water: A North American Survey, 2001-2002. Proceedings from AWWA Annual Conference and Exposition.

Rook, J.J., 1974. Formation of Haloforms during Chlorination of Natural Waters," Water Treatment and Examination, Vol. 23, pp. 234-243.

Schreiber, I.M., Mitch, W.A., 2006. Nitrosamine Formation Pathway Revisited: The Importance of Chloramine Speciation and Dissolved Oxygen. *Envir. Sci. & Technol.*, 39:3811.

USEPA, 1998. National Primary Drinking Water Regulations; Disinfectants and Disinfection Byproducts; Final Rule. 63 FR 69390. December 16, 1998.

USEPA, 2006. National Primary Drinking Water Regulations: Stage 2 Disinfectants and Disinfection Byproducts Rule; Final Rule. 71 FR 388. January 4, 2006.

USEPA, 2007. Unregulated Contaminant Monitoring Regulation (UCMR) for Public Water Systems Revisions. 72 FR 367, February 2007.

USEPA, 2017. National Primary Drinking Water Regulations; Announcement of the Results of EPA's Review of Existing Drinking Water Standards and Request for Public Comment and/or Information on Related Issues; Proposed Rules. 82 FR 3551. January 11, 2017.

EVALUATION OF OVER 100 DISINFECTION TRACER STUDIES IN WASHINGTON STATE

Russell L. Porter, P.E., Gray & Osborne, Inc.

Keith P. Stewart, P.E., Gray & Osborne, Inc.

ABSTRACT

This paper summarizes tracer studies performed at 32 water treatment plants in Washington State ranging from 5-5,400 gpm with contact basins from 200 gallons to 1.5 million gallons. Multiple tracer studies were performed at each plant. Tracer studies indicated that most facilities historically used baffling efficiency (T_{10}/T) values higher than what the tracer studies indicated. The majority of the basins were rectangular, baffled basins. Tracer studies indicated standard design parameters such as L/W ratio were not as accurate in predicting basin performance compared with calculating baffled channel cross sectional velocity. This effect was pronounced where cross sectional flow velocity was below the Reynolds laminar flow threshold. For circular, baffled basins, the cross sectional velocity and BE relationship was also evident.

KEYWORDS

Tracer Studies, Baffling Efficiency, T_{10}/T, Disinfection Contact Time

INTRODUCTION

Disinfection is a critical part of surface water treatment. Disinfection effectiveness against pathogenic organisms depends on disinfectant concentration, time and water quality parameters like pH and temperature. Much of disinfection is prescribed by the requirements of the Surface Water Treatment Rule and the use of disinfectants and the design of disinfection equipment is well understood in the water industry. In contrast, disinfection contact time and its design is likely the least understood part of disinfection and the design of disinfection contact basins is usually performed using a limited number of resources. Furthermore, contact time can be the largest single source of error in determining pathogen inactivation (Deem and Feagin, 2016).

Historically, the hydraulic efficiency of disinfection contact basins have been assigned based upon an assessment of baffling efficiency (BE), or T_{10}/T, usually per the USEPA ***Surface Water Treatment Manual*** (USEPA 1991) (Manual). The Manual assigns baffling efficiency values based on a qualitative assessment with categories of:

unbaffled (mixed flow) (BE $\leq$ 0.1),

poor (BE = 0.3),

average (BE = 0.5),

superior (BE = 0.7), and

plug flow (BE = 1.0).

The Manual gives some examples of each classification but the final assignment of a baffling efficiency for a particular design is subject to a perceived level of baffling rather than a set of design principles. Bishop et al. (1993) further refined the baffling categories with full scale and modeling results comparing BE values with other quantifiable parameters such as length to width (L/W) ratios for baffled channels.

Another resource that is often used for designing disinfection contact basins is ***Improving Clearwell Design for CT Compliance*** (AWWARF 1999). The document includes laboratory pilot studies and computer modeling, along with full scale verification, to develop relationships for correlating baffling efficiency and design parameters including the L/W ratio for a rectangular clearwell with parallel baffles. The L/W ratio is calculated by taking the total length of travel through the flow channels created by the baffles and dividing by the average width of each channel. The pilot data provided a logarithmic equation to allow for a quantitative prediction of baffling efficiency based on L/W ratio. Similar relationships are included for baffled, circular tanks and clearwells with perforated walls.

Computational Fluid Dynamic (CFD) modeling can be used to simulate clearwell performance during design and has been shown to match tracer studies in some studies (Zhang, 2011 and Kattnig and Venayagamoorthy, 2015). CFD modeling can be used as a design tool to predict BE. CFD modeling may not be available or can be too expensive for small systems so many small system designers still rely on historical resources for predicting hydraulic behavior for disinfection contact tanks.

For most plants, the BE of the constructed disinfection contact tank is determined by onsite tracer testing. Usually this onsite tracer determination is what is used for subsequent regulatory compliance calculations.

This work was prompted by the findings of an earlier study of disinfection data and practice by the Washington Department of Health (WDOH), which found that eight of the nine systems with a BE above 0.1 were using empirically-derived values not verified by a tracer study (Deem and Feagin, 2016). This group included two utilities with unbaffled chlorine contact tanks that were using an unusually high BE of 0.3. For the current study, WDOH staff screened all 135 operating surface water treatment plants in the state to identify plants meeting one of the following three criteria: 1) The plants used an empirical BE greater than 0.1 that had not been verified with a tracer study; 2) had a previous tracer study that included some questionable data due to interruptions of flow or tracer chemical during the study; or 3) had made system upgrades and changes which made their previous tracer study obsolete. This process identified 34 water treatment plants across the state that needed further evaluation. WDOH contacted the water utilities operating these plants and gave them the option to allow WDOH to perform tracer studies, perform their own tracer study, or provide an engineering analysis for their T_{10}/T.

Tracer studies were performed at 32 surface water treatment plants throughout Washington State. The characteristics of the water treatment plants are summarized below.

- The capacity of the treatment plants ranged from 5.0 gpm to 5,400 gpm.
- The smallest contact vessel tested was a 200 gallon ozone contact pressure vessel with concentric interior baffles. The largest contact vessel tested was a 1.5 MG circular concrete reservoir with interior baffles.
- The treatment plants in the study included 19 rapid sand filtration plants, five slow sand filtration plants, four bag filter plants, three membrane treatment plants, and one diatomaceous earth treatment plant.
- 31 of the plants in the study used chlorine for primary disinfection while one plant used ozone.
- 21 of the plants in the study had rectangular, baffled basins including some plants with parallel basins. Three of the plants had baffled circular tanks. Two of the plants had sequential, unbaffled, rectangular tanks. Two plants had an unbaffled, rectangular tank followed by a baffled, circular tank. Two of the plants had a rectangular, unbaffled tank followed by a circular, unbaffled reservoir. One plant had a single, unbaffled tank and one plant had an interior pipe coiled within a circular reservoir.

METHOD AND MATERIALS

The tracer studies were performed using standard step dosing strategies as outlined in ***Tracer Studies in Water Treatment Facilities: A Protocol and Case Studies*** (AWWARF, 1996). For step tests, a known concentration of tracer chemical is added continuously to a constant flow entering the basin. The outlet concentration of the basin is then monitored over time. Given a constant flow rate into and out of the basin and a known volume, the time at which the concentration leaving the basin is 10 percent of step dose is termed T_{10}. When the T_{10} is divided by the hydraulic residence time (T), calculated by basin volume divided by the flow rate, a T_{10}/T value is obtained that is equal to the BE.

Tracer Chemical Selection. The tracer chemical that was used in tracer studies needed to meet certain criteria including the following.

- The tracer chemical must be conservative, i.e. it must not react with anything in the water during the tracer test.
- The tracer chemical must be certified under ANSI/NSF Standard 60.
- The tracer chemical must be easily measured.
- The tracer chemical must be easily prepared and injectable at drinking water plants.
- The tracer chemical must meet consumer acceptance by water system customers.

Three tracer chemicals were used in the tracer studies.

Sodium Fluoride

Powdered sodium fluoride was the preferred tracer chemical for the study because of its ease of use and suitability of analytical methods. A stock solution was prepared at the treatment plant and injected using chemical feed pumps usually at a target dose of 1.0 to 1.5 mg/L. This was to provide a good range for analysis while keeping the total fluoride in the water (including any in the raw water) below the 2.0 mg/L secondary maximum contaminant level (MCL). Ion specific electrode (ISE) fluoride analysis was used to measure the fluoride level during the tracer analysis with grab samples for raw water and contact basin inlet water and semi-continuous sampling after the contact basin. The ISE equipment was calibrated to known fluoride standards prior to each tracer and checked against standards periodically during the testing. For most tracer tests, a grab sample was provided to a commercial lab for a comparative check.

Potassium Chloride

Powdered potassium chloride was used for some tracer tests where fluoride was not politically acceptable for the water purveyor or in cases where sequential tracer tests were performed where fluoride from a previous tracer test was still present in the basin. A stock solution was prepared at the treatment plant and injected using chemical feed pumps usually at a target dose of 50 mg/L. Ion specific electrode (ISE) chloride analysis was used to measure the chloride level during the tracer analysis with grab samples for raw water and contact basin inlet water and semi-continuous sampling after the contact basin. The ISE equipment was calibrated to known chloride standards prior to each tracer and checked against standards periodically during the testing. For most tracer tests, a grab sample was provided to a commercial lab for a comparative check.

Free Chlorine

For three of the treatment plants, free chlorine was used as the tracer chemical. At these plants, fluoride was not politically acceptable and potassium chloride was not feasible due to the large flow rates and logistical issues with using large volumes of stock solution. For these plants, an initial investigation was completed to confirm that chlorine levels would be stable over the duration of the test. The procedure included measuring the chlorine concentration before and after the contact basin during normal operation, as well as bench-scale chlorine decay tests of water taken immediately after disinfection. If these analyses confirmed that the chlorine concentration declined by only a maximum of 10 percent across the contact basin, chlorine was considered to be acceptable for use as a tracer for the study.

At these plants, chlorine solution was fed using a chemical feed pump or the existing chlorination addition system. A target of 0.5 to 1.0 mg/L of additional chlorine above normal operation was used to provide enough chlorine addition to be detectable. At each plant, online chlorine analyzers provided the post contact tank monitoring while portable colorimeters using standard DPD methods were used for grab samples at the contact basin inlet.

Tracer Study Logistics. For most treatment plants, at least three tracer tests were performed at a variety of flow rates including average and peak flow rates to ensure an accurate description of T_{10}/T values. Each treatment plant presented unique challenges, some of which are discussed below, and the tracer test approach was often modified to adapt to each plant.

At plants with contact basins that had relatively short hydraulic residence times, the step test was conducted until the tracer concentration leaving the basin was equal to that entering the basin indicating that tracer concentration in the basin was uniform. At that point, the tracer dosing was stopped and second tracer study was conducted by monitoring the declining tracer concentration leaving the basin, termed a "lag" study (as opposed to a "lead" study for a normal step test). The T_{10} time for a lag test was determined when the outlet tracer concentration reached 90 percent of the initial concentration prior to stopping tracer feed. The lag test was conducted until the tracer chemical concentration was equal to the background concentration indicating that the tracer chemical had been evacuated from the contact basin. At that point, a new lead test could be initiated.

For most of the plants, the hydraulic residence time was on the order of a few hours and it was not feasible to continue all of the tracer tests to a point where the tracer concentration leaving the basin was equal to the dose entering the basin. For these plants, the tracer test was conducted until the outlet tracer concentration was approximately 90 percent of the step dose. Even for well baffled and efficient plants, it usually took three to four hydraulic residence times for the outlet concentration to equal the step dose and the addition data gathered after the outlet reached 90 percent of the step dose did not add to the understanding of basin hydraulics. Consequently, only lead tests were conducted. After a tracer test, the plant would be operated normally for a few days to allow the tracer chemical to evacuate the basin before the next test could be conducted. As an alternative, a different tracer could be used if the next tracer test was scheduled sooner.

At a handful of plants with large contact basins with long hydraulic times, on the order of days, it was not feasible to conduct a step test until the basin outlet concentration approached the step dose due to the long amount of time required. For these applications, the tracer test was conducted until the outlet tracer concentration was at least three times the concentration of the T_{10} point. For these plants, several days or even weeks were often required between tests to allow the tracer chemical to evacuate the basin and the outlet tracer concentration to return to background levels.

Many of the treatment plants in the study had parallel contact basins. Most of these plants are small and don't have metering to allow verification of the flow splitting between the basins. At these plants, the outlet of the each of the basins were monitored during the tracer to provide an analysis of each basin individually and allow for an estimate of the flow splitting between the two basins.

For the plants with sequential basins, the tracer tests were designed to assess the hydraulic characteristics of each basin to the maximum extent possible. For many of these types of plants, the outlet of the first basin was monitored to assess its hydraulic efficiency while the outlet of the final basin was also monitored to determine overall hydraulic efficiency of the sequence of basins. In some cases, the tracer chemical was injected between basins in the sequence to look at individual basins within or at the end of the sequence.

RESULTS

Overall Results. In general, the tracer studies revealed that the baffling efficiencies used at a majority of the water treatment plants were higher than measured results from this study as shown in Figure 1. In some cases, the difference between the design baffling efficiency that the plant was operating under and the actual measured efficiency was significant. For example, one plant that was using 0.7 for a baffling efficiency but the measured baffling efficiency from the tracer study turned out to be only 0.3.

In summary, the tracer studies of the 32 plants that were studied resulted in the following changes to the baffling efficiency and CT compliance calculations:

- 7 plants were able to increase their baffling efficiency
- 5 plants had no change resulting from the tracer analysis
- 19 plants had decreased baffling efficiency resulting from the tracer analysis
- 2 plants had significant hydraulic issues identified by the tracer that made determining a baffling efficiency difficult and operational changes were suggested until the clearwell operation could be investigated and modified. One of the plants was able to make clearwell piping modifications and was able to be retested as part of the study with a resultant decrease in baffling efficiency.

Results by Basin Geometry. The tracer studies investigated several types of basin geometries. As part of the review of tracer results, the tracer studies of basins of similar construction were compared in an effort to potentially understand the effect of basin geometry on baffling efficiency. A summary of the general trends and observations noted during the tracer studies is included below by basin geometry.

<u>Unbaffled</u>

One of the tracer studies investigated a single, horizontal, unbaffled, cylindrical tank used for disinfection contact time. The tank had originally been a stainless steel milk truck trailer that had been damaged but was then repurposed as a disinfection contact tank. The tank had been modified to include perforated inlet and outlet pipes at opposite end of the tank.

Historically, 0.3 had been used for the tank's baffling efficiency. The tracer study results for the tank were 0.12 and 0.13 for the T_{10}/T and a baffling efficiency of 0.1 was recommended for the tank. This result is not surprising since most design guidelines for contact tanks use 0.1 baffling efficiency for unbaffled tanks.

Sequential Basins

Another method of constructing contact time is to use sequential basins that are operating in series. There were several of these studied as part of the tracer study and each on each one was unique. Consequently, it is difficult to develop conclusions regarding sequential basin hydraulics from this study but some individual examples provide insight.

One small camp water system used three concrete tanks plumbed in series. Each pre-fabricated concrete tank was 3,000 gallons. The inlet and outlet piping for each tank was at diagonally opposite corners with the outlet of one tank flowing to the inlet of the next tank. The system had historically used a baffling efficiency of 0.5. Two tracer tests were performed with measured baffling efficiencies of 0.45 and 0.48 so a baffling efficiency of 0.45 was assigned for compliance calculations. While the baffling efficiency of each individual tank was not measured, it is assumed that it would be close to 0.1 since each is unbaffled. The tracer results imply that sequentially operating unbaffled tanks in series can produce a system with a baffling efficiency that is relatively high given that none of the tanks are baffled.

Another sequential basin studied included two concrete clearwells and a pump wet pit that were part of the treatment plant building structure. The first clearwell was a concrete basin with a small window opening near the ceiling which sets a minimum weir level in the basin. Water flows through the weir window into the second clearwell and then through a pipe into the pump wet pit. The levels in the second clearwell and the pump wet pit vary depending upon demand in the system.

Because of supply issues with the system source, only one tracer was conducted. The tracer measured the baffling efficiency of the first clearwell and the system as a whole. The baffling efficiency of the first basin was measured at 0.11, a value consistent with the standard 0.1 for an unbaffled system. The baffling efficiency for the entire system was measured at 0.48 and a value of 0.4 was determined to be appropriate for compliance calculations. This was the value that had been historically used. As with the first example, the sequential use of unbaffled basins, each with a low baffling efficiency, produced a system with a surprising high efficiency.

Another example of a sequential basin was a two basin system of an L-shaped concrete basin with a weir wall controlling its level. At the start of the basin just after the inlet piping, there is a perforated cedar wall that is designed to provide uniform flow. The water flows from the first basin and over the weir wall into the second basin. The second basin also has a perforated cedar baffle wall prior the pump chamber. The level of the second basin varies depending upon pump operation.

Five tracer analyses produced values of 0.50 to 0.72 with an average of 0.62. A baffling efficiency of 0.6 was assigned to the two basin system, an improvement over what had been used in the past for compliance calculations. Historically, the system had calculated CT using a baffling efficiency of 0.5 for the first basin and 0.1 for the second. It is possible that the use of perforated cedar baffles enhanced the baffling efficiency in addition to the increase in efficiency seen in sequential unbaffled basins.

Baffled Rectangular

The single most prevalent geometry analyzed during the tracer studies was the baffled rectangular basin, 21 of which were included in the tracer project. Generally, baffled rectangular basins include a concrete structure with concrete or wood baffle walls or curtain baffles made of hypalon or other non-permeable fabric.

In addition to the EPA Guidance Manual guidelines concerning expected baffling efficiencies, there are other design resources that are used for the design of rectangular, baffled basins. A 1999 AWWARF publication, *Improving Clearwell Design for CT Compliance,* compiled utility survey data and summarized pilot study results to produce design parameters for predicting baffling efficiency. The document noted a relationship between length to width (L/W) ratio and baffling efficiency for rectangular basins ($T_{10}/T = 0.1984*Ln(L/W) - 0.0019$).

The results of the tracer studies indicated that the L/W ratio may not be the best indicator of baffling efficiency. Figure 2 shows the L/W ratio for the tracer studies performed during this project graphed relative to the relationship noted in the 1999 AWWARF publication. As can be seen in Figure 2, the relationship between L/W ratio and baffling efficiency is not pronounced and is widely variable. For example, several of the basins studied had baffling efficiencies of 0.6 even though the L/W ratios for these same basins varied from 10 to 110.

The pilot study work in the 1999 AWWARF publication was conducted in an 8-foot by 8-foot study basin with a 24 inch depth of water. The authors studied the L/W ratio by modifying the basin by adding baffle walls to create two to ten baffled channels ranging in width from four feet to 0.8 feet. The effect of depth was not noted. For some of the configurations with few baffle walls, the depth of the channel was greater than the width of the channel. For the configurations with more baffle walls and channels, the depth was greater than the channel width. For the most baffled condition with 10 channels of 0.8 feet each, the depth was 2.5 times the width.

For most full-scale basins, the depth of a channel is generally much greater than its width. For the 22 basins in this study, the average depth to width ratio of the baffled channel was 2.4 with the largest depth to width ratio at 4.3. Conceptually, it seems logical that if two basins of similar L/W ratio are compared, the basin that is deeper would be more like to experience short circuiting and have a lower baffling efficiency. It also seems reasonable the larger of the width or depth might be a controlling factor in baffling efficiency. Consequently, it seems that the smaller of the L/W ratio or the length to depth (L/D) ratio might be a better indicator than just L/W alone. Figure 3 shows the comparison of the smaller of the L/W and L/D ratios and the baffling efficiency. The data from the tracer studies match the relationship from the 1999

AWWARF publication better when the smaller of the L/W or L/D ratios are considered but there is still not a clear relationship that could be used to predict baffling efficiency based upon either L/W or L/D ratios.

One aspect that was investigated during the tracer studies is the velocity of the water in the baffled channels, calculated by taking the volumetric flow rate and dividing by the cross sectional area in a typical straight reach in the baffled channels. Figure 4 shows the relationship between the cross sectional velocity for each tracer study and the resulting measured baffling efficiency. There is still some variability between the cross sectional velocity and the BE for similar basins that is likely due to other basin factors such in inlet and outlet structures and the geometry of the turns between the baffled channels.

The data in Figure 4 suggest that the cross sectional velocity in a baffled channel is a better predictor than L/W or L/D ratio. This is a significant finding since this parameter is rarely considered during basin design.

One example of the significance of the velocity parameter was evident with one particular contact basin in the study. The disinfection contact basin for a small slow sand plant appeared to be well baffled (L/W=66) and had been assigned a T_{10}/T of 0.7 at the time of its design in line with industry practice. Tracer tests conducted to two plant flows showed T_{10}/T values for the lower and higher flows of 0.3 and 0.4, respectively and the basins T_{10}/T value for compliance was changed to 0.3. Upon calculating the flow velocity in the baffled channel, it was found that the flow velocities were very low. In fact, the Reynolds numbers in the baffled channel for the low and high flow were approximately 700 and 1,000, respectively.

When the Reynolds numbers in the baffled channels for all of the tracers studies were analyzed, all of the tracer studies with T_{10}/T values below 0.4 were conducted with flow velocities where the Reynolds number was below 1,900 indicating laminar flow (Re<2000). This is reasonable since Re>2000 corresponds to turbulent flows and one characteristic of turbulent flow is that the velocity across the water profile in a channel or pipe is constant. For laminar flow, it is possible that the cross sectional flow is not uniform and more short-circuiting within a channel is possible. Barnett and Venayagamoorthy (2014) noted a similar increase in baffling factor efficiency between turbulent and laminar flows in a study of a contact tank in Colorado.

Circular, Baffled Tanks

Four of the tracer studies included individual circular, baffled tanks. The four tanks were fairly typical of circular, baffled tanks in that they had parallel curtain baffles that create parallel channels.

Historically, when circular, baffled reservoirs are designed, they are general thought to be well-baffled under USEPA Guidance Manual guidelines and are assumed to have baffling efficiencies of 0.3 to 0.7. For the circular, baffled reservoirs analyzed with the tracer studies, the baffling efficiencies ranged between 0.16 to 0.47 for individual tracer studies and the baffling efficiency for future compliance assigned one reservoir a 0.2, two of the reservoirs a 0.3, while the fourth tank was assigned a baffling efficiency of 0.4.

Similar to rectangular, baffled reservoirs, the 1999 AWWARF publication included laboratory studies using an eight-foot circular reservoir into which one, three, five, and seven longitudinal baffles were added to create two, four, six, and eight baffled channel configurations. As with rectangular basins, the 1999 AWWARF publication found a relationship between T_{10}/T and L/W ratio. Figure 5 compares the L/W ratio and the T_{10}/T for the reservoirs in the study with the 1999 AWWARF publication relationship. The L/W ratio does not appear to be a good predictor for the T_{10}/T for the reservoirs in this study.

As with the rectangular basins, the cross-sectional flow velocity appears to be a better predictor of $T_{10}//T$ for the baffled circular basins. Figure 6 shows the relationship between the velocity in the baffled channel and the T_{10}/T. The logarithmic, best-fit relationship for circular basins is similar to that seen for rectangular basins in Figure 6 although with only four circular basins, the data are too limited to make a definitive conclusion.

CONCLUSIONS

The tracer analyses in this study produced the following conclusions.

> For the water treatment plants in this study, the T_{10}/T values that were assigned to the plant based upon industry standards or literature at the time of design in lieu of tracer studies at startup were generally higher than the values that were obtained by tracer analysis.
>
> For rectangular baffled contact basins, the cross sectional velocity of each channel appears to be a better predictor of T_{10}/T values than previous tools such as L/W ratio. The tracer study also appears to indicate that the cross sectional velocity can provide an estimate of T_{10}/T for circular basins with longitudinal baffles. This is especially true for Reynolds numbers in the laminar flow range.
>
> For the water plants with sequential basins in this study, no predictive relationship was apparent because each basin was unique and comparisons among the various plants was not possible. In general, sequential basins appear to provide significantly better baffling efficiency than the efficiency of each single basin. Full scale tracer analysis is always recommended for determining T_{10}/T values for compliance.
>
> Even with the velocity relationships shown in this study, there is variation in the T_{10}/T values for basins with similar cross sectional velocities. Consequently, full scale tracer studies are recommended to determine T_{10}/T values for compliance.

The tracer study also provided several lessons for designing and operating disinfection contact basins.

> Contact basin design should include provisions for tracer chemical injection and also location to monitor the tracer chemical including after injection and mixing prior to the contact basin and immediately after existing the contact basin.

Sequential basins should include sample locations between each basin to provide access for tracer analysis for each basin.

Incorporating an overflow weir wall structure prior to a pump chamber or basin outlet can ensure that the basin will maintain a constant volume. Accessibility to the weir for tracer sampling is encouraged.

Disinfection basin level sensing instrumentation should be checked annually to ensure that accurate basin volumes are recorded for compliance calculations.

REFERENCES

Barnett, T. C. and Venayagamoorthy, S. K., 2014 Laminar and turbulent regime changes in drinking water contact tanks. ***Journal of the American Water Works Association***, 106:12.

Bishop, M. M., Morgan, J. M., Cornwell, B., and Jamison, D. K., Improving the Disinfection Detention Time of a Water Plant Clearwell. 1993 ***Journal of the American Water Works Association***, 85:3.

Crozes, G. F., Hagstrom, J.P., Clark, M. M., Ducoste, J., and Burns, C., 1999. ***Improving Clearwell Design for CT Compliance***. AWWA Research Foundation, Denver, CO.

Deem, S. and Feagin, N., 2016. Disinfection Data Integrity in Washington State. ***Journal of the American Water Works Association***, 108:10.

Kattnig, J. J. and Venayagamoorthy, S. K., 2015 A Hybrid Approach for Increasing Baffling Factors in Contact Tanks. ***Journal of the American Water Works Association***, 107:12.

Teefy, S., 1996. ***Tracer Studies in Water Treatment Facilities: A Protocol and Case Studies***. AWWA Research Foundation, Denver, CO.

USEPA (U.S. Environmental Protection Agency), 1991 Edition. ***Guidance Manual for Compliance with the Filtration and Disinfection Requirements for Public Water Systems Using Surface Water Sources***. Washington, D.C.

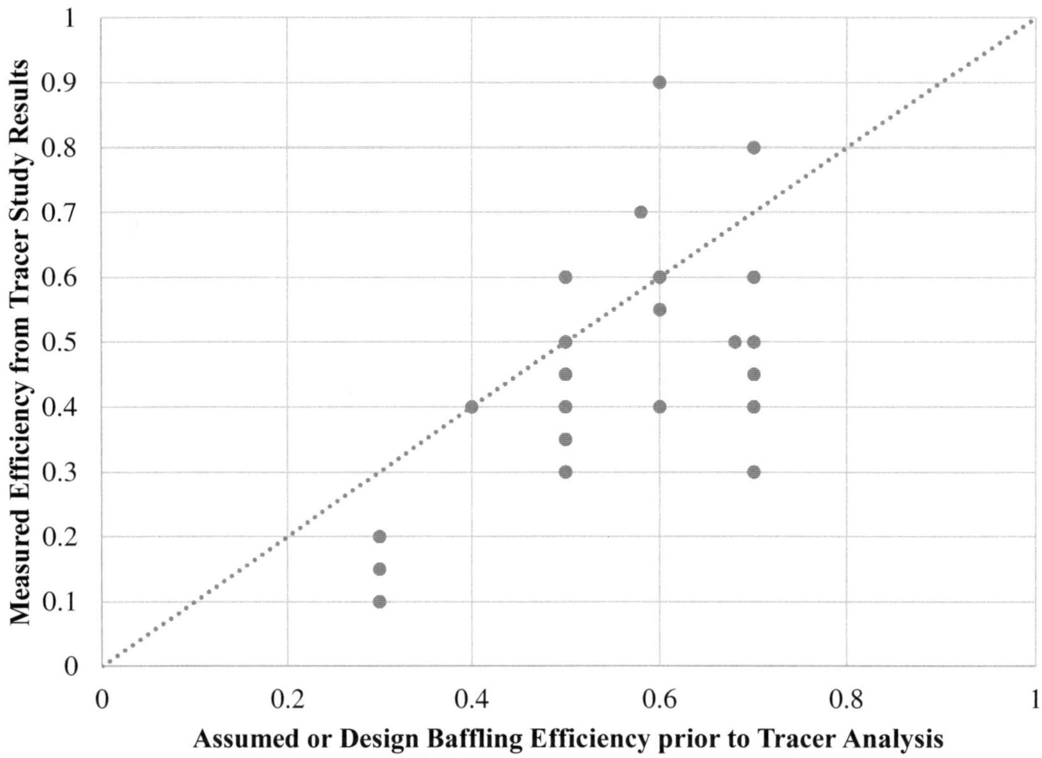

FIGURE 1 Comparison of historical baffling efficiency (BE) used at the water treatment plants in the study and the actual BE determined by the tracer analysis

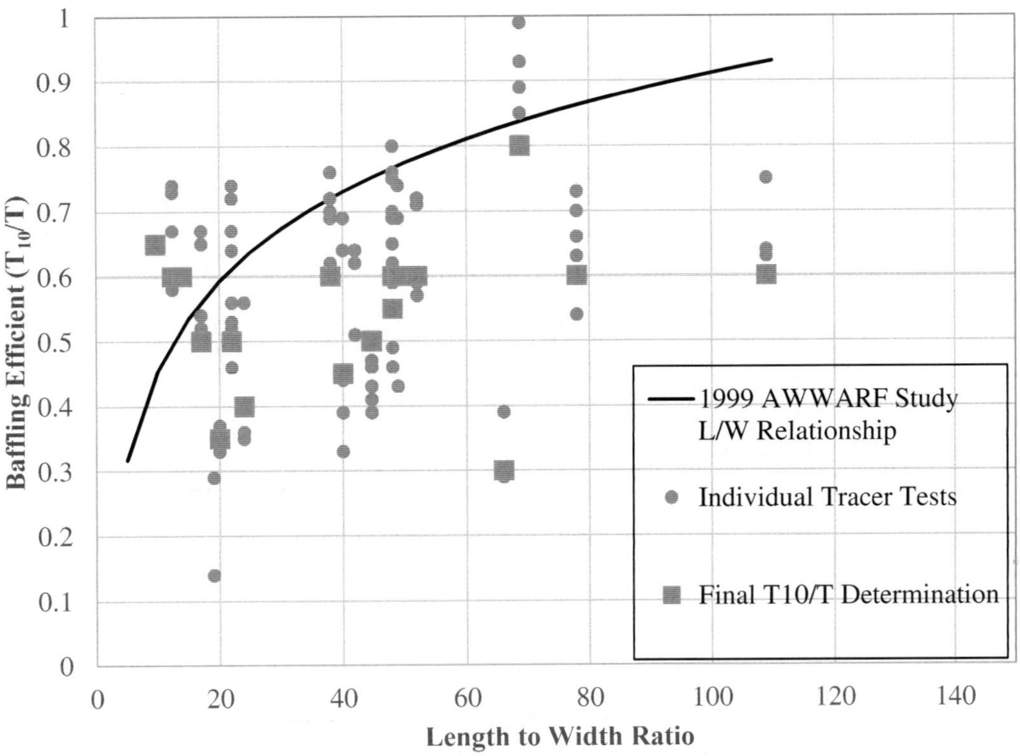

FIGURE 2 Comparison of baffled clearwell length to width ratios and baffling efficiencies (BE) as determined by tracer analysis for rectangular basins

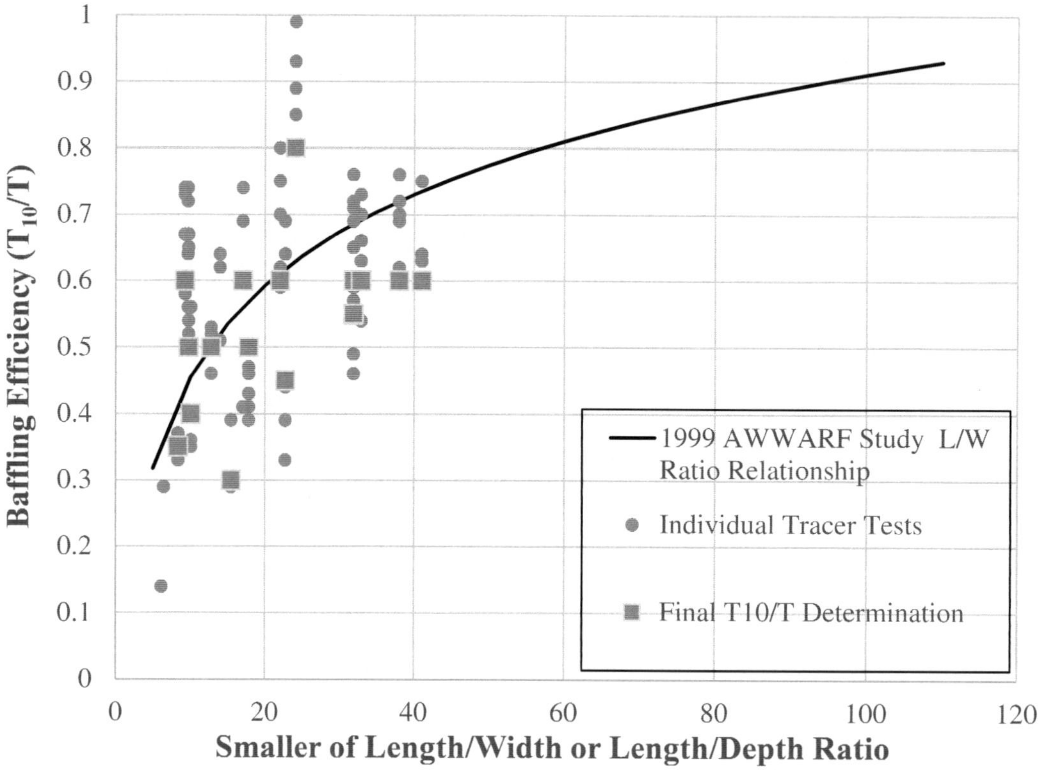

FIGURE 3 Comparison the smaller of baffled clearwell length to width ratios or length to depth ratios and baffling efficiencies (BE) as determined by tracer analysis

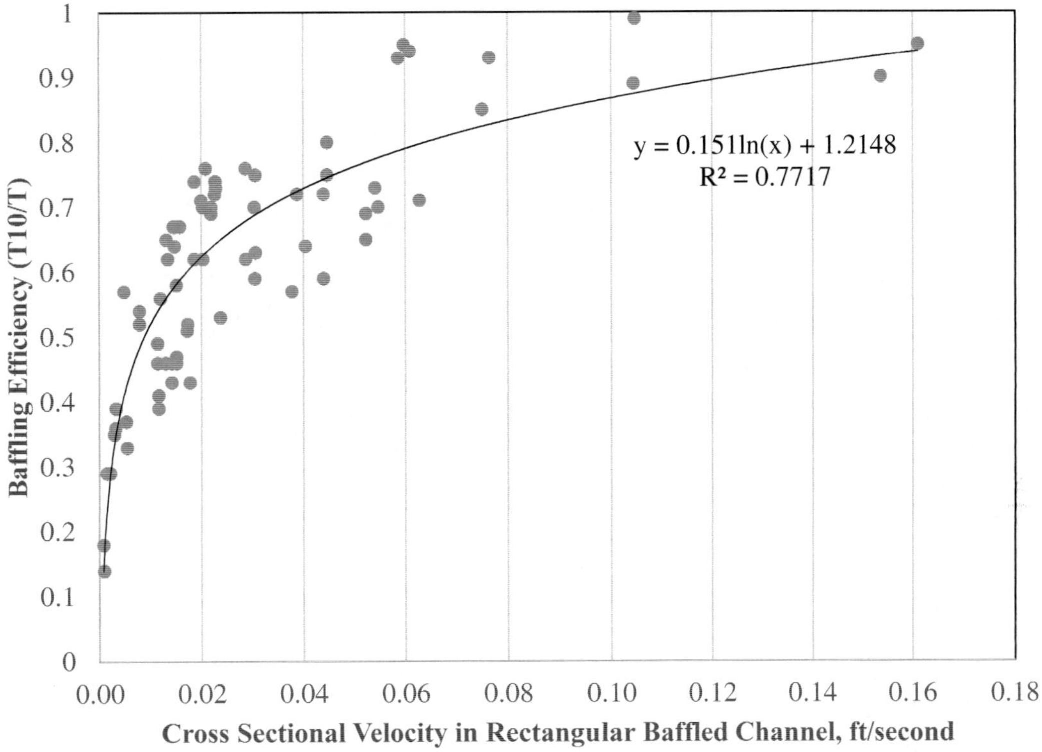

FIGURE 4 Comparison of cross sectional flow velocity in a rectangular, baffled channel with baffling efficiency

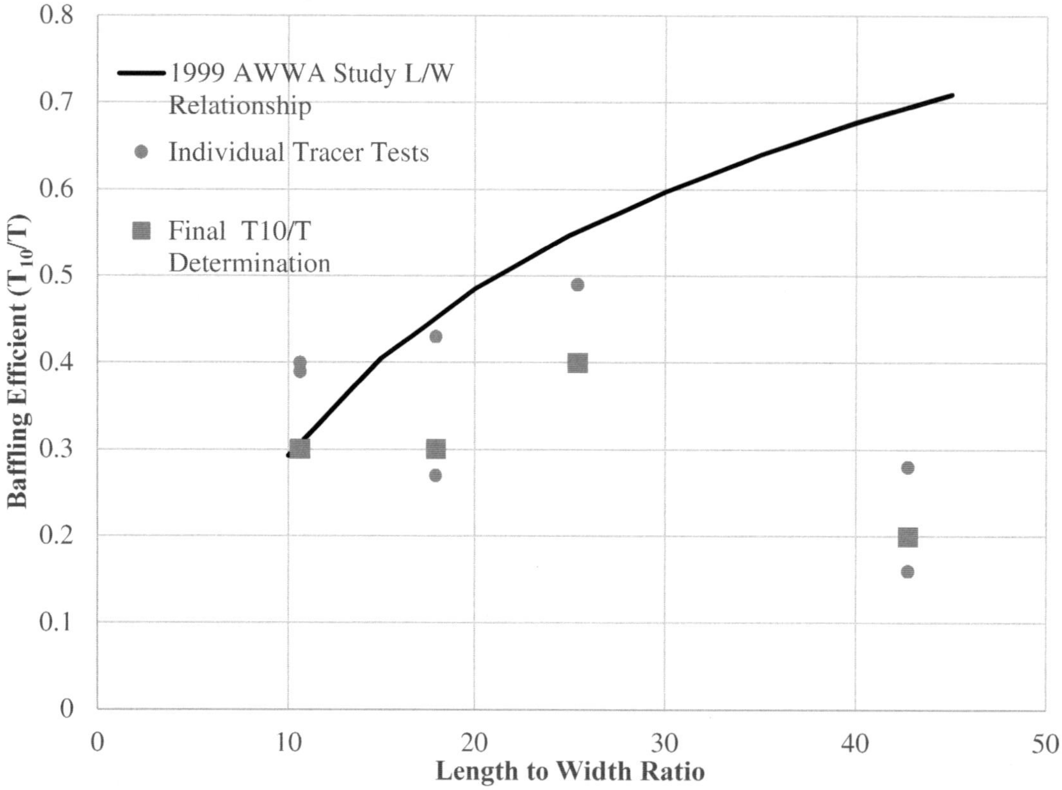

FIGURE 5 Comparison of baffled clearwell length to width ratios and baffling efficiencies as determined by tracer analysis for circular basins

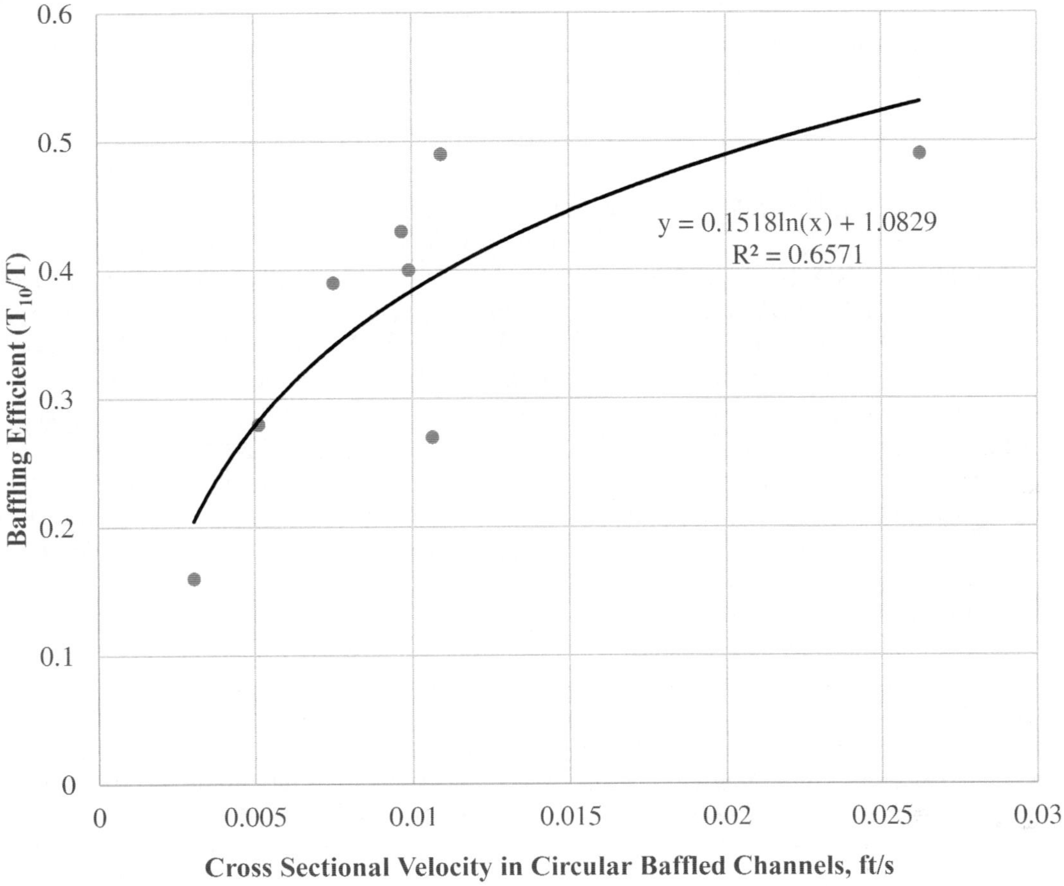

FIGURE 6 Comparison of cross sectional flow velocity in a circular, baffled channel with baffling efficiency

Ranitidine – A Potential Significant NDMA Precursor for Potable Reuse

Erica Marti, University of Nevada, Las Vegas
Caitlin M. Glover, Southern Nevada Water Authority
Eric R. V. Dickenson, Southern Nevada Water Authority

ABSTRACT
N-nitrosodimethylamine (NDMA) is an unregulated disinfection byproduct, but it is a top concern for potable water reuse. Treated wastewater is known to contain many different precursors that form NDMA upon chloramination or ozonation. This study focuses on the fate of one chloramine-reactive NDMA precursor, ranitidine (an antacid drug commonly consumed by humans), through two wastewater treatment plants (WWTPs) and along an urban wash that is dominated by the WWTP effluents. The urban wash feeds directly into the reservoir that supplies drinking water for the same community that produced the wastewater. Samples were collected after various wastewater treatment processes at two wastewater treatment plants and at well-studied locations along the urban wash. The samples were analyzed for ranitidine and NDMA formation potential. Ranitidine was present at significant concentrations in the primary effluent (2.2 μg/L), the UV-disinfected final effluent (0.66 μg/L), and throughout the urban wash (0.084-0.12 μg/L). Ranitidine accounted for 12-24% of the NDMA formation potential in the wastewater samples (722-3222 ng/L) and 5-6% of the NDMA formation potential in the urban wash (256-356 ng/L). Considering the substantial number of known NDMA precursors, this single compound accounts for a sizeable fraction of the NDMA formation potential at this site. Ranitidine was not present in the treated wastewater effluent for two other wastewater treatment plants (chlorination disinfection) located along the same urban wash. An implication of this study is that changing from UV to chlorination or ozonation would result in removing this high-yield precursor from the urban wash, which would reduce the potential for NDMA in the community's drinking water.

KEYWORDS
Disinfection byproducts, DBPs, NDMA, ranitidine, potable reuse

INTRODUCTION
Unregulated disinfection byproducts (DBPs) have received much attention in recent years owing to the detection of more than 500 compounds (Richardson, 1998) and the higher toxicity of many compounds compared to regulated DBPs (Muellner et al., 2015; Plewa et al., 2008). One notable unregulated DBP is *N*-nitrosodimethylamine (NDMA), which is classified as a probable human carcinogen (US EPA, 2014). NDMA is formed as a DBP due to chloramination of drinking water, chlorination of wastewater that contains ammonia, and ozonation of wastewaters containing NDMA precursors (Krasner et al., 2013).

A wide range of compounds have been demonstrated as chloramine NDMA precursors. The list includes pesticides, pharmaceuticals, and personal care products with amine moieties (Chen and Young, 2008; Selbes et al., 2013; Le Roux et al., 2011; Shen and Andrews, 2011a). On average, the precursors found in wastewater have molar yields of 2% (i.e., 100 mol/L of compound X results in 2 mol/L of NDMA). However, a select set of tertiary amines with α-aryl-N,N-dimethylamine substituents have notably higher yields (>50%) (Le Roux et al., 2011; Shen and Andrews, 2011b; Shen and Andrews, 2013). One example is ranitidine, which is a frequently used, over-the-counter antacid for treating heartburn and stomach ulcers (GlaxoSmithKline, 2015). The molar yield for ranitidine ranges from 50 to 95% depending on the matrix (i.e., pH, DO, nitrite, and dissolved organic matter) (Shen and Andrews, 2011a; Shen and Andrews, 2013; Seid et al., 2018). Between 30-70% of the compound is excreted from the body (GlaxoSmithKline, 2015) and ends up in wastewater where it is only partially removed during biological treatment (Radjenovic et al., 2009; Jelic et al., 2011). Attenuation in the environment by sorption and biodegradation is nominal; ranitidine is anticipated to have low sorption potential and a biodegradation half-life of 3.54 days (EPA Chemistry Dashboard, 2018). Solar photolysis may occur and result in ranitidine degradation (half-life <0.2 days), but this only holds true for surface water with no wastewater impact and no light hindrance (i.e., turbidity) (Latch et al., 2013). Thus, ranitidine is found at a significant concentration in the environment, especially in water sources influenced by treated wastewater.

This study examined the occurrence of ranitidine in a previously studied wastewater-dominated urban wash (Woods & Dickenson, 2016) and the relative importance of this precursor for NDMA formation. In addition, ranitidine was measured after several stages of wastewater treatment, including final disinfection of the effluent. As with many locations, this urban wash is upstream of a drinking water source. The results provide valuable insight on this NDMA precursor and the implications of final disinfection treatments on reducing NDMA formation potential in potable reuse situations.

METHODOLOGY

Samples were collected in early summer (two sampling campaigns) from two wastewater treatment plants (P2 and P3, Figure 1) and the Las Vegas Wash (LVW), which is an urban stream consisting of mostly treated wastewater effluent (~90%). Dry- and wet-weather runoff and shallow groundwater intrusion make up less than 10% of the LVW (Stave, 2001; Holdren and Turner, 2011). LVW samples were collected upstream and downstream of the wastewater treatment plants as seen in Figure 1.

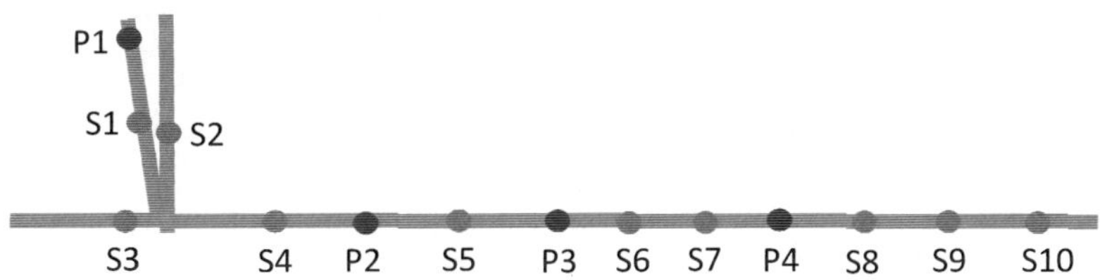

Figure 1. Not-to-scale depiction of locations for sample sites (●) and wastewater treatment plants (●) along the Las Vegas Wash.

Analyses were performed for ambient NDMA, NDMA chloramine maximum formation potential (NDMA Max FP), and ambient ranitidine. Formation potential tests involved preformed monochloramine. The Max FP 10-day tests were conducted at 140 mg/L as Cl_2 at pH 7 with phosphate buffer (Mitch et al., 2003). NDMA was analyzed by solid phase extraction followed by gas chromatography tandem mass spectrometry using isotope dilution (Holady et al., 2012). Ranitidine was analyzed without extraction (only 0.2 µm filtration with glass fiber filter) by liquid chromatography tandem mass spectrometry (Marti et al., 2017).

RESULTS

Key findings:

- Only P3 contributed ranitidine to the LVW in its UV-disinfected effluent.
- Treatment trains using chlorination/dechlorination or ozone resulted in no detectable ranitidine (<0.050 µg/L) in the effluent.
- Ranitidine was removed/transformed the most by ozonation (>97%) and chlorination (>90%). Secondary treatment was also effective (55-86%), and UV was moderately effective (33%).
- Ranitidine accounts for 5-6% of the NDMA Max FP in the LVW.
- Ranitidine accounts for 12-27% of the NDMA Max FP in wastewater samples at P3.

Table 1. Ranitidine and NDMA formation potential results for locations on the Las Vegas Wash and after various wastewater treatment processes.

Site ID	**Ranitidine in µg/L (St dev)**	**NDMA Max FP in ng/L (St dev)**	**Site ID**	**Ranitidine in µg/L (St dev)**	**NDMA Max FP in ng/L (St dev)**
S1	<0.050	989 (15.6)	P2 effluent before chlorination	0.53	933
S2	<0.050	52.5 (0.7)	P2 effluent after chlorination	<0.050	104
S3	<0.050	19	P2 effluent after dechlorination	<0.050	78
S4	<0.050	77	P3 primary effluent	2.2	3222
S5	<0.050	233	P3 secondary effluent	1.0	878 (15.6)
S6	0.43	233	P3 UV effluent	0.66 (0.03)	722
S7	0.12 (0.01)	356 (0.0)	P3 ozonated effluent	<0.050	133
S8	0.089	311	P3 chlorinated effluent	<0.050	61.5 (2.1)
S9	0.089	278			
S10	0.089	256			

Table 2. Fraction of NDMA Max FP attributed to ranitidine for Las Vegas Wash samples and wastewater samples based on 74% molar yield for ranitidine. Dilution of Las Vegas Wash samples by side streams was taken into account.

Site ID	Fraction attributed to ranitidine	Site ID	Fraction attributed to ranitidine
S6	32%	P2 effluent before chlorination	10%
S7	6%	P3 primary effluent	12%
S8	5%	P3 secondary effluent	20%
S9	5%	P3 UV effluent	27%
S10	6%		

DISCUSSION

Previous work (Woods & Dickenson, 2016) indicated the presence of NDMA precursors in the LVW. The precursors underwent natural attention through biodegradation (32-88%), photolysis by sunlight (52%), and sorption to soil (4%) as determined with bench-scale microcosm and photolysis experiments. However, the sources of the precursors were not determined. During a recent biofiltration study (Marti et al., 2017), ranitidine was quantified at a significant concentration (2 µg/L) in the tertiary filtered effluent at P3. Thus, it was hypothesized that ranitidine may be present in the discharged effluent at P3 and other wastewater treatment plants along the LVW.

Ranitidine Concentrations in the Las Vegas Wash

Results for ranitidine are shown in Table 1 and the sample locations are shown in Figure 1. Ranitidine was not present in the LVW ahead of P3. Neither P1 nor P2 appear to contribute ranitidine to the LVW in their effluent. In addition, P4 discharges just downstream of S7 and ranitidine did not increase after the discharge point. Both P1 and P4 use chlorination/ dechlorination as their final disinfection step. Chlorination is known to transform ranitidine (Jeon et al., 2016). P2 also uses chlorination/dechlorination, and ranitidine was below the method reporting limit (<0.050 µg/L) after chlorination. Only P3 is known to contribute ranitidine to the LVW via the treatment train with UV-disinfected effluent. P3 also discharges 25-25% of its total flow via a treatment train with membrane filtration and ozonation; however, ozonation transforms ranitidine (Jeon et al., 2016). Ranitidine was not present in runoff at S2 or S3. Also, S7 had a lower ranitidine concentration (72% decrease) after a tributary joined the LVW, which suggests no ranitidine input from this stream. However, this tributary was not specifically analyzed for ranitidine. Wash water for these tributaries (S2, S3, and S7) consists of urban runoff and is not expected to contain any treated wastewater. Thus, it is presumed that all of the ranitidine is due to the UV-treated effluent discharged by P3 to the LVW.

The concentration of ranitidine decreased along the LVW. After the initial input from P3, ranitidine decreased by 72% between S6 and S7, decreased by 26% from S7 to S8, and then was relatively consistent at 89 ng/L along the rest of the LVW. Part of the decrease could be explained by dilution as tributaries join the LVW. Dilution factors from Woods & Dickenson (2016) for inputs to the Las Vegas Wash were used to adjust the ranitidine concentration. After adjusting for dilution, ranitidine decreased only 16% from S7 to S10. Possible explanations include sorption to sediment, plant uptake, and biodegradation. Direct photolysis is also a known mechanism for ranitidine removal (Latch et al., 2003; Bergheim et al., 2012; Rivas et al., 2010) and could explain the decrease along the LVW. The flow-adjusted concentrations did not change the 72% decrease between S6 and S7. Possible reasons for the large difference include poor mixing at S6 (just downstream from the outfall at P3) and diurnal variability. Grab samples along the LVW were taken over a four hour period in the morning, and they are not hydraulically linked. Diurnal variability could partly account for the difference in ranitidine concentration at sampling locations.

Ranitidine Removal During Wastewater Treatment

Ranitidine removal varied for the different wastewater treatment processes. Removal for each treatment was determined by taking the difference between the initial concentration before the treatment step and the concentration after the treatment step.

Removal for P2:

- Secondary treatment = 86%
- Chlorination = >90%
- Overall treatment train = >98%

Removal for P3:

- Secondary treatment = 55%
- Tertiary filtration = 2%
- UV = 33%
- Ozone = >97%
- Chlorination = >95%
- Overall treatment train with UV = 70%

In general, biological treatment and oxidation by chlorine and ozone were the main routes for ranitidine transformation in the wastewater treatment plants.

NDMA FP Concentrations in the Las Vegas Wash

Results for NDMA Max FP are shown in Table 1. Although ranitidine was not detected, P1 has substantial NDMA precursors (989 ng/L) in its treated effluent. The precursors are unknown, but it is possible that the treatment plant uses polymers. P2 was low (78 ng/L) in NDMA precursors compared to P3 (572.5 ng/L). The concentration of NDMA precursors decreased (22 ng/L, 7%) between S7 and S8 where P4 discharges, which suggests a low NDMA precursor concentration for that treatment plant.

After adjusting for dilution from tributaries, there is relatively little change (19%) in the concentration of NDMA precursors along the LVW from S7 to S10. This is in agreement with the minimal change (16%) for ranitidine concentration along the same stretch.

NDMA FP Removal During Wastewater Treatment
NDMA precursors were removed at different rates for various wastewater treatment processes. The same equation for ranitidine removal was used to calculate removal for NDMA FP.

Removal for P2:
- Chlorination = 89%

Removal for P3:
- Secondary treatment = 73%
- Tertiary filtration = 18%
- Microfiltration = 5%
- UV = 21%
- Ozone = 84%
- Chlorination = 91%
- Overall treatment train with UV = 82%

This matches well with previous research on NDMA precursor removal through biodegradation (secondary treatment) and oxidation (ozonation, chlorination). UV was not as effective at destroying NDMA precursors, which agrees with the low ranitidine removal by UV for CCWRD.

Ranitidine Contribution to NDMA Max FP Concentration
Ranitidine was detected in all LVW samples downstream from P3. With a high molar yield for NDMA formation, it could be a significant precursor in the LVW. Based on an experimentally-determined molar yield for the Max FP test using LVW water (74%), the percentage of NDMA Max FP attributed to ranitidine was calculated (Table 2). Ranitidine accounts for 5-6% of the NDMA formation potential in the LVW. While this is a significant fraction, there are certainly many other NDMA precursors present in the LVW.

Ranitidine accounts for a higher fraction of NDMA Max FP in treated wastewater samples as compared to the LVW samples. For primary, secondary, tertiary filtration and membrane filtration, ranitidine accounts for 12-24% of the NDMA Max FP at P3. Ranitidine was not present after ozonation and chlorination, but it was present after UV irradiation and accounted for 27% of the NDMA FP. Ranitidine accounted for 10% of the NDMA Max FP in the effluent at P2 just prior to chlorination.

CONCLUSIONS
This follow-up study on NDMA precursors in the LVW demonstrated that ranitidine is a significant precursor in this urban wash. It accounts for 5-6% of the NDMA formation potential. Based on available information, ranitidine is discharged solely by one wastewater treatment plant in its UV-disinfected effluent. Three other treatment plants along the LVW use chlorination, which transforms ranitidine. In potable reuse situations, changing from UV disinfection to chlorination or ozonation should result in reduced NDMA formation potential from ranitidine, as well as other unidentified NDMA precursors.

REFERENCES

Chen, W. H., Young, T. M., 2008. NDMA formation during chlorination and chloramination of aqueous diuron solutions. Environ. Sci. Technol., 42, (4), 1072-1077.

GlaxoSmithKline, 2015. Product Monograph: Zantac®: Histamine H2-receptor antagonist. https://ca.gsk.com/media/592906/zantac.pdf (accessed 12.03.18).

Holady, J. C., Trenholm, R. A., Snyder, S. A., 2012. Use of automated solid-phase extraction and GC-MS/MS to evaluate nitrosamines in water matrices. American Laboratory, 44, (3), 25-30.

Holdren, G.C., Turner, K., 2010. Characteristics of Lake Mead, Arizona-Nevada. Lake and Reservoir Management, 26 (4), 230-239.

Jelic, A., Gros, M., Gindebreda, A., Cespedes-Sanchez, R., Ventura, F., Petrovic, M., Barcelo, D., 2011. Occurrence, partition and removal of pharmaceuticals in sewage water and sludge during wastewater treatment. Water Res., 45 (3), 1165-1176.

Krasner, S. W., Mitch, W. A., McCurry, D. L., Hanigan, D., & Westerhoff, P., 2013. Formation, precursors, control, and occurrence of nitrosamines in drinking water: a review. Water Res., 47 (13), 4433-4450.

Le Roux, J., Gallard, H., Croué, J.P., 2011. Chloramination of nitrogenous contaminants (pharmaceuticals and pesticides): NDMA and halogenated DBPs formation. Water Res. 45 (10), 3164-3174.

Marti, E. J., Dickenson, E. R., Trenholm, R. A., & Batista, J. R., 2017. Treatment of Specific NDMA Precursors by Biofiltration. Journal-American Water Works Association, 109 (6), E273-E286.

Mitch, W. A., Gerecke, A. C., Sedlak, D. L., 2003. A N-nitrosodimethylamine (NDMA) precursor analysis for chlorination of water and wastewater. Water Res., 37 (15), 3733-3741.

Muellner, M. G., Wagner, E. D., McCalla, K., Richardson, S. D., Woo, Y. T., Plewa, M. J., 2007. Haloacetonitriles vs. regulated haloacetic acids: are nitrogen-containing DBPs more toxic? Environ. Sci. Technol., 41 (2), 645-651.

Plewa, M. J., Wagner, E. D., Muellner, M. G., Hsu, K. M., Richardson, S. D., 2008. Comparative mammalian cell toxicity of N-DBPs and C-DBPs, In: Disinfection By-Products in Drinking Water, Vol. 995, Karanfil, T., Krasner, S., Westerhoff, P. and Xie, Y (Eds.), American Chemical Society, pp. 36-50.

Radjenovic, J., Petrovic, M., Barcelo, D., 2009. Fate and distribution of pharmaceuticals in wastewater and sewage sludge of the conventional activated sludge (CAS) and advanced membrane bioreactor (MBR) treatment. Water Res., 43 (3), 831-841.

Richardson, S.D., 1998. Drinking water disinfection by-products, In: The Encyclopedia of Environmental Analysis and Remediation, Vol. 3, Meyers, R.A. (Ed.), John Wiley & Sons, pp. 1398-1421.

Seid, M.G., Cho, K., Lee, C., Park, H.-M., Hong, S.W., 2018. Nitrite ion mitigates the formation of N-nitrosodimethylamine (NDMA) during chloramination of ranitidine. Sci. Total Environ., 633 (15), 352-359.

Selbes, M., Kim, D., Ates, N., Karanfil, T., 2013. The roles of tertiary amine structure, background organic matter and chloramine species on NDMA formation. Water Res., 47 (2), 945-953.

Shen, R., Andrews, S.A., 2011a. Demonstration of 20 pharmaceuticals and personal care products (PPCPs) as nitrosamine precursors during chloramine disinfection. Water Res. 45 (2), 944-952.

Shen, R., Andrews, S.A., 2011b. NDMA formation kinetics from three pharmaceuticals in four water matrices. Water Res., 45 (17), 5687-5694.

Shen, R., Andrews, S.A., 2013. Formation of NDMA from ranitidine and sumatriptan: the role of pH. Water Res., 47 (2), 802-810.

Stave, A.K., 2001. Dynamics of Wetland Development and Resource Management in Las Vegas Wash, Nevada. Journal of the American Water Resources Association, 37 (5), 1369-1379.

U.S. Environmental Protection Agency (U.S. EPA), 2014. Technical Fact Sheet – N-nitroso-dimethylamine (NDMA). https://www.epa.gov/sites/production/files/2014-03/documents/ffrrofactsheet_contaminant_ndma_january2014_final.pdf. (accessed 17.04.17).

U.S. Environmental Protection Agency Chemistry Dashboard, 2018. https://comptox.epa.gov/dashboard/dsstoxdb/results?search=DTXSID0052835#biodeg_hl (accessed 03.04.2018).

Woods, G.C., Dickenson, E.R.V., 2016. Natural attenuation of NDMA precursors in an urban, wastewater-dominated wash. Water Res., 89 (1), 293 – 300.

GETTING A JUMP ON DISINFECTION BYPRODUCTS: A PRO-ACTIVE APPROACH FOR OPTIMIZING RECLAIMED WATER DISINFECTION

ABSTRACT

As communities grow across the United States, public utilities are faced with unique challenges in meeting future water demands on the front-end and higher level treatment of wastewater on the back-end. The traditional view of water and wastewater as separate and distinct resources is giving way to a more holistic approach, particularly as potable reuse is beginning to dominate the conversation of potential alternative water supplies. One critical component of this holistic approach is the optimization of wastewater treatment and production of high quality reclaimed water effluent. Providing high-level chlorine disinfection to protect against pathogen exposure is a key treatment step directly impacting reclaimed water quality. However, the formation of disinfection byproducts (DBPs) is an unintended consequence of disinfection with free chlorine. With regulatory agencies beginning to scrutinize and enforce DBPs limits in reclaimed waters, utilities are tasked with managing their treatment and operational processes to meet these limits. In this study, DBP formation potential testing of reclaimed water from a full-scale water reclamation facility was performed to assess the impacts of converting from UV to chlorine disinfection as part of a 20 MGD plant expansion. The DBP formation potential results were empirically modeled to develop an operations support tool for predicting the formation of DBPs as a function of process flows and the chlorine contact time. The DBP and modeling evaluation results demonstrated that dechlorination in conjunction with careful control of the chlorine contact chamber operation were vital for meeting regulated DBP limits. Thus, the developed modeling tool provides treatment operators a practical, proactive strategy for optimizing their disinfection process and improve their reuse water quality by helping maintain lower DBP levels.

KEYWORDS

Reclaimed water, disinfection, disinfection byproducts, optimization, modeling

INTRODUCTION

In wastewater treatment, disinfection is required for the inactivation of disease-causing microorganism prior to reclaimed water discharge or reuse. Disinfection may be accomplished by using gas or liquid chlorine, combined chlorine (chloramines), ultraviolet light (UV), peracetic acid, ozone, or chlorine dioxide. Of these, chlorination is by far the most widely applied disinfection strategy. Chlorine disinfection effectively protects against microbial pathogen exposure; however, it also reacts with the organic and inorganic matter in reclaimed water to form undesirable disinfection byproducts (DBPs). Regulatory guidelines for discharges from wastewater treatment facilities will continue to tighten and consider regulatory limits for chlorine DBPs, particularly trihalomethanes (THMs) and haloacetic acids (HAAs). This regulatory shift is of particular interest for the majority of facilities that use chlorination as well as those facilities that are looking to convert from an alternative method, such as UV, to free chlorine. While this study will focus on a specific subset of THMs, including bromodichloromethane and dibromochloromethane, it is important to note that the THMs and HAAs regulated under the drinking water rules would need to also be considered in potable reuse applications. As such, it is critical for wastewater utilities to understand the factors that impact DBP formation and the options available to limit their formation.

Formation of DBPs depends primarily on organic content, disinfectant type and dosing, contact time, pH, temperature, and the presence of halide ions (Reckhow & Singer, 2011; Krasner, 2009). The highest DBP formations tend to occur when chlorine is applied as the disinfectant (Bougeard et al., 2009; Hua & Reckhow, 2007). Typically, DBP levels increase with increasing organic precursor material, chlorine dose and residual, and temperature (Liang & Singer, 2003; Reckhow & Singer, 2011; Zhang et al., 2012). While THMs generally increase with pH levels and contact time with chlorine, HAAs are favored under more acidic pH values and are less predictable with chlorine contact time due to possible biodegradation at low chlorine residuals (Liang & Singer, 2003; Reckhow & Singer, 2011). The presence of bromide in water results in a higher fraction of brominated DBPs (Obolensky & Singer, 2005).

Based on these general factors that govern DBP formation, the following techniques can be employed for reducing DBP formation in reclaimed water:

- Reducing the chlorine dosage and/or contact time
- Using alternate disinfectants (chloramines, UV, peracetic acid, ozone, or chlorine dioxide)
- Removing organic precursors (granular activated carbon, biofiltration, ion exchange, or nanofiltration or reverse osmosis membranes)
- Removing THMs and/or HAAs after their formation (aeration for THMs and biofiltration for HAAs)

The applicability of a DBP control strategy is site specific and should be carefully evaluated in terms of the discharge or reuse application (non-potable vs potable), regulatory requirements and limits, operational requirements, ease of integration into an existing system with fixed contactor volumes, and costs. Of the DBP control strategies listed, this study focused on treatment and operation optimization through the management of chlorine dosage and contact time for a proposed 20 MGD water reclamation facility expansion in Hillsborough County, Florida.

Hillsborough County is currently expanding their Northwest Regional Water Reclamation Facility (WRF) from its original treatment capacity of 10 MGD to 30 MGD on an annual average daily flow (AADF) basis. The facility expansion included the conversion from UV disinfection to chlorine disinfection. The chlorine disinfection system was designed to provide high-level disinfection to allow reuse for public access irrigation. In addition to supplying the reclaimed water distribution system, reclaimed water flows can be discharged to the Tampa Bay through surface water discharge. The facility's surface water discharge permit sets limits on the quantity of DBPs, specifically bromodichloromethane ($CHBrCl_2$) and dibromochloromethane ($CHBr_2Cl$), that can be discharged to the receiving water body. These regulated species are part of the THM group of chlorinated DBPs, for which the regulated maximum levels are summarized in **Table 1**.

Table 1. Northwest Regional WRF Surface Water Discharge Limits for DBPs

Contaminant	Limit
Bromodichloromethane	22.0 µg/L
Dibromochloromethane	34.0 µg/L

With UV disinfection, regulated DBP formation is not a concern as UV disinfection does not produce THMs. However, the conversion to chlorine disinfection increases the formation potential of these DBPs. Because of this, the purpose of this study was to assess the impacts of converting to chlorine disinfection on DBP formation. This was accomplished by implementing a three-step approach. First, the chlorine contact dose required for high-level disinfection was determined by analyzing fecal coliform monitoring results. Second, DBP formation potential testing was performed to gauge the potential for exceeding the regulatory DBP limits. Third, the DBP formation results were used to develop an operations support tool for optimizing the disinfection process by allowing operations to predict the formation of DBPs for varying flows and chlorine contact chamber sections in operation. The DBP modeling tool was preliminarily validated with recent full-scale DBP monitoring data. Overall, the results of this study offer a case study framework for implementing a practical, proactive strategy for optimizing the disinfection process and improve reclaimed water quality relative to regulated DBPs.

PERTINENT REGULATORY REQUIREMENTS FOR DISINFECTION

Optimization of the wastewater treatment process relative to minimizing disinfection byproducts begins with establishing the lowest disinfection level that upholds public and environmental safety. In Florida, the trade-off between minimizing exposure to pathogens and high chemical costs and DBP formation is balanced through a minimum required total chlorine residual and tiered disinfection contact time (CT) level based on fecal coliform counts.

Reclaimed water facilities that supply public access reuse systems must comply with the high-level disinfection requirements outlined in Chapter 62-600.440 of the Florida Administrative Code (FAC). According to Chapter 62-600.440, the minimum requirements for achieving high-level disinfection include a total chlorine residual of at least 1.0 mg/L and contact time of at least 15 minutes at peak hourly flow. Additionally, sub-paragraph 62-600.440(5)(c) sets tiered design criteria for the product of the total chlorine residual and contact time (CT) based on the level of fecal coliforms in the reclaimed water effluent prior to disinfection. The tiered CT requirements are summarized in **Table 2**.

Table 2. Tiered Criteria for the Product of Total Chlorine Residual and Contact Time

Fecal Coliforms (CFUs/100 mL)	CT Required (mg/L-min)
<1,000	25
1,000 – 10,000	40
>10,000	120

METHODOLOGY

Fecal Coliform Monitoring Program for High-Level Disinfection

To assess the CT required for high-level disinfection, a fecal monitoring plan was conducted over seven consecutive days to characterize the level of fecal coliforms in the Northwest Regional WRF's reclaimed water prior to disinfection. From July 27, 2016 through August 2, 2016, eight (8) grab samples were collected each day from the filter effluent before disinfection or chemical addition. Based on the reviewed historical influent flows presented in **Figure 1** and operations input, the grab samples were collected hourly from 7 am to 10 am and from 6 pm to 9 pm. These sampling times represent estimated times of peak influent flows to the facility. An overview of the fecal monitoring plan is provided in **Table 3**. The samples were transported to a certified analytical laboratory and analyzed for fecal coliforms within 8 hours of collection.

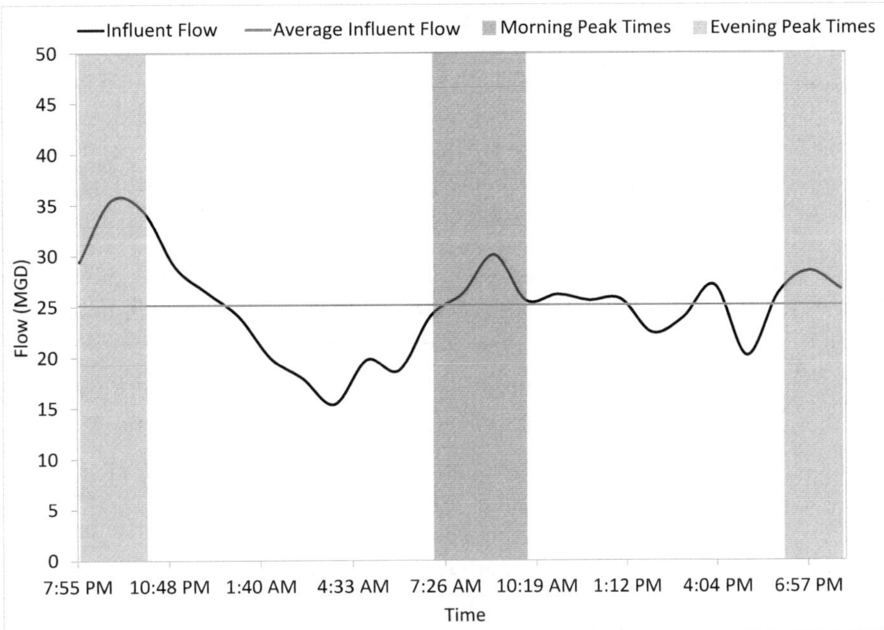

Figure 1. Diurnal Influent Flows

Table 3. Fecal Coliform Monitoring Plan

Sampling Parameter	Value
Number of Consecutive Sampling Days	7
Estimated Morning Peak Sampling Times	7:00 AM 8:00 AM 9:00 AM 10:00 AM
Estimated Evening Peak Sampling Times	6:00 PM 7:00 PM 8:00 PM 9:00 PM
Number of Fecal Coliform Samples Collected per Day	8
Total Number of Fecal Coliform Samples	56

The fecal coliforms in the NWRWRF filter effluent samples were measured according to Standard Method 9222D. This standard method applies the membrane filter technique to determine fecal coliform bacterial densities. The fecal coliform membrane filter method uses an enriched lactose medium and incubation temperature of 44.5 ± 0.2 °C for selectivity and provides 93% accuracy according to the method (9222D). Colonies produced by fecal coliform bacteria are counted and reported as colony forming units (CFUs) per 100 mL.

Disinfection Byproduct Formation Potential Analysis

To determine the impact of converting from UV to chlorine disinfection, a DBP formation potential investigation was performed to simulate levels of bromodichloromethane and dibromochloromethane formed in chlorinated reclaimed water effluent from the Northwest WRF. On June 14, 2016, approximately twelve (12) gallons of treated effluent were collected from the NWRWRF's combined filter effluent channel prior to UV disinfection or chemical addition. The NWRWRF filter effluent sample was brought to Tetra Tech's in-house laboratory for effluent water quality, chlorine demand, and DBP formation potential testing.

Filter Effluent Water Quality

The background inorganic and organic water quality of the filter effluent was characterized by measuring the pH, temperature, turbidity, color, and ultraviolet absorbance at 254 nanometer wavelength (UV-254). The water quality results are provided in **Table 4**. The measured pH was near neutral, and the temperature was near room temperature. The water had a relatively low filtered water turbidity of 0.3 NTU. The observed color and UV-254 levels, which are surrogate measures of organic matter, were in the moderate range comparable to UV 254 levels observed in surface waters (0.113 to 0.67 1/cm) (Edzwald et al., 1985). Generally, higher THM formation is favored under high organic matter, pH, and temperature conditions. Therefore, based on the measured water quality, extreme values of bromodichloromethane and dibromochloromethane formation are not anticipated.

Table 4. Filter Effluent Water Quality

Parameter	Unit	Value
pH	--	6.81
Temperature	°C	23.1
Turbidity	NTU	0.3
UV-254	1/cm	0.2
Color	Color Units (Pt-Co)	30

Chlorine Demand Determination

The chlorine demand of the NWRWRF filter effluent sample was estimated and modeled to determine the chlorine dose required to achieve total chlorine residual targets of 1.0 mg/L (minimum required by Chapter 62-600, FAC) and 2.0 mg/L (typical minimum operating residual). Nine batches of filter effluent samples were dosed with increasing doses of sodium hypochlorite ranging from 2 to 8 mg/L as Cl_2. Sodium hypochlorite solutions typically contain about 12% to 13% of active, available chlorine. For each batch, the chlorine residual was measured after 30 minutes of contact time. The chlorine demand results are summarized in **Table 5**. The 2.0 and 2.5 mg/L chlorine doses did not produce a stable chlorine residual and were excluded from the chlorine demand modeling. A chlorine demand curve was developed by plotting the measured chlorine residual versus applied chlorine dose. The scatter-plot data, shown in **Figure 2**, was modeled as a linear regression and used to calculate the chlorine doses necessary to reach the target chlorine residuals. The chlorine demand analysis revealed that a 4.41 and 5.88 mg/L sodium hypochlorite dose is needed to target a 1.0 and 2.0 mg/L chlorine residual, respectively, for the DBP formation potential testing.

Table 5. Chlorine Demand Results

Sequence	Target Cl_2 Dose (mg/L)	Contact Time (mins)	Cl_2 Residual (mg/L)
1	2	30	0.03
2	2.5	30	0.05
3	3	30	0.1
4	3.5	30	0.31
5	4	30	0.71
6	5	30	1.38
7	6	30	2.01
8	7	30	2.96
9	8	30	3.35

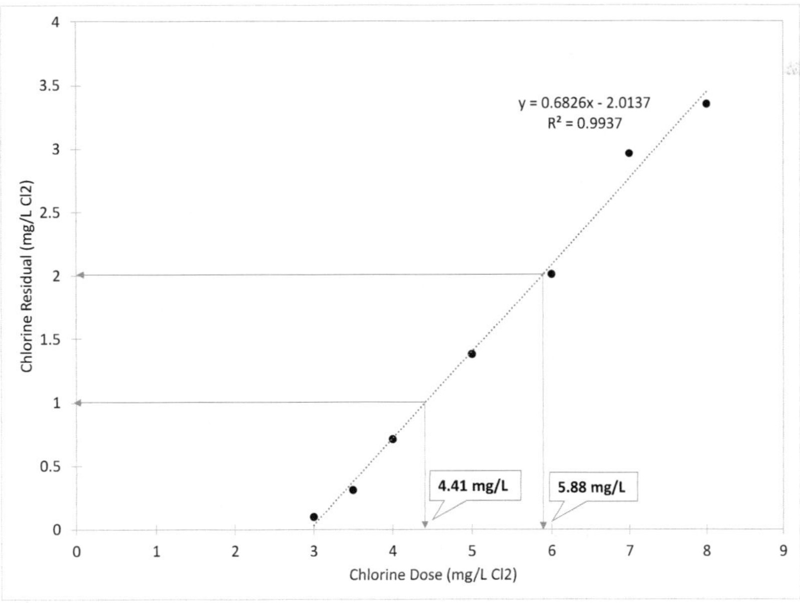

Figure 2. Chlorine Residual versus Applied Chlorine Dose Modeling

DBP Formation Potential Testing

DBP formation potential testing was performed on the filter effluent to simulate the expected levels of bromodichloromethane and dibromochloromethane formation. Two separate bulk batches of filter effluent samples were dosed with 4.41 and 5.88 mg/L of sodium hypochlorite to target total chlorine residuals of 1.0 and 2.0 mg/L, respectively, after 30 minutes of contact time. After disinfection, the chlorinated samples were incubated in borosilicate glass bottles at ambient temperature. The samples were contacted with the chlorine for up to six hours, and THM formation was monitored every 15 minutes in the first hour, every 30 minutes in the second hour, and every hour in the remaining four hours.
At each contact time, the sample was analyzed for chlorine residual then the sample was quenched or dechlorinated for THM analysis. At the end of the 6-hour contact time, the collected THM samples were shipped on ice to a certified analytical laboratory for THM analysis. The concentration of the four major THM species, including bromodichloromethane and dibromochloromethane, were analyzed according to EPA Method 624. A summary of the DBP formation potential testing matrix is presented in **Table 6**.

Table 6. DBP Formation Potential Testing Matrix

Experimental Parameters	Disinfection Scenarios	
Target Chlorine Residual:	1 mg/L Cl_2 at 30 minutes	2 mg/L Cl_2 at 30 minutes
Target Chlorine Dose:	4.41 mg/L Cl_2	5.88 mg/L Cl_2
DBP Water Quality Monitoring:	Chlorine Residual Bromodichloromethane Dibromochloromethane	
Experimental Contact Times:	T1 = 15 minutes T2 = 30 minutes T3 = 45 minutes T4 = 60 minutes T5 = 90 minutes T6 = 2 hours T7 = 3 hours T8 = 4 hours T9 = 5 hours T10 = 6 hours	

Optimization Tool Development

The chlorine demand and DBP formation results were used to develop an operations support tool for predicting the formation of DBPs as a function of process flows and the proposed chlorine contact chamber sections in operation. The operations support tool was based on the empirical relationships between the DBP levels and contact times for both the 1.0 mg/L and 2.0 mg/L chlorine residual scenarios. The relationships were modeled by plotting the DBP concentration versus contact time results. The scatter-plot data was then fit to a logarithmic equation using the curve regression function in Excel.

The DBP formation potential data for the 1.0 and 2.0 mg/L residual scenarios were extrapolated to estimate the DBP formation at a higher 4.0 mg/L chlorine residual after 30 minutes of contact time. The extrapolated data was also fit to a logarithmic equation to provide an estimate of DBP levels during disinfection at a higher total chlorine residual.

RESULTS AND DICUSSION

Fecal Coliform Monitoring Program Results

These fecal coliform results were organized into morning, evening, and daily sampling and analyzed using descriptive statistics, confidence intervals, and box-plot distributions. The average, minimum, and maximum fecal coliforms are presented in **Table 7**. The corresponding 95 percent confidence intervals for each average is also presented in **Table 7**. The 95 percent confidence interval provides a range in which the average is expected to fall.

Throughout the seven days of consecutive sampling, the fecal coliforms ranged from 10 to 700 CFUs/100 mL in the treated wastewater filter effluent. On average, the observed fecal coliforms were approximately 36% higher during the evening sampling as compared to the morning sampling. Based on the historical hydrograph shown in **Figure 1**, the higher evening average may have resulted from greater flow loadings to the plant during the evening hours.

The daily average was 177 CFUs/100 mL. At a 95 percent confidence level of ±34, the average fecal coliforms are expected to fall within 143 and 211 CFUs/100 mL. Similarly, the morning and evening averages are expected to reach up to 159 and 278 CFUs/100 mL, respectively. Based on these 95 percent confidence intervals, the average fecal coliforms are expected to remain below the 1,000 CFUs/100 mL tier for a required CT of 25 mg/L-min.

To further assess the variability in the fecal coliform results, box-plots were constructed using the morning, evening, and daily sampling results. A comparison of the box-plots is presented in **Figure 3**. For each sampling category, the box-plot shows the median (50th percentile) as the center bar and the quartiles (25th and 75th percentiles) as the box. The box includes the middle 50% of the data and is known as the inner-quartile range. The "whiskers" or bars represent the minimum and maximum fecal coliforms. The box-plots emphasize that the majority of the fecal coliforms were distributed between 100 and 300 CFUs/mL. Furthermore, extreme coliform values remained below 1,000 CFUs/100 mL.

Table 7. Fecal Coliform Results (July 27th to August 2nd 2016)

Fecal Coliform Monitoring	Number of Samples	Average	Minimum	Maximum	95% Confidence Level
Morning Sampling	28	137	10	300	±25
Evening Sampling	28	217	40	700	±61
Daily Sampling	56	177	10	700	±34

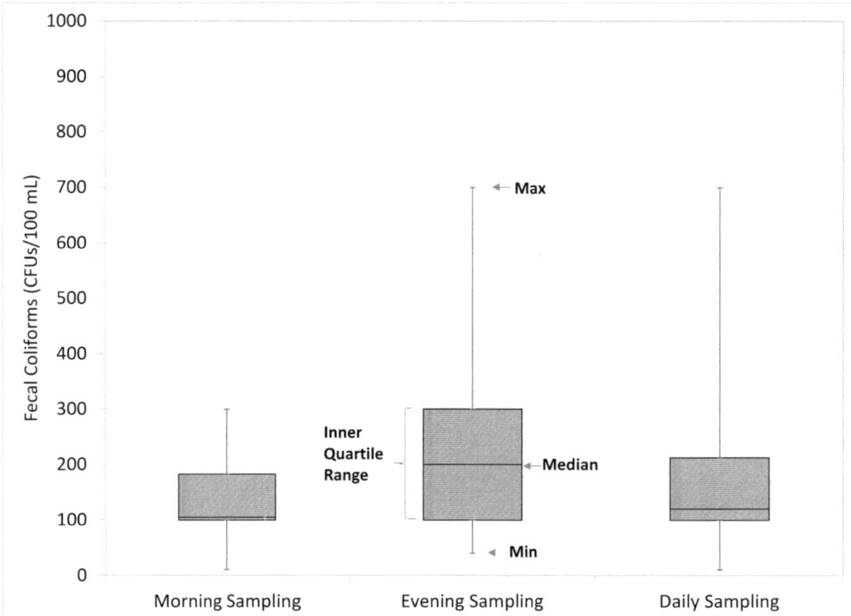

Figure 3. Fecal Coliform Box-Plot Comparison

Proposed Chlorine Contact Time (CT) Design Criteria

The fecal coliform monitoring results revealed that the Northwest Regional WRF produced reclaimed filter effluent containing 177 CFUs/100 mL of fecal coliforms on average over a seven-day, peak-hour sampling period. Furthermore, maximum fecal coliform levels remained less than 1,000 CFUs/100 mL. Therefore, the fecal coliform data supports a CT design criteria of 25 mg/L-min, as required in Chapter 62-600(5)(c), FAC.

As outlined in the regulatory summary in **Table 2**, a CT criteria 25 mg/L-min is the lowest disinfection tier allowed for public access reuse in Florida. Hence, this fecal coliform monitoring program demonstrated the ability of the Northwest Regional WRF to produce a high-quality filter effluent that requires the lowest CT. This lower CT requirement allows the facility to operate a lower total chlorine disinfectant residual, which in turn can be applied to lower DBP formation.

Disinfection Byproduct Formation Potential Results

The chlorine residual, bromodichloromethane, and dibromochloromethane concentrations for the 1.0 mg/L and 2.0 mg/L chlorine residual targets are presented in **Table 8** and **Table 9**, respectively.

Table 8. DBP Experimental Results for 1 mg/L Cl_2 Residual Target

Contact Time (minutes)	Chlorine Residual (mg/L)	Bromodichloromethane (µg/L)	Dibromochloromethane (µg/L)
15	1.35	8.6	1.9
30	1	11	2.2
45	0.81	13	2.5
60	0.6	14	2.6
90	0.44	16	3
120	0.23	18	3.1
180	0.11	20	3.7
240	0.06	21	3.5
300	0.04	21	3.6
360	0.04	22	3.6

Table 9. DBP Experimental Results for 2 mg/L Cl_2 Residual Target

Contact Time (minutes)	Chlorine Residual (mg/L)	Bromodichloromethane (µg/L)	Dibromochloromethane (µg/L)
15	2.52	9.1	1.9
30	2.08	12	2.4
45	1.96	14	2.6
60	1.75	15	2.9
90	1.4	17	3.2
120	1.18	19	3.3
180	0.87	22	3.5
240	0.66	23	3.9
300	0.49	24	3.8
360	0.36	25	4

Chlorine decay curves, shown in **Figure 4**, were developed by plotting chlorine residual results against the experimental contact time. The chlorine decay curves reveal a negative logarithmic relationship between the remaining chlorine concentration and contact time. Initially, chlorine reacted rapidly with the reduced inorganic and organic compounds present in the reclaimed water, then the reaction kinetics slowed after about two hours of contact time. For the 1.0 mg/L chorine residual, the chlorine residual was nearly depleted by the end of the six-hour period.

The corresponding DBP formation potential curves for the bromodichloromethane and dibromochloromethane species are presented in **Figure 5** and **Figure 6**, respectively. The bromodichloromethane formation potential curve exhibits a positive logarithmic shape, corresponding to the initial rapid reaction of organics with chlorine to form DBPs. During the six-hour contact time, bromodichloromethane levels reached up to 25 µg/L. The bromodichloromethane levels exceed the 22 µg/L limit after three hours of contact time for the 2 mg/L chlorine residual target, and after about 5 hours for the 1 mg/L chlorine residual target. The dibromochloromethane formation potential curve has a less pronounced logarithmic shape, which is likely due to the overall low formation. Based on the data, dibromochloromethane formation was well below the 34 µg/L limit and should not be a concern for possible exceedance of the limit.

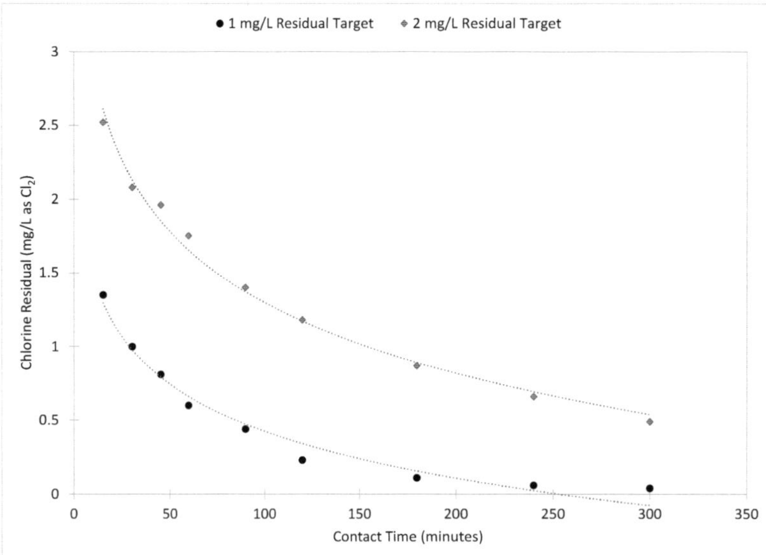

Figure 4. Chlorine Decay Curves

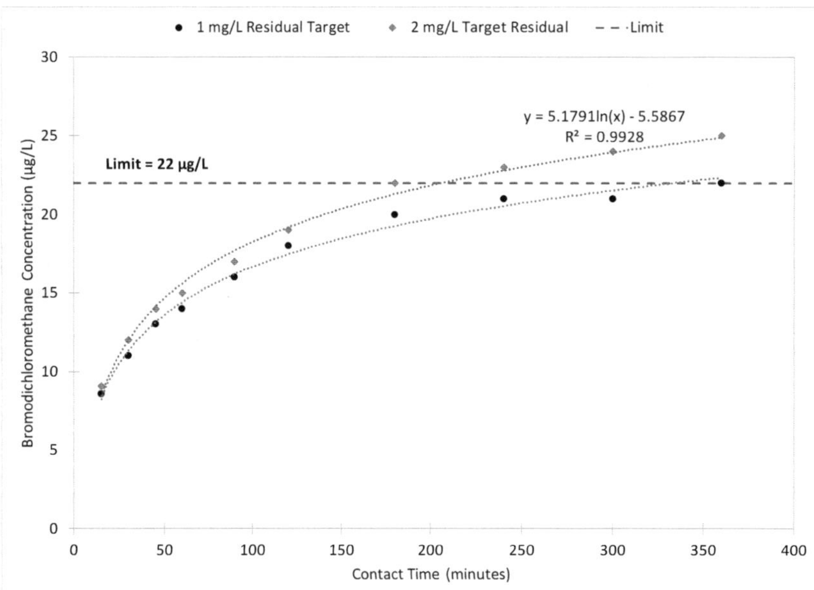

Figure 5. Bromodichloromethane Formation Potential Curves

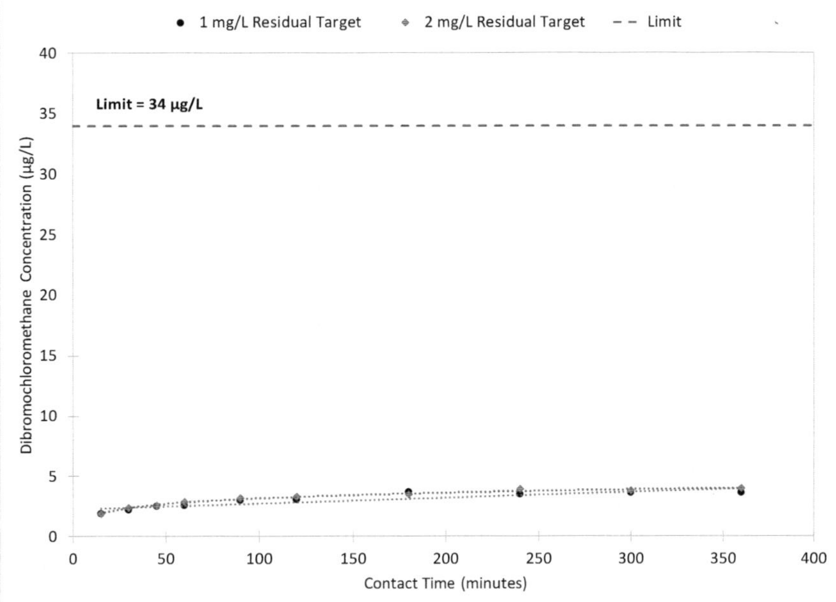

Figure 6. Dibromochloromethane Formation Potential Curves

Dechlorination Requirements

Based on the bromodichloromethane formation potential results, dechlorination will be necessary to maintain the levels below the maximum limit of 22 µg/L. **Figure 7** presents the formation potential data and provides a suggested dechlorination or quench time based on the water quality and chlorination conditions tested. Bromodichloromethane is not expected to be a concern as long as dechlorination is provided at a contact time of around 2.5 hours.

Since the dibromochloromethane formation potential was found to be well below the maximum limit of 34 µg/L, dechlorination for dibromochloromethane control is not a critical concern (see **Figure 6**).

Operations Support Tool for Chlorine Contact Operation

Chlorine contact chamber volumes are designed to meet the required minimum contact time (CT = Hydraulic Contact Time x Total Chlorine Residual) at peak hour flow. However, actual hydraulic residence times are generally higher and vary with varying diurnal flows into the facility. Therefore, careful management of the hydraulic chlorine contact time and operating chlorine residual is critical for controlling DBP formation.

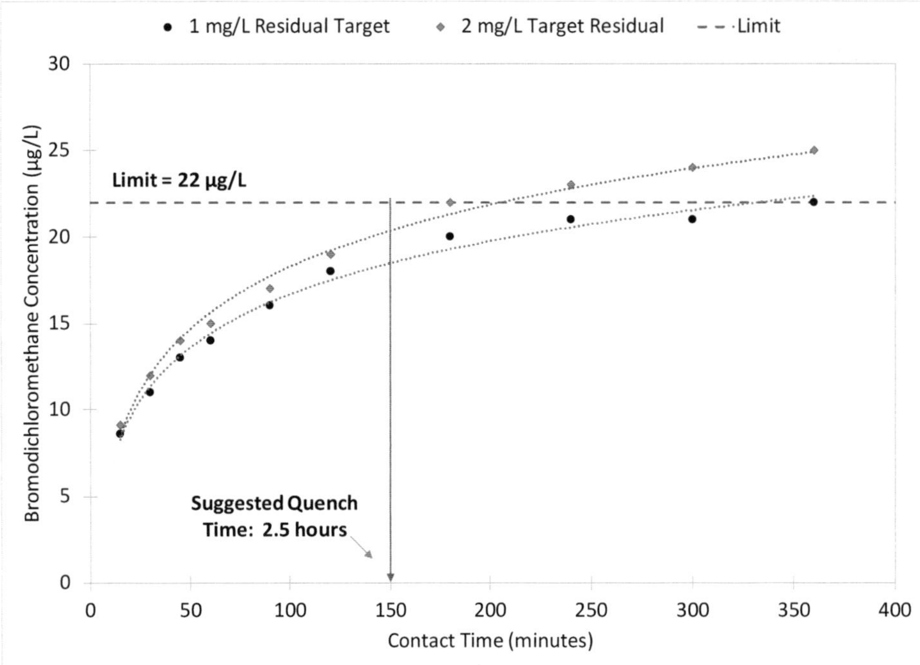

Figure 7. Suggested Dechlorination Time for Controlling Bromodichloromethane Levels

The Northwest Regional WRF currently has two 0.181 MG chlorine contact chamber sections that can be placed in or out of service. Additionally, the disinfected reclaimed water passes through an additional 0.125 MG of contact volume in the effluent wet well prior to dechlorination. The facility expansion will add two larger 0.395 MG chlorine contact chamber sections. The combined future chlorine contact volume will provide 30 minutes of contact time at the facility's design peak hour flow of 60 MGD. At lower flows, this total contact volume would provide chlorine contact in excess of 2.5 hours. Consequently, a DBP formation modeling tool was developed to assist operations with maintaining DBP compliance prior to dechlorination.

The operations support tool, which is presented in **Figure 8**, was developed by modelling DBP formation as an empirical function of chlorine contact time. The support tool allows the operator to input the plant process flow and select which chlorine contact chamber section is in operation and. Based on these inputs, the support tool automatically calculates the chlorine contact time and corresponding bromodichloromethane concentration at the end of the chlorine contact chamber. If the predicted bromodichloromethane concentration is greater than 20 µg/L, the cell is automatically highlighted red to alert the operator that there is a potential for exceeding the limit under the selected operating scheme.

OPERATIONS SUPPORT TOOL FOR CHLORINE CONTACT OPERATION

Chlorine Contact Chamber (CCC) Design Parameters and Assumptions:

CCC Volumes (MG):

Existing CCC Section No. 1	0.181
Existing CCC Section No. 2	0.181
New CCC Sections No. 3 & 4	0.395
Existing CCC Effluent Wet Well	0.125
Total CCC Volume	1.277

Reclaimed Water Chlorine Demand[1]:

Chlorine Demand (Dose - Residual) ≈ 4 mg/L

6 mg/L Dose ≈ 2 mg/L Residual

8 mg/L Dose ≈ 4 mg/L Residual

Bromodichloromethane ($CHBrCl_2$) Formation Potential Modeling[2]:

$CHBrCl_2$ @ 2 mg/L Cl_2 Residual (µg/L)	$CHBrCl_2 = 5.18 \times ln(CT) - 5.59$
$CHBrCl_2$ @ 4 mg/L Cl_2 Residual (µg/L)[3]	$CHBrCl_2 = 6.68 \times ln(CT) - 9.32$
Sugested Quench Time @ 2 mg/L (mins)	150
Sugested Quench Time @ 4 mg/L (mins)	90

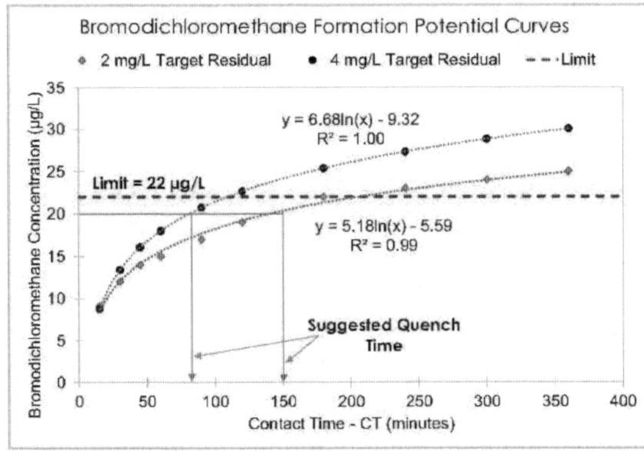

CCC Operations Tool Description and Instructions:

- The CCC Operations Tool helps the user determine CCC sequencing under varying process flow scenarios to maintain bromodichloromethane levels below the maximum limit of 22 µg/L for surface water discharge requirements.
- The tool includes anticipated and designed-based process flow scenarios.
- The 'User Input Scenario' allows user to input an alternative process flow (MGD) in red font cell.
- In yellow highlighted cells under 'CCC Section No. 1 - 4', input 1 to bring a CCC section online or input 0 to place a CCC section off line.
- The tool calculates the hydraulic contact time based on the volume of the CCC Section(s) online and process flow.
- Using the calcualted contact time and $CHBrCl_2$ equations, the tool projects the $CHBrCl_2$ concentrations at target chlorine resuduals of 2 and 4 mg/L.
- $CHBrCl_2$ cells highlighted in red indicate that there is a potential for exceeding the maximum limit and that the number of CCC Sections online should be reduced.

Process Flow Scenario	Process Flows (MGD)	CCC Section No. 1 (Y=1; N=0)	CCC Section No. 2 (Y=1; N=0)	CCC Section No. 3 (Y=1; N=0)	CCC Section No. 4 (Y=1; N=0)	CCC Volume (MG)	Contact Time (mins)	$CHBrCl_2$ @ 2 mg/L Cl_2 Residual (µg/L)	$CHBrCl_2$ @ 4 mg/L Cl_2 Residual (µg/L)
Minimum	6	1	0	0	0	0.306	73	16.7	19.4
nterim	12	1	1	0	0	0.487	58	15.5	17.8
Start-up AADF	15	1	1	0	0	0.487	47	14.3	16.4
Design AADF	30	0	0	1	1	0.915	44	14.0	15.9
Max Month	35	1	1	1	1	1.277	53	15.0	17.2
Max Day	45	1	1	1	1	1.277	41	13.6	15.5
Peak Hour	60	1	1	1	1	1.277	31	12.1	13.5
User Input Scenario	3.0	1	0	0	0	0.306	147	20.3	24.0

Notes:

(1) Chlorine demand is based on bench-scale chlorine dosing and residual testing of reclaimed water from the Northwest Regional Water Reclamation Facility (NWRWRF). Sodium hypochlorite dosing ranged from 2 to 8 mg/L Cl_2.

(2) Bromodichloromethane formation potential modeling is based on disinfection byproduct (DBP) testing of reclaimed water from the NWRWRF. The DBP formation potential was monitored from 15 minutes to 6 hours of contact time.

(3) $CHBrCl_2$ formation potential was extrapolated from the bench-scale DBP formation potentials.

Figure 8. Operations Support Tool for Chlorine Contact Operation

To correspond with the full-scale target chlorine residual, the operations support tool was set-up to predict bromodichloromethane concentrations at target chlorine residuals of 2.0 and 4.0 mg/L after 30 minutes of contact time. **Table 10** summarizes the bench-scale DBP results for the 1.0 and 2.0 mg/L chlorine residual scenarios. As also presented in **Table 10**, the results from the 1.0 and 2.0 mg/L scenarios were extrapolated to estimate the DBP formation potential at a 4.0 mg/L chlorine residual. The DBP versus contact time scatter-plots and corresponding best-fit logarithmic equations used in the operations support tool are presented in **Figure 9**.

Table 10. Bench-Scale DBP Formation Modeling Data

Time (mins)	$CHCl_2Br$ @ 1 mg/L Cl_2 Residual (µg/L)	$CHCl_2Br$ @ 2 mg/L Cl2 Residual (µg/L)	Δ $CHCl_2Br$ (µg/L)	$CHCl_2Br$ @ 4 mg/L Cl_2 Residual (µg/L)
15	8.3	8.4	0.2	8.8
30	11.3	12.0	0.7	13.4
45	13.1	14.1	1.0	16.1
60	14.4	15.6	1.2	18.0
90	16.2	17.7	1.5	20.7
120	17.5	19.2	1.7	22.7
180	19.3	21.3	2.0	25.4
240	20.6	22.8	2.2	27.3
300	21.5	24.0	2.4	28.8
360	22.4	24.9	2.5	30.0

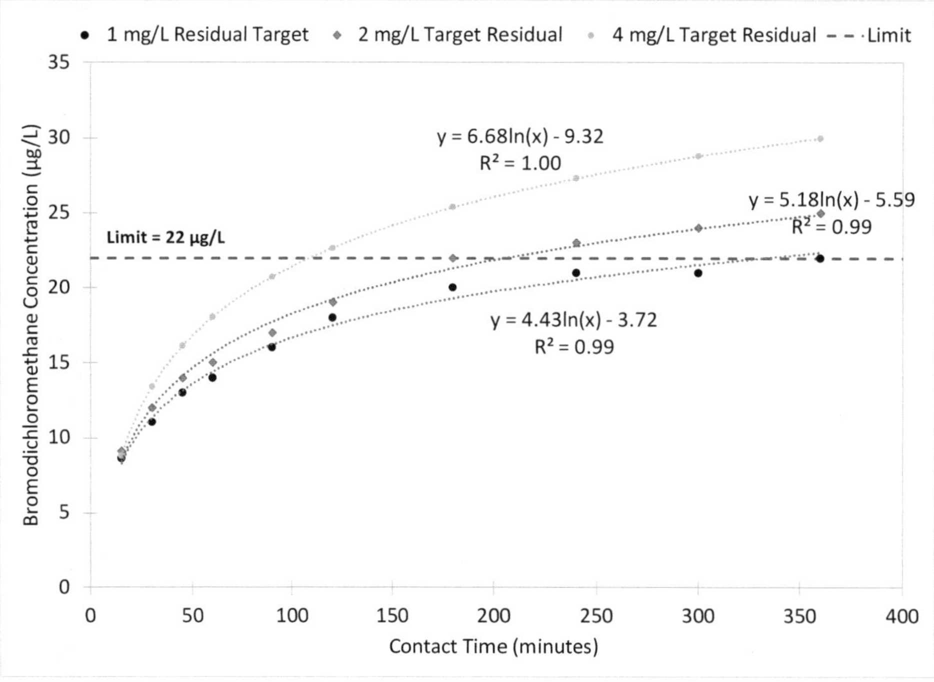

Figure 9. DBP Formation versus Contact Time Scatterplots

Preliminary Model Validation

As part of an interim plant expansion from 10 MGD to 12 MGD AADF, the Northwest Regional WRF was recently converted from UV disinfection to free chlorine disinfection. Since the conversion, facility operations staff have been collecting bromodichloromethane data from the chlorine contact chamber effluent. A summary of this preliminary data is presented in **Table 11**. To compare predicted to measured DBP levels, the operations support tool was used to estimate the bromodichloromethane formation at the full-scale conditions and simulated 4.0 mg/L target residual. These predicted bromodichloromethane concentrations are included in **Table 11**.

The predicted versus measured bromodichloromethane scatterplot is presented in **Figure 10**. The scatterplot reveals a relatively linear fit with an r-squared value of 0.63. While the modeled DBP values slightly underpredicted the measured levels for three of the four samples, the predicted versus measured analysis supports that the operations support tool provides an adequate estimate of the full scale performance. Additional full-scale data is necessary to further validate the model, particularly at chlorine residuals near 3.0 and 4.0 mg/L. Nevertheless, the operations support tool is expected to provide a practical, pro-active approach for optimizing the operation of the chlorine contact chamber sections and minimizing DBPs prior to discharge.

Table 11. Bromodichloromethane Results from the Interim Full-Scale Chlorine Conversion

Date	Plant Flow (MGD)	CCC Volume (MG)	Additional Wet Well Volume (MG)	Chlorine Residual (mg/L)	Measured $CHCl_2Br$ (µg/L)	Predicted $CHCl_2Br$ (µg/L)
22-Feb	8.50	0.181	0.125	2.6.	20.7	17.1
27-Mar	13.60	0.181	0.125	3.12	12.0	13.9
10-Apr	12.79	0.181	0.125	3.32	18.3	14.3
15-May	12.33	0.181	0.125	3.6	16.0	14.6

N.A. = Not Available.

CONCLUSIONS

Based on the results from the fecal coliform monitoring, DBP formation potential, and operations support tool development and evaluations, the following conclusions are offered:

- The Northwest Regional WRF produces a high-quality effluent with fecal coliforms averaging near 180 CFUs/100 mL and maximum levels remaining below 1,000 CFUs/100 mL over a seven-day, peak-hour sampling period. Based on these fecal coliform results, the required CT for high-level disinfection is the lowest treatment tier of 25 minutes. A lower CT requirement would allow the WRF to operate at lower total chlorine residuals, and in turn lower the potential for DBP formation.
- With the conversion to chlorine disinfection, there is a potential for exceedance of the bromodichloromethane limit of 22 µg/L at contact times greater than 2.5 hours with 2.0 mg/L of chlorine residual. Therefore, dechlorination through the addition of sodium bisulfite is necessary after the chlorine contact chamber.

- Dibromochloromethane formation should not be a concern for exceedance due to its low formation potential.
- The developed operations support tool provides a simple, yet practical method for operations to estimate the levels of bromodichloromethane in the chlorine contact chamber effluent based on the process flows and selected contact chamber sections on-line. By understanding the potential for bromodichloromethane formation, operators can make decisions on how best to operate their chlorine contact chamber section for minimizing DBP formation.
- Additional bench-scale and full-scale DBP formation testing is recommended to further refine and validate the support tool model. Additional bench-scale DBP formation testing should include formations at varying chlorine doses and residuals.
- The DBP optimization strategy and operations support tool developed for the Northwest Regional WRF can serve as a be tailored to other WRFs at other Based on

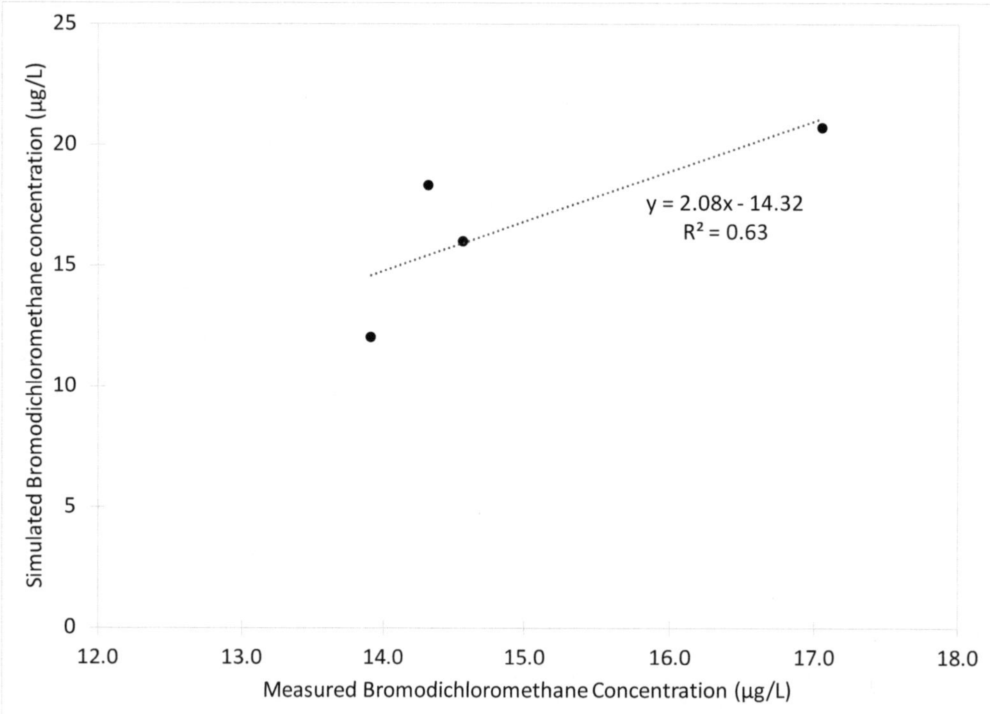

Figure 10. Predicted versus Measured Bromodichloromethane

REFERENCES

Bougeard, C. M. M., Goslan, E. H., Jefferson, B., & Parsons, S. A. (2010). Comparison of the disinfection by-product formation potential of treated waters exposed to chlorine and monochloramine. *Water Research, 44*(3), 729-740. doi:10.1016/j.watres.2009.10.008

Edzwald, J., Becker, W., & Wattier, K. (1985). Surrogate Parameters for Monitoring Organic Matter and THM Precursors. *Journal American Water Works Association*, 77(4), 122-132. Retrieved from http://www.jstor.org/stable/41272390.

Hua, G., & Reckhow, D. A. (2007). Comparison of disinfection byproduct formation from chlorine and alternative disinfectants. *Water Research, 41*(8), 1667-1678. doi:10.1016/j.watres.2007.01.032

Krasner, S. W. (2009). The formation and control of emerging disinfection by-products of health concern. *Philosophical Transactions of the Royal Society A: Mathematical, Physical and Engineering Sciences, 367*(1904), 4077-4095. doi:10.1098/rsta.2009.0108

Liang, L., & Singer, P. C. (2003). Factors influencing the formation and relative distribution of haloacetic acids and trihalomethanes in drinking water. *Environmental Science and Technology, 37*(13), 2920-2928.

Metcalf & Eddy (2003). *Wastewater Engineering Treatment and Reuse*. New York: McGraw-Hill.

Reckhow, D. A., & Singer, P. C. (2011). Chapter 19: Formation and control of disinfection by-products. In J. K. Edzwald (Ed.), *Water quality and treatment: A handbook on drinking water* (Sixth ed., pp. 19.1) McGraw-Hill.

Zhang, X., Yang, H., Wang, X., Fu, J., & Xie, Y. F. (2012). Formation of disinfection by-products: Effect of temperature and kinetic modeling. *Chemosphere.*

Experiences and Successful Strategies for Controlling Reuse Water Chorine Residual

Bing Lin, EPCOR Water Services; Alfredo Suarez, EPCOR Water Services

ABSTRACT

A small portion of the final effluent from EPCOR Gold Bar Wastewater Treatment Plant (GB WWTP) is filtered by membrane process and sent to customers for industrial reuse. Chlorine is added to the product water to maintain required chlorine residual and prevent biological growth in the transportation pipeline as well as the water storage tank on the customer's site. It was a challenge maintaining the chlorine residual consistently within the required 0.5 to 1.0 mg/L range during the first few years of operation. After trials and improvements, the following changes have been made to the chemical dosing control strategies in order to maintain the chlorine residual within the required 0.5 to 1.0 mg/L range:

- Changed chlorine contact tanks operation from in series to in parallel;
- Changed to the First Good (FG) value control from the higher value control;
- Introduced product water flowrate feedforward control; and
- Prevented Clean in Place (CIP) solution entering the membrane product water.

Since implementing the above changes, the chorine residual in the reuse water has continued to remain stable and meet the customer's requirements.

KEY WORDS

Chlorine Residual, Process Control, Water Reuse, Membrane Filtration

INTRODUCTION

EPCOR, through its subsidiaries, builds, owns and operates electrical, natural gas and water transmission and distribution networks, water and wastewater treatment facilities, sanitary and storm water systems, and infrastructure in Canada and the United States. The company also provides electricity, natural gas and water products and services to residential and commercial customers. EPCOR, headquartered in Edmonton, is an Alberta Top 70 employer. EPCOR's website address is www.epcor.com.

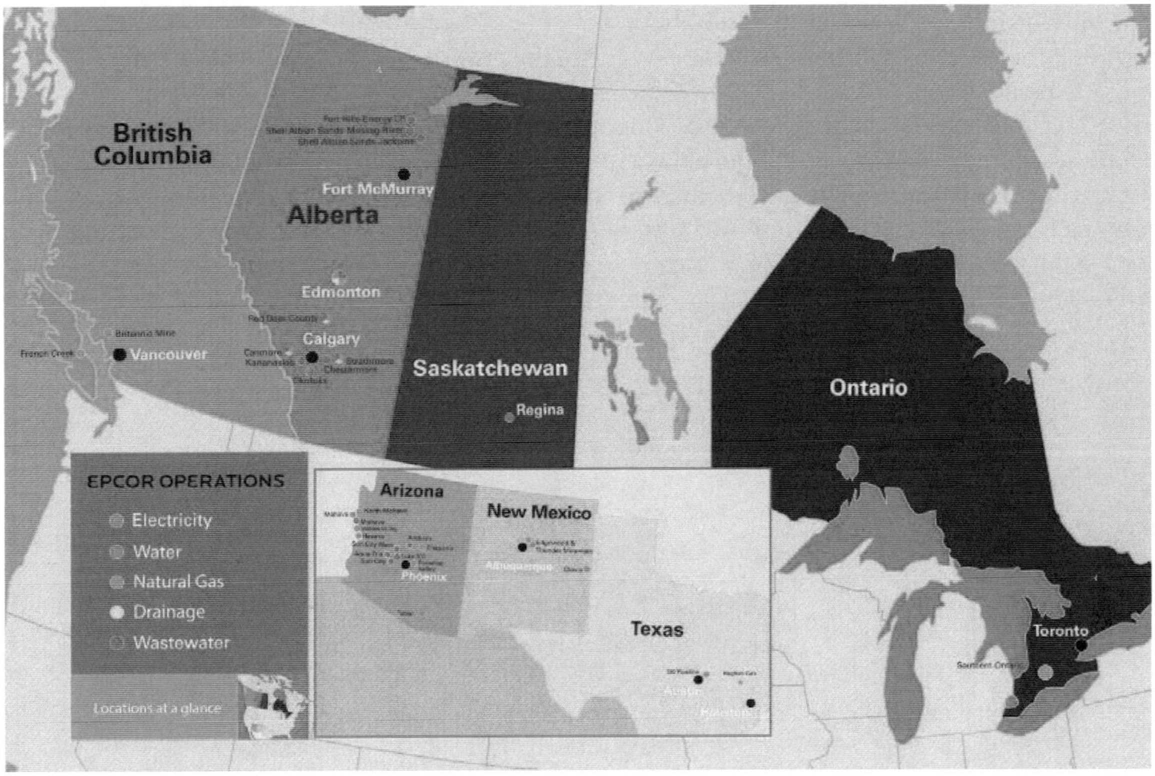

Fig.1 EPCOR Business Geographical Areas

In 2005, a refinery company teamed with GB WWTP and built the largest membrane-based water reuse project in Canada at the time. A small portion of the final effluent from the Biological Nutrient Removal (BNR) treatment process from EPCOR GB WWTP is treated by Ultra Filtration (UF) hollow fiber membrane. Sodium hypochlorite is added for chlorination, and the membrane product water is then sent to two customers through a 5.5 km pipeline for industrial use. Two online chlorine analyzers control the chemical dosing to maintain chlorine residual between 0.5 to 1.0 ppm. Fig.2 provides an overview of the membrane production process.

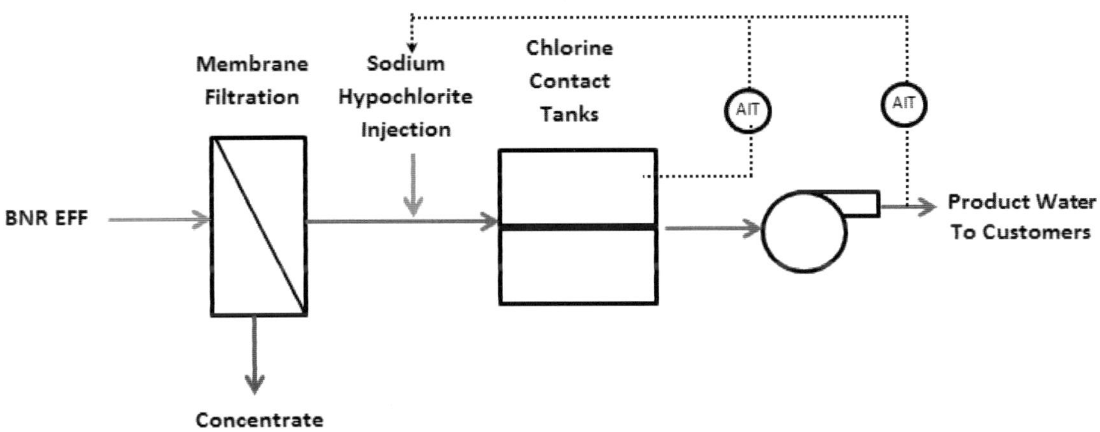

Fig.2 Simplized Reuse Water Production Process

Controlling the residual chlorine concertation within the required range is critical to the membrane facility operation since one of the customer feeds the reuse water directly to a Reverse Osmosis (RO) membrane system, which is very sensitive and subject to damage by high chlorine concentration. Maintaining the chorine residual in the membrane product water was a challenge due to inadequate chemical dosing strategies. Fig.3 shows a typical example of membrane product water chlorine residual moving between 0.4 and 3.2 mg/L when targeting to achieve chlorine residual of 1.0 mg/L during a five-day production period, when the lowest flowrate was 7.7 MLD and the highest flowrate was at 15.4 MLD.

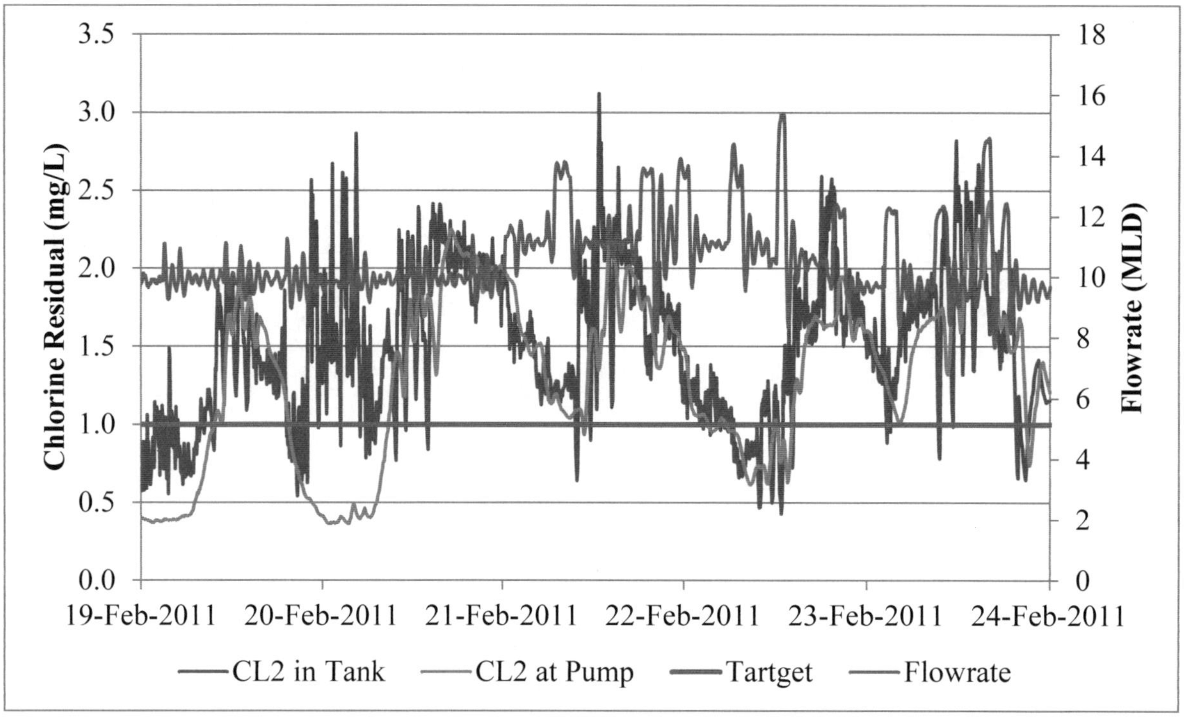

Fig.3 Membrane Product Water Chlorine Residual Swung Widely in a Five Day Production Cycle

METHODOLOGY AND STRATEGIES FOR CHLORINE RESIDUAL CONTROL

During more than 10 years of operating the membrane filtration facility, GB WWTP experienced various challenges stabilizing chlorine residual in the product water. Respective chlorine control strategies were developed and implemented based on trials and learnings.

Change Contact Tanks Operation From In Series to in Parallel

There are two chlorine contact tanks after sodium hypochlorite injection and before pumping the product water to customers to provide chlorine contact time, with the flexibility to operate two tanks in series or in parallel with manual valve control. Initially, the two contact tanks were operated in series, as shown in Fig.4, to provide longer contact time. Two chlorine online

analyzers were used to control chemical dosing, and the higher value of the two measurements was selected by the program to prevent chemical overdosing. When the second analyzer was selected, it caused longer reaction time and bigger swings in chlorine residual concentration. It also took more time for the chlorine concentrations to stabilize, as shown in previous Fig.3.
A change was made to operate the two tanks in parallel, as shown in Fig. 5. With this change, the control reaction time reduced to half and stabilized the chlorine concentration sooner. Although the chlorine contact time was reduced, the 5.5 km of pipeline ensures sufficient chlorine contact time.

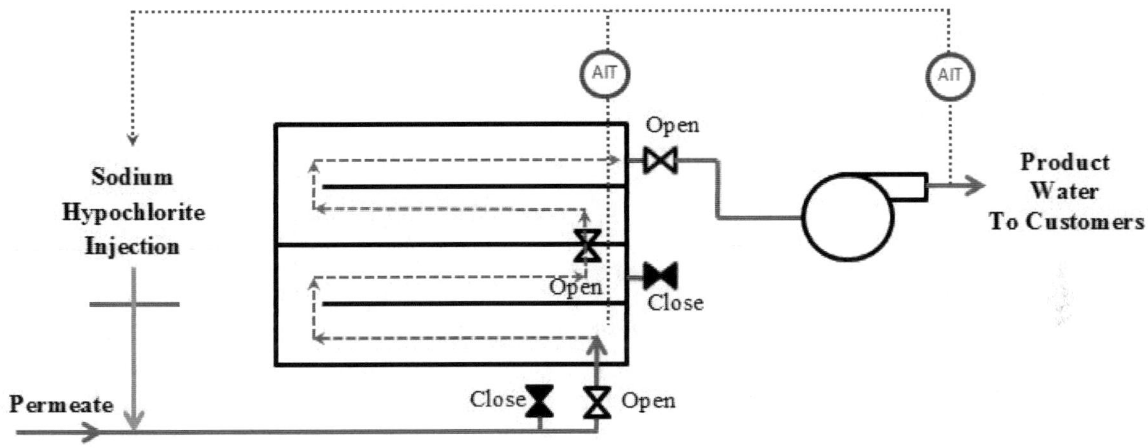

Fig. 4 Chlorine Contact Tank Operation in Series and Chemical Dosing Control with Flow Path

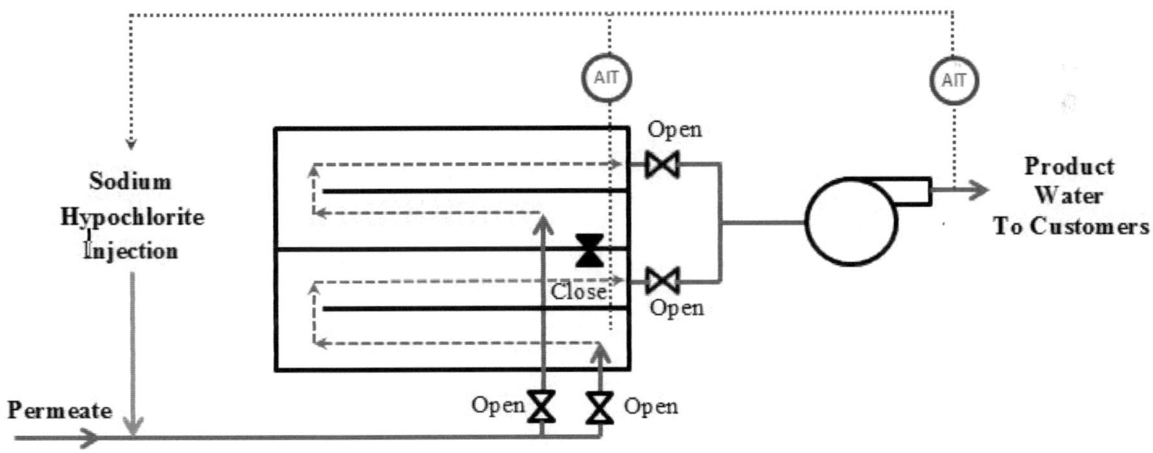

Fig. 5 Chlorine Contact Tank Operation in Parallel and Chemical Dosing Control with Flow Path

Change Chemical Dose Control from the Higher Value to the First Good Value

There are two chlorine online analyzers that monitor and control sodium hypochlorite dosing. The higher value of the two measurements was used to control chemical dose and resulted in big swings of chlorine residual in the product water due to the process configuration. A change was made to the program to use the First Good (FG) value to control chemical dosing. This normally involves the instrument closer to the chemical injecting point controlling the chemical dosing, which would lessen the swings and shortens the chlorine concentration stabilization time in the product water.

Product Water Flowrate Feedforward Control Chemical Dosing

Because reuse water demand varies between EPCOR's two downstream customers, the membrane product water flowrate can vary from 5 MLD to 15 MLD frequently throughout of the day, which contributed to the chlorine residual variation significantly. A flowrate feed ward control was introduced to the chemical dosing control to fine tune the chemical dosing. This resulted in the control system reacting sooner and achieving more stable chlorine residual in reuse water.

Prevent Clean in Place (CIP) Solution Entering Product Water

Membranes need periodical chemical cleans to maintain their permeability, and sodium hypochlorite has been used to recover membranes from biological fouling. On a few occasions, a membrane tank inlet gate malfunctioned and failed to close during the chemical cleaning process. This resulted in highly concentrated chlorine solution entering the common feed channel and caused high chlorine residual in the product water, which creates a risk that customers' RO membranes are damaged. Based on these learnings, a procedure was developed to instruct the operator to monitor membrane tank inlet gate movement from the HMI screen during critical steps and stop chemical recovery clean process when necessary to prevent chemical clean solution entering the product water.

RESULT

After changes were made to the chlorine residual control strategies, the chorine residual in the reuse water has been stabilized and meets the control target most of the time during membrane filtration process operation, as illustrated in Fig.6. EPCOR's customer receives the expected chlorine residual of 0.5 to 1.0 mg/L in the membrane product water when the water arrives at their site, as demonstrated in Fig.7

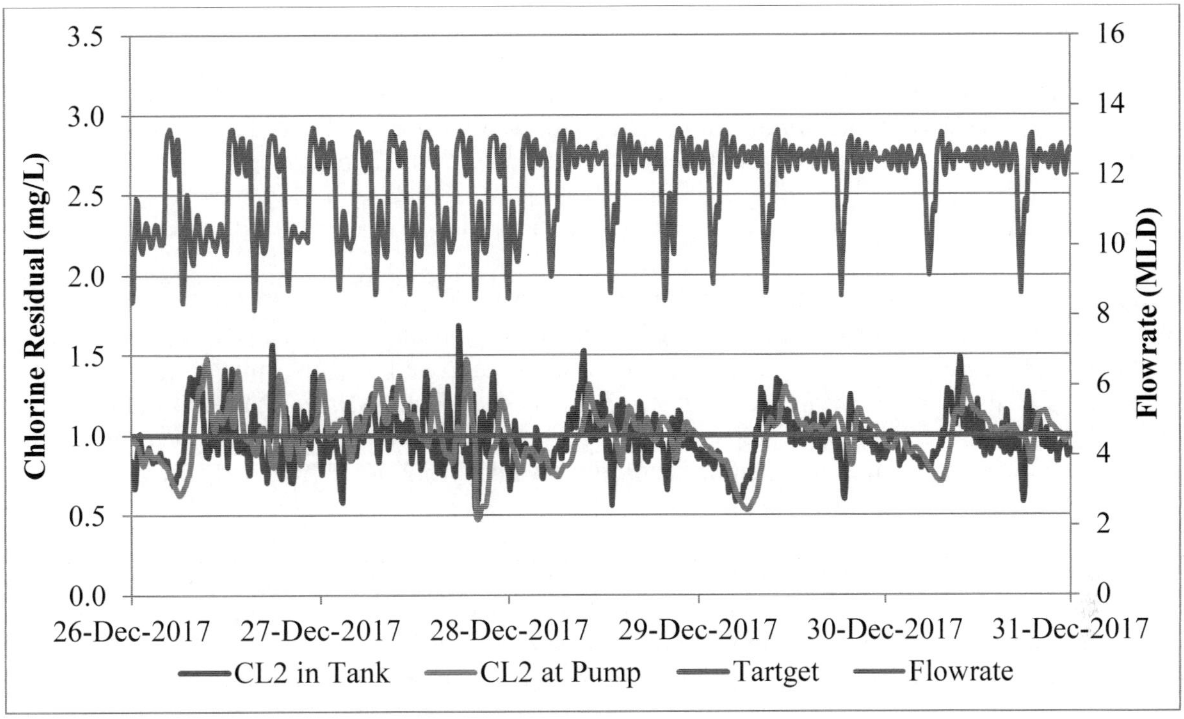

Fig.6 Improved Chlorine Residual Control in Reuse Water and Target

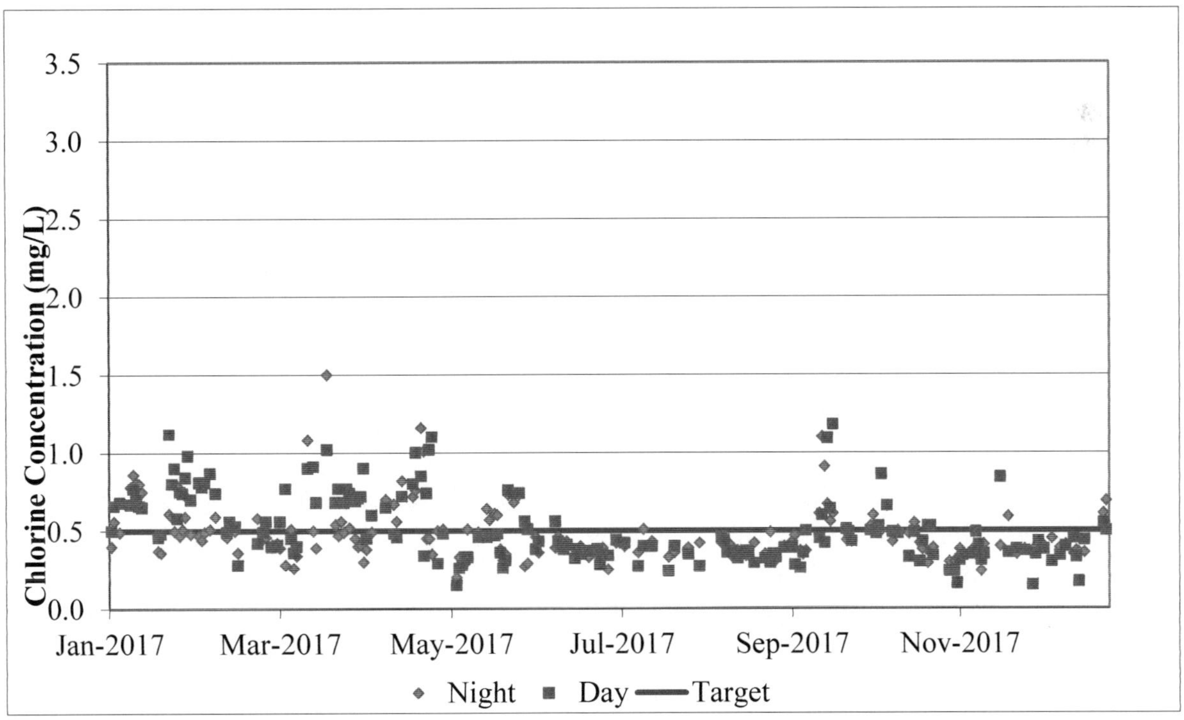

Fig.7 Customer Tested Chlorine Residual in Reuse Water and Target

DISCUSSION

The experience and learnings for the chlorine residual control in the membrane product water obtained from GB WWTP membrane facility may not be site specific; it can be adapted for similar scenarios and facilities in order to improve chlorine residual control.

CONCLUSION

Many factors can contribute to the challenges of reuse water chlorine residual control. At GB WWTP, it was identified that process configuration and operation, chlorine residual monitoring and control strategies, and other operational issues were significant factors impacting the chlorine residual in the membrane product water. It has been demonstrated that the following changes to the chemical dosing strategies and process management procedures significantly stabilized chlorine residual in the membrane product water for industrial reuse at EPCOR GB WWTP:

- Changing chlorine contact tanks operation from in series to in parallel;
- Changing to the First Good value control from the higher value control;
- Introducing product water flowrate feedforward control; and
- Preventing the Clean in Place solution from entering the membrane product water.

PureWaterSF: PureWaterSF: Direct Potable Reuse Demonstration in San Francisco

Manisha Kothari*, Paula Kehoe**, Andrea F. Corral, Nicola Fontaine, Mark Millan, Andrew Salveson

*Water Resources Division - San Francisco Public Utilities Commission - 525 Golden Gate Avenue, 10th floor - San Francisco, CA 94102. Email: mkothari@sfwater.org

** Water Resources Division - San Francisco Public Utilities Commission - 525 Golden Gate Avenue, 10th floor - San Francisco, CA 94102. Email: pkehoe@sfwater.org

Abstract
The San Francisco Public Utilities Commission (SFPUC) is a leader in the innovative and sustainable use of water in an urban setting. Currently, SFPUC uses a constructed wetland system to treat the wastewater generated in its headquarters building for non-potable reuse. PureWaterSF will add to the existing system a direct potable reuse (DPR) building-scale demonstration system that includes ultrafiltration, reverse osmosis, and an ultraviolet advanced oxidation process (UF/RO/UVAOP) to purify the tertiary recycled water effluent from the wetland system. The treatment train was designed and selected to have a small footprint and produce high-quality water that meet every regulated drinking water standard and standards detailed in state and national DPR guidance documents, providing treatment for CECs and "unknown unknowns". During the study, the treated water will be redirected to the non-potable reuse system.

PureWaterSF focuses on the latest concerns and research challenges for direct potable reuse in California. The State Water Resources Control Board expert panel recommended the use of non-target analysis (NTA) and bioassays to better understand the significance of the "unknown unknowns." During 9 months of extensive testing, PureWaterSF will continually monitor the process using precise, on-line real time monitoring tools. This project will leverage the ability to document on-line real time precise and accurate data and the use of advanced analytics to understand the impact of the "unknown unknowns" and well as look at surrogates for assessing treatment performance.

The results from the PureWaterSF project will build on a body of knowledge and will provide valuable information about the reliability of advanced water treatment systems for potable water reuse, in particular for Direct Potable Reuse. Finally, the viability of DPR at a building scale will be demonstrated while educating the public on the importance and safety of potable water reuse.

Keywords: potable reuse, non-target analysis, bioassays, monitoring, decentralized.

Introduction

Advanced treatment of wastewater for direct potable reuse (DPR) is operational at one facility in the United States, the Colorado River Municipal Water District's Raw Water Production Facility in Big Spring Texas. Ongoing research of that facility is demonstrating the production of a high quality water that is protective of public health (Steinle-Darling et al., 2015). These results demonstrated the effective use of multiple barriers for reduction of trace pollutants and pathogens. While providing high quality water, the "Big Spring" facility relies upon monitoring systems designed for indirect potable reuse (IPR) applications. Nationally, the National Water Research Institute (NWRI) recently published a 173-page "how to" document on DPR, titled Framework for Direct Potable Reuse (NWRI, 2015). Central to this document was the use of precise and accurate monitoring technologies for public health protection in DPR applications. Within California, an extensive research program (>$6M), the California DPR Initiative, has been undertaken to define the necessary level of treatment for a DPR project in California, and inform the discussion of DPR nationally. The Division of Drinking Water (DDW) is part of this Initiative, providing third party review of all research as they consider the possibility of regulating DPR in California. Even with the success of "Big Spring," with the development of clear guidelines for safe DPR implementation, and with extensive funding for research, the public and regulatory concern over "unknown unknowns" remains. What is that next pollutant? How do we find it? Are trace levels of pollutants harmful? The State Water Resources Control Board recently conducted an expert workshop to lay the groundwork for tracking down these questions (SWRCB, 2015). The expert workshop team recommended the use of non-target analysis (NTA) and bioassays to better grasp the significance of the "unknown unknowns." These key research needs, the ability to document real time precise and accurate monitoring technologies and the use of advanced analytics to understand the impact of the "unknown unknowns," are the primary objectives of this proposed research project. There is a secondary value of this project, which is the integration of DPR methodologies into building-scale treatment. The proposed project would use the existing constructed wetlands with tertiary treatment that harvests wastewater from the building and treats it to non-potable water reuse standards, and then purify the water to potable standards.

In total, the goals of the demonstration are:

- Demonstrate innovative building-scale treatment of wastewater for DPR.
- Procure purification processes that produce potable water in accordance with health criteria established in National documents (NWRI, 2015).
- Use leading edge online analytical techniques to demonstrate the performance of each treatment process.
- Use advanced analytical monitoring to understand the potential impact of unknown trace level pollutants.
- Clearly document the costs of a potential future DPR system for utilities in California.
- Educate regulators and community members about the safety of properly engineered potable water reuse treatment systems.

This ambitious project will span 9 months, and includes a substantial work effort, which is supported by funding from The Water Research Foundation, US Bureau of Reclamation, San Francisco Public Utilities Commission (SFPUC), and Carollo Engineers, Inc.

Research Approach

Building-Scale Treatment for Non-Potable Water Reuse

This project starts with raw wastewater, harvested from the 13-story, 900 employee SFPUC headquarters building. The advanced, ecologically based tertiary treatment system currently collects and treats wastewater for non-potable reuse inside the structure. The tertiary treatment system consists of a two-stage, recirculating, engineered wetland system with subsequent filtration and disinfection units (collectively called a tertiary treatment system henceforth in this paper) and is housed in landscaped planters on the interior and exterior of the structure. The tertiary treatment system can treat a maximum flow of 5,000 gallons per day. The system consists of primary treatment and flow equalization followed by a wetland system, denitrification, polishing and disinfection and a reclaimed water reservoir. The system has proven capable of treating raw wastewater with a small physical footprint, appropriate to an urban setting.

The value of decentralized wastewater treatment cannot be overstated. Water can be treated and used within one watershed, eliminating the need for sewers, pump stations, and wasted conveyance energy. Demonstrating advanced purification of the reclaimed water to potable water standards is possible and safe may lead to a radical revolution in the water industry.

Purification Processes for Potable Water Reuse

There are numerous treatment trains that could be used for potable water reuse. Within California, the particular processes that could be employed for this type of project are more limited (CDPH, 2014). In particular, IPR projects in California that include 100 percent purified water (no dilution) and do not benefit from surface spreading (soil aquifer treatment), must have reverse osmosis (RO) and advanced oxidation processes (AOP) within the treatment train. Using these two processes as a starting point, and relying upon the NWRI Framework for Direct Potable Reuse (NWRI, 2015), the purification process proposed for this treatment train are ultrafiltration (UF), RO, ultraviolet light (UV) AOP (Figure 1). These processes will provide multiple barriers to both pollutants and pathogens, as shown in Table 1.

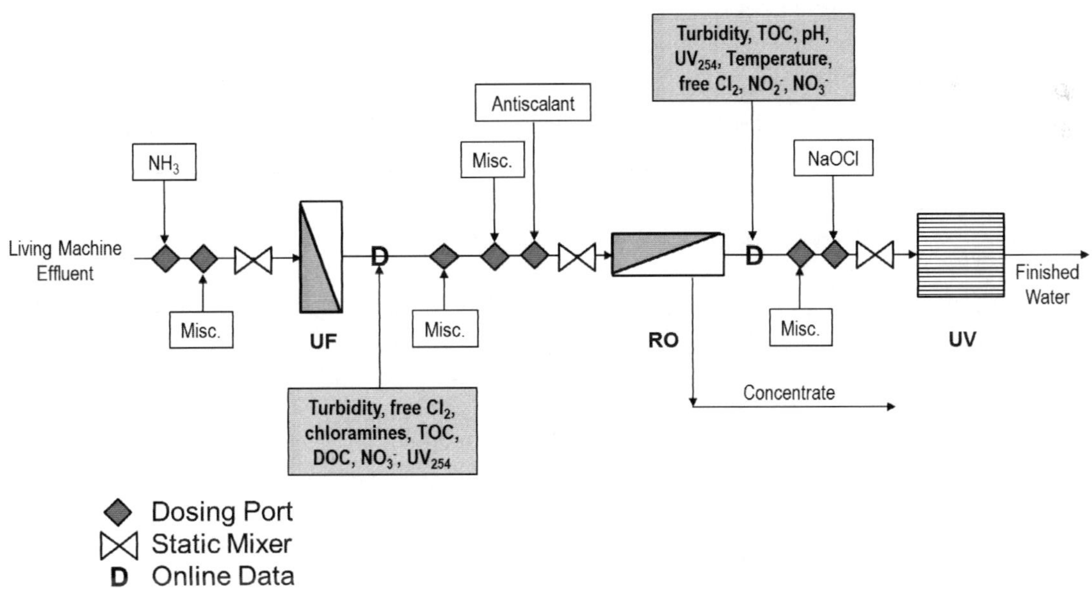

Figure 1 Advanced Water Treatment Train with Chemical Feed. Sampling Ports are Part of the Treatment Systems

Table 1 Use of Multiple Barriers for Purification

Process	Bulk Organic Removal	Trace Organic Removal	Virus Removal	Protozoa Removal	Bacteria Removal
Primary, Secondary, and Tertiary Treatment	●	●	●	●	●
UF	●	–	–	●	●
RO	–	●	●	●	●
UV AOP	–	●	●	●	●

Monitoring Technologies

Potable reuse trains such as the long-standing treatment process at the Orange County Water District and others (CWS, 2014; Trussell, 2013), have repeatedly demonstrated the ability to meet Environmental Protection Agency (USEPA) drinking water and California DDW Title 22 standards. To expand the body of knowledge from the existing demonstration studies, the SFPUC advanced water treatment train will monitor for chemical constituents across the treatment process. Chemical classes to be monitored for include:

- Pharmaceuticals and personal care products (PPCPs)
- Perfluorinated compounds (PFCs)
- N-Nitrosodimethylamine (NDMA), NDMA formation potential (FP), and NDMA Simulated Distribution System (SDS)
- Total N-Nitrosamine (TONO)
- TONO SDS
- TONO max FP
- Disinfection byproducts (DBPs) Trihalomethanes (THM) and Haloacetic Acids (HAAs) FP
- THM and HAA FP and uniform formation conditions (UFC)
- Bioassays and non-target analysis (NTA)

Pathogen monitoring will be performed by collecting grab samples at various points in advanced water treatment train (UF feed and effluent and RO permeate), thus allowing for an evaluation of pathogen reduction through the purification processes. Pathogens (and surrogate organisms) to be evaluated include:

- Bacteria: Male specific phage
- Protozoa: Giardia and Cryptosporidium
- Virus: enterovirus, norovirus, and Adenovirus (all via qPCR)

In addition to the above parameters, and relevant to this TM, an on-line system will monitor several parameters in the RO feed (UF effluent) and the RO permeate. The final effluent will be sampled for drinking water maximum contaminant levels (MCLs).

The ability for these processes to produce high quality water in accordance with regulations is not in question. What this project looks to define is the ability to continuously monitor the performance of the advanced treatment systems in real time. This will be done through the use of precise and accurate metering of the critical control points in the purification process. To that end, two s::can micro::stations will be used to perform such monitoring, as shown in Table 2. These parameters will be used to demonstrate process-by-process performance as follows:

- UF - UF filtrate turbidity and *E.* coli concentrations will closely track UF performance. These continuous measurements will be paired with daily pressure decay test (PDT) results to provide real-time confidence in protozoa and bacteria removal performance.
- RO - TOC values collected pre and post RO allow for clear determination of a conservative surrogate for pathogen removal by RO as well as consistent reduction in TOC. TOC values will be paired with online electrical conductivity (EC) to verify TOC performance values.
- UV AOP - Destruction of total chlorine across UV systems has now been shown to correlate directly with UV dose, which then correlates directly to pathogen removal and destruction of pollutants such as NDMA. Free chlorine measurements and UV absorbance (UVA) can be used to develop a "chlorine weighted UV dose," which has recently been shown to correlate directly with destruction of trace pollutants by UV AOP.

Table 2 Parameters Measured by s::can micro::stations at RO Feed and Permeate Locations

Parameters	RO Feed	RO Permeate
Chloramines	✓✓	
Free Chlorine	✓	✓
TOC	✓	✓
Dissolved Organic Carbon (DOC)	✓	
Turbidity	✓	✓
UVA at 254 nanometers	✓	✓
Nitrate	✓	✓
Nitrite		✓
pH		✓
Temperature		✓

The information from the s::can micro::stations will be logged for the duration of the 9-month demonstration and used to evaluate overall reliability in system performance. These values will also be used to monitor system performance remotely, available 24 hours / 5 days per week. The research will take one further step, the investigation of the "unknown unknowns." While hundreds of chemicals have been detected in water, thousands more likely occur at very low concentrations but have not yet been detected. Chemical surrogates and indicators are often used to gauge the efficacy and efficiency of a particular treatment process and/or multi-barrier train (Yu et al., 2015; Merel et al., 2015; Anumol et al., 2015; Gerrity et al., 2012). However, these measures do not provide any reference to biological effects and thus do not account for the potential additive or synergistic effects of chemical mixtures. Bioassay-based monitoring complements chemical analysis by providing a comprehensive assessment of the mixture of substances present in a particular water sample (Escher et al., 2014). A limitation of bioassays is the ability to determine what substance, or substances, were responsible for the bioactivity observed. Therefore, non-targeted analysis (NTA) will also be performed using high-resolution mass spectrometry (HRMS) with both gas chromatography (GC) and liquid chromatography (LC) interfaces for volatile and non-volatile organic compounds, respectively. National experts convened in California recently to examine two promising techniques for such investigation (SWRCB, 2015). In that two-day workshop, the expert group concluded that these two methods, non-target analysis (NTA) and bioassays should be paired.

Data Analysis

Three distinct sets of data will be collected. What those data are, and how they will be utilized, is defined below:

- **Online Data** - online data will be logged and performance probability distribution functions (PDFs) will be created, which document the statistical reliability of each process to provide the desired results (for pathogen and pollutant reduction)
- **Grab Sample Data** - trace pollutant data will be collected and compared against industry standards, and then used to compare pollutant levels with the results from the advanced analytics. Pathogen data will be used to set a baseline of pathogen levels in the purification feed water, and then document the levels of reduction of those pathogens to the new potable water supply, clearly documenting compliance (or lack thereof) with published health standards (CDPH, 2014; NWRI, 2015).
- **Advanced Analytics** - NTAs and bioassays will be paired together and compared/contrasted with the trace pollutant data.

Results

Operation of the demonstration DPR system started mid-June 2018. Results will be available after a month of operation.

References

Anumol, T., Sgroi, M., Park, M., Roccaro, P., Snyder, S.A. (2015). *Predicting Trace Organic Compound Breakthrough in Granular Activated Carbon using Fluorescence and UV Absorbance as Surrogates*. *Water Research*, 76:76-87.

CDPH (2014). *Groundwater Replenishment Using Recycled Water (Water Recycling Criteria. Title 22, Division 4, Chapter 3, California Code of Regulations)*. California State Water Resources Control Board Division of Drinking Water.

Clean Water Services (CWS) (2014). *Direct Potable Reuse Water Reuse Demonstration*. Carollo Engineers for Clean Water Services, Oregon.

Escher, B.I., Allinson, M. Altenburger, R., Bain, P.A., Balaguer, P., Busch, W., Crago, J., Denslow, N.D., Dopp, E., Hilscherova, K. *et al.* (2014). *Benchmarking Organic Micropollutants in Wastewater, Recycled Water and Drinking Water with In Vitro Bioassays*. *Environmental Science and Technology*, 48(3):1940-1956.

Gerrity, D., Gamage, S., Jones, D., Korshin, G.V., Lee, Y., Pisarenko, A., Trenholm, R.A., von Gunten, U., Wert, E.C., Snyder, S.A. (2012). *Development of Surrogate Correlation Models to Predict Trace Organic Contaminant Oxidation and Microbial Inactivation during Ozonation*. *WaterResearch*, 46(19):6257-6272.

Merel, S., Anumol, T., Park, M., Snyder, S.A. (2015). *Application of Surrogates, Indicators, and High-Resolution Mass Spectrometry to Evaluate the Efficacy of UV Processes for Attenuation of Emerging Contaminants in Water*. *Journal of Hazardous Materials*, 282:75-85. 2

NWRI (2015). *Report of an NWRI Expert Panel for Developing a Direct Potable Reuse Framework.* Alexandria, VA. Co-sponsored by American Water Works Association and the Water Environment Foundation.

Steinle-Darling, E., Juby, G.; Salveson, A. (2015). *Direct Potable Reuse - Full-Scale Demonstrations Testing Results from Big Spring, Texas*. Proceedings from California Water Environment Association Conference.

SWRCB (2015). *California Water Boards: Recycled Water Research Workshop*. Costa Mesa CA.

Trussell, R. R.; Salveson, A.; Snyder, S. A.; Trussell, R. S.; Gerrity, D.; Pecson, B. M. (2013). *Potable Reuse: State of the Science Report and Equivalency Criteria for Treatment Trains*. WateReuse Research Foundation, Alexandria, VA.

Yu, H-W., Anumol, T., Park, M., Pepper, I., Scheideler, J., Snyder, S.A. 2015. On-line Sensor Monitoring for Chemical Contaminant Attenuation During UV/H2O2 Advanced Oxidation Process. *Water Research*, 81(0):250-260.

Quantification and Prediction of Degradation and Deactivation Kinetics of Antibiotic Resistance Genes during Disinfection/Oxidation Processes: From Bench-scale to Full-scale

Huan He[†,*], Peiran Zhou,[†] Kyle K. Shimabuku[†], Xuzhi Fang[†], Annika Anderson[§], Michael C. Dodd[†]
[†]Department of Civil and Environmental Engineering, University of Washington, Seattle, WA
[§]Washington University in St. Louis
*Corresponding author e-mail address: huanhe@uw.edu

KEY WORDS

Antibiotic Resistance Genes; Disinfection; Deactivation; Degradation

ABSTRACT

This work determined kinetics models and rate constants for DNA reactions with free available chlorine (FAC), monochloramine (NH_2Cl), ozone (O_3), chlorine dioxide (ClO_2), ultraviolet (UV) light and hydroxyl radical ($^{\bullet}OH$), and developed a framework for *a priori* prediction of fundamental sequence-specific rate constants governing antibiotic resistance gene (ARG) degradation/deactivation during selected bench-scale and full-scale disinfection processes. A model ARG *blt*, carried by chromosomal DNA of multidrug-resistant *Bacillus subtilis* strain 1A189, was used to study kinetics of ARG degradation and deactivation by each disinfectant, as assessed by qPCR assays targeting a set of varying-length *blt* amplicons and a culture-based transformation assay, respectively. The observed rate constant for degradation of a given amplicon was found to depend on its length and nucleotide content in a predictable manner, with reactivities for the chemical disinfectants declining in the order $HO^{\bullet} \gg O_3 > HOCl \gg ClO_2 > NH_2Cl$. During treatment with each disinfectant (except $^{\bullet}OH$), ARG deactivation tracked with loss of qPCR amplicons having length and nucleotide content equivalent to a ~900 bp region of DNA flanking the point mutation inducing *blt* activity, consistent with the *excluded length* of DNA required for homologous recombination in *B. subtilis*. Through deriving the fundamental second-order rate constants for DNA reactivity toward each disinfectant, a promising quantitative framework has been developed to predict ARG degradation/deactivation kinetics during disinfection processes under a range of conditions. Model predictions agreed well with measured changes in concentrations of clinically-relevant ARGs including *tet*A, *van*A, NDM-1 and KPC-2 during bench- and full-scale chlorination of non-nitrified secondary effluent from a local WWTP, where the dominant disinfectant was NH_2Cl rather than FAC due to the presence of excess NH_3.

INTRODUCTION

Antibiotic resistant bacteria and resistance genes (ARB & ARGs) have been widely detected in natural and engineered aquatic systems, which may represent an increasingly significant concern in public health, particularly in light of increasing water reuse. Although disinfection processes in (waste)water treatment can potentially act as a crucial barrier to ARB/ARG transport by prodiving effective means of bacterial inactivation, the ARG-containing DNA within cell debris may still remain biologically active after disinfection, which could be able to spread resistance to downstream bacteria (even to those temporally or spatially distant from donor cells) via horizontal gene transfers such as natural transformation (Lorenz and Wackernagel 1994). Increasing attention has been paid to the efficiencies of commonly-used disinfection/oxidation processes for ARB/ARG removal, though limited information is available on the fundamental kinetics parameters governing ARG elimination and consequent impacts on their biological activities during these processes (Chang et al. 2017, Czekalski et al. 2016, Yoon et al. 2017). In this context, the current work was undertaken to (1) characterize kinetics and mechanisms of ARG *degradation* (i.e., detectable decrease in ARG copy numbers) and *deactivation* (i.e., loss of ARG transforming activity) during treatment by free available chlorine (FAC), monochloramine (NH_2Cl), ozone (O_3), chlorine dioxide (ClO_2), ultraviolet (UV) light and hydroxyl radical ($^{\bullet}OH$), (2) develop a *quantitative framework* for predicting the rate constant for *degradation* of a given ARG sequence by each disinfectant/oxidant, and (3) enable prediction of ARG *degradation* and/or *deactivation* kinetics during bench-scale and selected full-scale disinfection processes.

METHODOLOGY

Chemicals and Materials. Molecular-biology-grade or at least reagent-grade chemicals and culture media were purchased from commercial suppliers. Working stocks of the chemicals and culture media (except disinfectants) were sterilized through either autoclave or 0.2 μm filtration if non-autoclavable. A commercially-available aqueous sodium hypochlorite (NaOCl) stock (w/w 5%) was diluted as the working stock of FAC. Aqueous stocks of NH_2Cl, O_3, and ClO_2 were prepared according to Kumar et al. (1986), Bader and Hoigné (1981), and Gates (1997), respectively. 254 nm UV light and $^{\bullet}OH$ were generated using a 15-W, low-pressure Hg lamp, and via irradiation of H_2O_2 by 290-400 nm wavelengths, respectively.

Determination of Disinfectant Exposures. Residual disinfectant concentrations ([Disinfectant], in units of mg/L) were measured by means of colorimetric methods for FAC and NH_2Cl (APHA 2005), O_3 (Bader and Hoigné 1981), and ClO_2 (Pinkernell et al. 2000). Intensity (I, in units of mW/cm^2) and the average path length (in units of cm) of 254 nm UV light were quantified by means of chemical actinometry (Rahn 1997a, b, Zepp 1978). Disinfectant exposures (or CT values, in units of mg/L·min) or fluences (or IT values, in units of mJ/cm^2) were calculated as $\int_0^t [\text{Disinfectant}]\,dt$ or the product of intensity and time ($I \times t$), respectively. Exposure of $^{\bullet}OH$ was quantified according to Elovitz and von Gunten (1999).

Bacterial Strains and Associated ARGs. *Bacillus subtilis* strains 1A189 (multidrug-resistant) and 1A1 (non-resistant) were purchased from the Bacillus Genetic Stock Center (BGSC; Ohio State University), and used as a DNA donor-recipient pair in natural transformation assays. Chromosomal DNA of 1A189 carries a model ARG *blt* with point mutation *acfA*, which encodes efflux pump mediated resistance to a variety of antibiotics. *Escherichia coli* ECE228, an NEB 5-α strain cloned with an artificially-ligated *pCFPamy* plasmid, was also purchased from BGSC, and applied together with *B. subtilis* 1A189 as two internal standards in DNA extraction of wastewater samples. Pathogenic ARB strains methicillin-resistant *Staphylococcus aureus*

(MRSA), vancomycin-resistant *Enterococci* (VRE), *E. coli* SMS-3-5, *Pseudomonas aeruginosa* 01 (PA01), and *Klebsiella pneumoniae* strains 1.53 and BAA-1705, each carrying clinically-relevant ARG *mec*A, *van*A, *tet*A, *amp*C, NDM-1, and KPC-2, respectively, were either purchased from American Type Culture Collection (ATCC) or donated by Dr. Marilyn Roberts at UW. The above six pathogenic ARB and their associated ARGs were applied as model disinfection targets in bench-scale wastewater disinfection experiments and/or in preparation of ARG standards for quantitative polymerase chain reaction (qPCR) assays.

Disinfection Treatment. **Bench-scale Disinfection in Phosphate Buffer.** To determine ARG degradation/deactivation kinetics, chromosomal DNA of *B. subtilis* 1A189 was treated with each disinfectant over *CT* or *IT* ranges typical of (waste)water treatment. Disinfection was conducted by treating either extracellular 1A189 DNA pre-isolated from viable cells via phenol-chloroform extraction (Saito and Miura 1963) or its intracellular form (as vegetative cells) in phosphate buffer (PB) at pH 7 (for FAC, ClO_2, O_3, UV, and $^{\bullet}OH$) or 8 (for NH_2Cl, to minimize NH_2Cl self-decomposition). FAC, NH_2Cl, ClO_2, and O_3 were dosed in $\geq$ 20-fold molar excess of DNA to ensure less than 10% decay during the whole period of treatment, while UV light and $^{\bullet}OH$ were generated as described above at a constant intensity or in situ concentration. At targeted *CT* or *IT* values, reaction solutions of 1A189 DNA or cells diluted in PB were sampled and if necessary, quenched with thiosulfate (for FAC, NH_2Cl, O_3, ClO_2) or bovine liver catalase (for H_2O_2 used to generate $^{\bullet}OH$). For intracellular experiments, the treated cells were recovered on membrane filters and processed with phenol-chloroform extraction as above. For collected DNA samples, qPCR amplicon copy numbers, *blt* transforming activities, and DNA fragment size distributions were analyzed as described below. Cell viability loss was also measured for intracellular samples prior to extraction. **Full-scale Disinfection in Wastewaters.** Pre and post-disinfection secondary effluent (SE) samples were collected from a wastewater treatment plant (WWTP) located in Seattle, Washington, which uses chlorine as disinfectant. SE samples were spiked with the internal standards 1A189 and ECE228, then concentrated through vacuum filtration (0.2 μm), and finally processed with DNA extraction using a FastDNA Spin Kit (MP Biomedicals). DNA extracts were analyzed with qPCR assays to determine copy numbers of the above six clinically-relevant ARGs in native bacterial DNA, as well as the two internal standard genes (*blt* and *pCFPamy*) to determine DNA extraction recoveries. **Bench-scale Disinfection in Wastewaters.** To simulate full-scale chlorination, NaOCl was dosed to 0.2 μm filtered pre-disinfection SE at pH 8.0, and then vegetative cells of several ARB strains (*B. subtilis* 1A189, MRSA, VRE, *E. coli* SMS-3-5, and PA01), altogether or individually, were dosed to the solution to initiate disinfection. At pre-determined time intervals, samples were collected and quenched with excess $Na_2S_2O_3$. The treated cells were pelleted and processed with DNA extraction as described for full-scale SE samples. ARG copy numbers were analyzed for DNA extracts, while associated ARB viabilities were measured for samples prior to extraction.

Analytical Approaches. **qPCR assays.** qPCR primers were either designed in this study or obtained from literature (Luo et al. 2013, McKinney and Pruden 2012). For the model ARG *blt*, four sets of qPCR primers were applied to target varying-length amplicons (266, 832, 870, and 1017 bp) encompassing or adjacent to the *acfA* mutation site in the *blt* gene locus. For the two internal standard genes, qPCR primers target the 1017-bp *blt* amplicon (as above), and a 663-bp amplicon on the *pCFPamy* plasmid, encompassing the ligation site of *cfp* and *amyE* genes. For the six pathogen-related ARGs, qPCR primers generally target long amplicons of ~ 800-1100 bp. **Natural transformation assays.** Transforming activity of the model ARG *blt* was analyzed by a culture-based natural transformation assay using the wild-type non-resistant *B. subtilis* 1A1 as recipient and acriflavine as the selective agent for transformants (Anagnostopoulos and Spizizen 1961, Bott and Wilson 1967). **Cell viability measurements.** Viabilities of ARB cells before and after disinfection treatment were determined by direct spread-plating or spot-titering (Beck et al. 2009) of serially-diluted sample aliquots on nutrient-rich agar with corresponding selective antibiotic if required (acriflavine for *B. subtilis* 1A189 and tetracycline for PA01). **Gel electrophoresis.** Agarose gel electrophoresis (AGE) was applied to verify DNA fragment size of the qPCR amplified products, while pulsed-field gel electrophoresis (PFGE) was used to analyze that of chromosomal 1A189 DNA before and after disinfection treatment.

RESULTS & DISCUSSION

Kinetics and Mechanisms of ARG Degradation/Deactivation. According to qPCR results of bench-scale disinfection in phosphate buffer, *degradation* of the 266, 832, 871 and 1017 bp *blt* amplicons in the vicinity of the *acfA* point mutation generally exhibited second-order (or fluence-normalized first-order) kinetics during treatment with each disinfectant, with reactivities of a given *blt* amplicon toward the chemical disinfectants declining in the order of $^{\bullet}OH$ >> O_3 > FAC >> ClO_2 >> NH_2Cl.

The measured rate constants were found to positively depend on amplicon lengths ($R^2 > 0.93$), implying that a larger number of nucleotides per amplicon results in a higher probability of an amplicon sustaining damage that prevents qPCR amplification. Moreover, the measured rate constants correlated particularly strongly with the contents of specific interstrand base pairs (bps) or intrastrand doublets per amplicon, implying that many of these disinfectants selectively attack specific sites on a given DNA strand with distinctive mechanisms. Briefly, FAC and NH_2Cl appear to preferentially attack GC and AT bps, respectively, in the first step of a two-step sequential pathway (N-chlorination followed by halogenation), while O_3 exhibits equal preference in attacking GC and AT bps. ClO_2 preferentially reacts with guanosine doublets (5'-GG-3'), while UV primarily targets thymine doublets (5'-TT-3') to form thymine dimers. For the above five disinfectants, the *deactivation* of *blt* exihibited similar rates as *degradation* of the 832, 871 and 1017 *blt* amplicons (but faster rates than for degradation of the 266 bp amplicon). In contrast with the selective modification of nucleobases by the above disinfectants, $^{\bullet}OH$ reacts non-selectively with DNA via extensive strand fragmentation (confirmed by PFGE analyses), leading to a significantly faster loss of transforming activity than of qPCR signals.

Kinetics of intracellular ARG deactivation generally tracked with those of extracellular ARGs for FAC, NH_2Cl, O_3, ClO_2, and UV (except at very high exposures for ClO_2 and UV). At disinfectant exposures representative of (waste)water practice, extracellular and intracellular *blt* genes were >90% degraded/deactivated by FAC, O_3, and UV, but recalcitrant to NH_2Cl and ClO_2. Degradation/deactivation of the *blt* gene always lagged behind 1A189 cell inactivation.

Development and Validation of a Quantitative Framework to Predict ARG Degradation/Deactivation. From the linear regressions of the measured rate constants for the *degradation* of *blt* amplicons versus specific nucleotide contents (AT, GC, or AT+GC bps, or 5'-GG-3' or 5'-TT-3' doublets), fundamental *sequence-independent* rate constants were derived for DNA reactivity toward each disinfectant. These fundamental rate constants enable prediction of the *degradation* kinetics of a given DNA segment during treatment with each disinfectant, as long as provided with its sequence. Further, this approach also enables prediction of the kinetics of ARG *deactivation* resulting from DNA damage for each disinfectant (except $^{\bullet}OH$), provided with the ARG's sequence and the requirements for its transformation to a given recipient organism.

According to reported requirements for natural transformation in *B. subtilis* (Dubnau 1993), a ~800-1000 bp homologous flanking sequence centered on the *acfA* point mutation in *B. subtilis* 1A189 DNA should be necessary to achieve efficient transformation of *blt*-mediated multiple antibiotic resistance to *B. subtilis* 1A1. Therefore, the kinetics of intracellular and extracellular 1A189 DNA deactivation by FAC, NH_2Cl, O_3, ClO_2, and UV can generally be predicted based on degradation kinetics of the requisite homologous flanking sequence, consistent with the measured results. The quantitative framework was also utilized to predict rate constants for degradation of a wide array of ARG amplicons during UV irradiation, yielding predicted values which are in excellent agreement with measured values obtained from prior studies (Chang et al. 2017, Yoon et al. 2017, Yoon et al. 2018). Further, model predictions obtained for degradation of ARGs by NH_2Cl also agreed well with observations of insignificant changes in measured levels of pathogen-related ARGs during bench- and full-scale wastewater chlorination. In this case, the predominant chlorine species in solution following FAC addition was confirmed to be NH_2Cl, rather than FAC itself, due to rapid conversion of dosed FAC to NH_2Cl upon reaction with excess ammonia (> 25 mg N/L) in the wastewater. The limited degree of ARG degradation observed here and in previous studies during wastewater chlorination is consistent with the very low reactivity of NH_2Cl toward ARGs (as noted above), and indicates that chlorination of non-nitrified effluent is generally not likely to provide an effective barrier to ARG dissemination into downstream waterbodies.

CONCLUSIONS

The current work illustrates that common (waste)water disinfectants can exhibit a wide range of reactivities and distinct DNA damage mechanisms in degrading and deactivating ARGs. Relatively high efficacies were observed for FAC, O_3, and UV light in ARG degradation/deactivation over ranges of disinfectant exposures typical of (waste)water treatment, while extremely low efficacies were observed for NH_2Cl and ClO_2. A promising quantitative framework has been developed to predict ARG degradation/deactivation kinetics during disinfection processes under a range of conditions, and also supports use of qPCR as a proxy for tracking ARG deactivation under certain circumstances. The applicability of this framework was confirmed through comparisons with results from bench- and full-scale studies of wastewater chlorination, as well as literature results on UV irradiation and wastewater chlorination. Ongoing work is being undertaken to assess the general applicability of this predictive framework to other bench- and full-scale disinfection processes.

ACKNOWLEDGEMENTS

This material is based upon work supported by the National Science Foundation under Grant Number CBET-1254929. Dr. Dan Zeigler (Bacillus Genetic Stock Center), Dr. Marilyn Roberts (School of Public Health, UW) and Dr. Scott Meschke (School of Public Health, UW) are greatly acknowledged for provision of *B. subtilis* strains, *Klebsiella pneumoniae* strains, and access to PFGE facilities, respectively.

REFERENCES

Anagnostopoulos, C. and Spizizen, J. (1961) Requirements for transformation in *Bacillus subtilis*. Journal of Bacteriology 81(5), 741.

APHA (2005) Standard Methods for the Examination of Water and Wastewater, APHA, AWWA, WPCF, Washington.

Bader, H. and Hoigné, J. (1981) Determination of ozone in water by the indigo method. Water Research 15, 449-456.

Beck, N., Callahan, K., Nappier, S., Kim, H., Sobsey, M. and Meschke, J. (2009) DEVELOPMENT OF A SPOT - TITER CULTURE ASSAY FOR QUANTIFYING BACTERIA AND VIRAL INDICATORS. Journal of Rapid Methods & Automation in Microbiology 17(4), 455-464.

Bott, K.F. and Wilson, G.A. (1967) Development of competence in *Bacillus subtilis* transformation system. Journal of Bacteriology 94(3), 562-570.

Chang, P.H., Juhrend, B., Olson, T.M., Marrs, C.F. and Wigginton, K.R. (2017) Degradation of extracellular antibiotic resistance genes with UV254 treatment. Environmental Science & Technology.

Czekalski, N., Imminger, S., Salhi, E., Veljkovic, M., Kleffel, K., Drissner, D., Hammes, F., m H. and Von Gunten, U. (2016) Inactivation of antibiotic resistant bacteria and resistance genes by ozone: from laboratory experiments to full-scale wastewater treatment. Environmental Science & Technology 50(21), 11862-11871.

Dubnau, D. (1993) Bacillus subtilis and other gram-positive bacteria: biochemistry, physiology, and molecular genetics. American Society for Microbiology. Sonenshein, A.L., Hoch, J.A. and Losick, R. (eds), pp. 555-584, American Society for Microbiology, Washington D.C.

Elovitz, M.S. and von Gunten, U. (1999) Hydroxyl radical/ozone ratios during ozonation processes. I. The Rct concept.

Gates, D.J. (1997) The Chlorine Dioxide Handbook. American Water Works Association, Denver, CO.

Kumar, K., Day, R.A. and Margerum, D.W. (1986) Atom-transfer redox kinetics: general-acid-assisted oxidation of iodide by chloramines and hypochlorite. Inorganic Chemistry 25(24), 4344-4350.

Lorenz, M.G. and Wackernagel, W. (1994) Bacterial Gene Transfer by Natural Genetic Transformation in the Environment. Microbiological Reviews 58(3), 563-602.

Luo, Y., Yang, F., Mathieu, J., Mao, D., Wang, Q. and Alvarez, P.J.J. (2013) Proliferation of Multidrug-Resistant New Delhi Metallo-β-lactamase Genes in Municipal Wastewater Treatment Plants in Northern China. Environmental Science & Technology Letters 1(1), 26-30.

McKinney, C.W. and Pruden, A. (2012) Ultraviolet disinfection of antibiotic resistant bacteria and their antibiotic resistance genes in water and wastewater. Environmental Science & Technology 46(24), 13393−13400.

Pinkernell, U., Nowack, B., Gallard, H. and Von Gunten, U. (2000) Methods for the photometric determination of reactive bromine and chlorine species with ABTS. Water Research 34(18), 4343-4350.

Rahn, R.O. (1997a) Potassium iodide as a chemical actinometer for 254 nm radiation: Use of iodate as an electron scavenger. Photochemistry and Photobiology 66(4), 450-455.

Rahn, R.O. (1997b) Potassium iodide as a chemical actinometer for 254 nm radiation: Use of iodate as an electron scavenger (vol. 66, pg. 450, 1997). Photochemistry and Photobiology 66(6), 885-885.

Saito, H. and Miura, K. (1963) Preparation of transforming deoxyribonucleic acid by phenol treatment. Biochimica et Biophysica Acta 72(4), 619-629.

Yoon, Y., Chung, H.J., Di, D.Y.W., Dodd, M.C., Hur, H.-G. and Lee, Y. (2017) Inactivation efficiency of plasmid-encoded antibiotic resistance genes during water treatment with chlorine, UV, and UV/H2O2. Water Research 123, 783-793.

Yoon, Y., Dodd, M.C. and Lee, Y. (2018) Elimination of transforming activity and gene degradation during UV and UV/H 2 O 2 treatment of plasmid-encoded antibiotic resistance genes. Environmental Science: Water Research & Technology.

Zepp, R.G. (1978) Quantum yields for reaction of pollutants in dilute aqueous solution. Environmental Science & Technology 12(3), 327-329.

Assessing chemical disinfectant alternatives using I-CT dose response kinetics for multiple microorganisms.

Roberta Maffettone[1,2], Siva Sarathy[1,2] , Paris Neofotistos[3], Domenico Santoro[1,2], Ronald Gehr[4]

[1]University of Western Ontario
[2]Trojan Technologies
[3]USP Technologies
[4]McGill University

ABSTRACT

Performic acid (PFA), a novel chemical disinfectant alternative, was compared against more common chemical (hypochlorite and peracetic acid (PAA)) and physical (ultraviolet (UV)) disinfectants using the Integrated-CT (I-CT) dose concept. The disinfectants were compared on primary and secondary effluent wastewaters for conventional fecal indicator bacteria and bacteriophages. Results showed that PFA required a low I-CT dose for inactivating fecal indicator bacteria. Higher inactivation of MS2 was achieved by UV than by PFA, PAA and NaOCl. T1 and T7 were hardly inactivated by PAA, whereas several log reductions were achieved by UV, PFA and NaOCl. The ability of PFA to achieve very high levels of bacterial inactivation suggests that further investigations in reuse applications should be pursued. Additionally, because of the very low I-CT, PFA can provide flexibility in short hydraulic retention time scenarios such as space-constrained plants, peak flow events, and combined sewer overflows.

KEYWORDS

Performic acid, Peracetic acid, Hypochlorite, Ultraviolet light, Disinfection, Wastewater, Dose, Bacteriophage, Fecal indicator bacteria

INTRODUCTION

In North America, the most widely used method for wastewater disinfection is chlorination, treating about 65% of the wastewater flow, followed by ultraviolet (UV) disinfection, accounting for approximately 35%. Other chemical alternatives include ozone and peracetic acid (PAA) with the latter gaining interest due to the ease of retrofitting existing chlorination operations. Performic acid (PFA) is a wide-spectrum disinfectant recently considered as an alternative for wastewater disinfection due to the low doses required for inactivation of typical indicator organisms and pathogens (Ragazzo et al. 2013, Chhetri et al. 2014, Ragazzo et al. 2017). Research on this innovative disinfectant has been performed in Europe for disinfecting municipal wastewater treatment plants (WWTPs) effluents and combined sewer overflows (CSOs) (Luukkonen & Pehkonen, 2017, Cheetri et al. 2014). However, a holistic approach to compare PFA with other disinfectants under the same experimental conditions is not included in the reported studies.

In this study, we compared PFA against sodium hypochlorite (NaOCl), PAA, and UV. The work evaluated the disinfection efficacy of the disinfectants in secondary and primary effluents against a wide range of test organisms, including fecal indicator bacteria (i.e. *E. coli*, enterococci, fecal coliforms, and total coliforms) and F-specific and somatic coliphages (i.e. MS2, T1, and T7). Preliminary technical comparisons were then conducted by calculating the initial concentration required by PAA and PFA at different HRTs (5, 10 and 25 minutes), and determining their concentration profiles under different scenarios.

METHODOLOGY

Secondary and primary effluents from a wastewater treatment plant in London, Ontario, Canada were collected in April and May 2016 and characterized for water quality parameters (Table 1). Batch chemical disinfection tests were performed in Trojan's lab (London, ON, Canada) using stock solutions of PFA, PAA and NaOCl, dosed at known concentrations. PAA and NaOCl were commercially available solutions, whereas PFA was produced at bench scale following the procedure described by Gehr et al. (2009).

UV disinfection tests were performed in Trojan's lab (London, ON) using a collimated beam (CB) apparatus. Viable indicator organisms (i.e. *E. coli*, fecal coliforms, enterococci, and total coliforms) were enumerated by standard filtration methods (i.e. 9222B, 9222D, 9222G and 9215D) (APHA et al., 2017). Bacteriophage MS2, T1, and T7 stock solutions were prepared as reported in Beck et al. (2015), and inoculated in the secondary effluent. Phage concentrations were determined by GAP EnviroMicrobial Services lab (London, ON), using the double agar layer method. Initial concentrations of the required chemical disinfectant (PFA, PAA) and resulting effluent concentrations were also calculated for three hydraulic retention times (HRTs) (5, 10, and 25 min) to determine PAA and PFA concentration profiles.

Dose-response kinetics of the chemical disinfectants were evaluated following the concept of "Integrated-CT" (ICT), as given by the equation 1 (Santoro et al., 2015):

$$\mathrm{ICT} = \frac{C_0(1-\alpha)}{k'}\left(1 - e^{-k't}\right) \qquad \text{[Equation 1]}$$

where: k'= disinfectant first-order disinfectant decay rate constant (min^{-1}); C_0 = initial disinfectant dose (mg/L); α = initial instantaneous demand of disinfectant at C_0 as a fraction of C_0; and t = time from start of experiment to time of sample (min).

Microbial inactivation curves were fitted with the double exponential model given in equation 2, and this model was used to determine the dose (I-CT or UV fluence) required to achieve different levels of disinfection (e.g., from 1 to 5 log inactivation).

$$N = N_d e^{-k_d (Dose)^m} + N_p e^{-k_p Dose} \qquad \text{[Equation 2]}$$

where N is the concentration of viable microorganisms (cfu or pfu / 100 mL); N_d is the concentration of particle-associated microorganisms (cfu or pfu / 100 mL); N_p is the concentration of dispersed microorganisms (cfu or pfu / 100 mL); k_d is the reaction rate constant for particle-associated microorganisms (L/mg·min); k_p is the reaction rate constant for dispersed microorganisms; (L/mg·min); m is a model parameter to account for shouldering effects (dimensionless, m=1 for UV only); "Dose" is I-CT (given by equation 1) for the chemical disinfectants (mg·min/L), or fluence for UV (mJ/cm^2).

Table 1: Water characteristics of primary effluent (PE) and secondary effluent (SE) samples collected in April and May 2016 for disinfection testing.

Sample	Ammonia as N (mg/L)	BOD (mg/L)	COD (mg/L)	pH	TSS (mg/L)	UVT (%)
SE – April 12	0.18	3	<10	7.2	2	77
SE – April 20	0.20	3	<10	NM	NM	75
SE – April 26	0.06	<2	124	7.2	6	79
PE – May 10	33.2	140	280	7.3	84	23
PE – May 17	29.5	357	351	7.5	80	15

NM- not measured

RESULTS

Fecal bacterial inactivation

Table 2 presents the calculated I-CT values for the three disinfectants acting on the four fecal bacteria indicator organisms, and at two inactivation levels for the secondary and primary effluents tested.

In the secondary effluent, PFA and hypochlorite required the lowest I-CT than PAA to achieve 2 and 3 log reductions of *E. coli*, enterococci, fecal coliforms, and total coliforms (respectively at initial counts in the order of 10^4, 10^4, 10^5, and 10^3 cfu / 100 mL). Interestingly, PAA needed an order of magnitude higher I-CT to achieve the same log reduction of the aforementioned fecal indicator bacteria. UV was able to achieve high levels of reduction for *E. coli*, fecal coliforms, and total coliforms at doses of less than 20 mJ/cm^2, whereas it could not achieve more than 1 log reduction of enterococci with a dose greater than 80 mJ/cm^2. PFA achieved 2.2 cfu/100 mL of total coliforms for reuse at I-CT of 5 –10 mg·min/L, and enterococci concentration down to the detection limit of the enumeration method at I-CT of 20 mg min/, while PAA could barely achieve 100 cfu/100 mL at 60 mg·min/L of I-CT. This suggests that further investigations in water reuse applications should be pursued on PFA.

For the primary effluent, PAA still required higher I-CT than PFA to achieve equivalent disinfection levels, but in contrast to the secondary effluent, the difference in I-CT between the two disinfectants was not as large. Compared to PFA, although NaOCl required a lower I-CT to inactivate the fecal indicator bacteria in secondary effluents, in primary effluent a higher I-CT was instead observed for NaOCl than PFA. This was due to the higher demand required for NaOCl in this water matrix, where all the added free chlorine was converted into combined chlorine species due to the relatively high concentration of ammonia (~30 mg/L) and organics (up to 350 mg/L COD). For PFA, even for long HRTs, an initial concentration as low as 0.25 mg/L could be dosed due to the very low demand of this disinfectant.
In the primary effluent, due to the high initial concentrations of the target indicators (i.e. *E. coli*, fecal coliforms, total coliforms, and enterococci in the order of 10^7, 10^7, 10^7, and 10^6 cfu / 100 mL, respectively) several log reductions could be achieved by PFA and PAA at lower I-CTs

compared to those calculated for the secondary effluent. This is due to the tailing observed at 2 log reduction for PAA and 3 log reduction for PFA in the secondary effluents with initial concentrations of target bacteria within a range up to four order of magnitude lower than the primary effluent (10^3-10^5 cfu/100 mL).

Table 2: Summary of I-CT (PFA, PAA, and NaOCl, mg·min/L) and IT (UV, mJ/cm^2) doses required to achieve 2, and 3 log reductions of various indicator organisms in secondary and primary effluents.

Sample	Microbial indicator	2 log reduction				3 log reduction			
		PFA	**PAA**	**NaOCl**	**UV**	**PFA**	**PAA**	**NaOCl**	**UV**
Secondary effluent	*E. coli*	3	13	1	8.5	9	17	3.4	13.6
	Fecal coliforms	2	44	0.5	10.5	4		2	
	Total coliforms	2	26	1.8	10.3	17	49	5.8	17.5
	Enterococci	19	32	6					
Primary effluent	*E. coli*	2	5	13	NM	4	5	20	NM
	Fecal coliforms	2	6	16	NM	4	7	24	NM
	Total coliforms	2	9	16	NM	3	19	26	NM
	Enterococci	4	22	37	NM	7	35	62	NM

NM: Not Measured

Coliphages inactivation

With respect to coliphage inactivation, PAA and PFA were not able to achieve greater than 1 log reduction of MS2 phage within I-CT of 30 mg·min/L for PFA, and 40 mg·min/L for PAA. NaOCl was more effective than peracids, but it could barely achieve 2 log reduction. In contrast, UV could achieve 4 log inactivation of MS2 for a UV dose greater than 80 mJ/cm^2 (Figure 1).

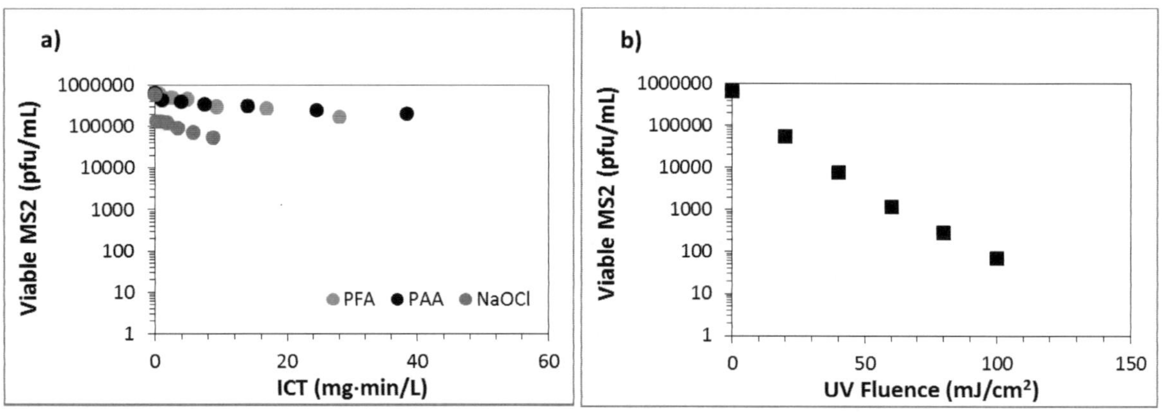

Figure 1: MS2 inactivation by PAA, PFA and NaOCl (a) and by UV (b).

T1 and T7 were inactivated easily by PFA and hypochlorite (Figures 2 and 3). PAA was effective against T7 but showed poor inactivation of T1. Finally, the bacteriophages were all susceptible to inactivation by UV, with MS2 being the most challenging phage tested.

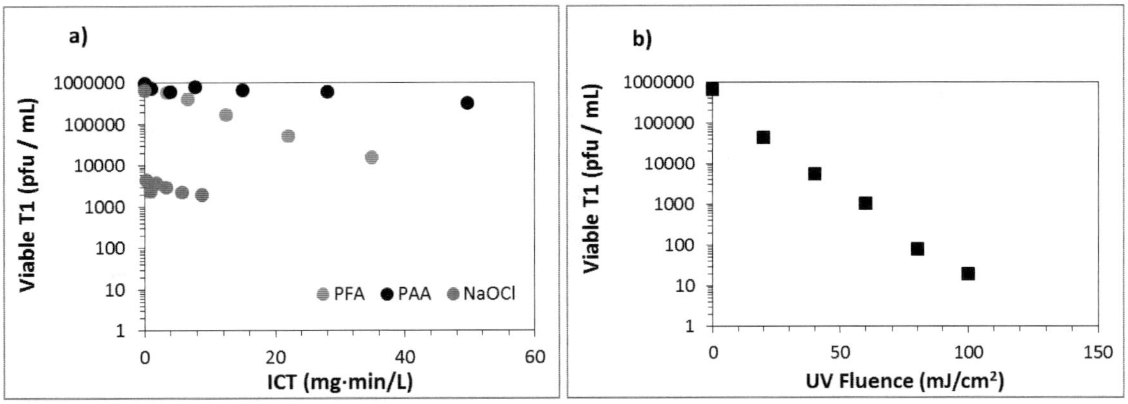

Figure 2: T1 inactivation by PAA, PFA and NaOCl (a) and by UV (b).

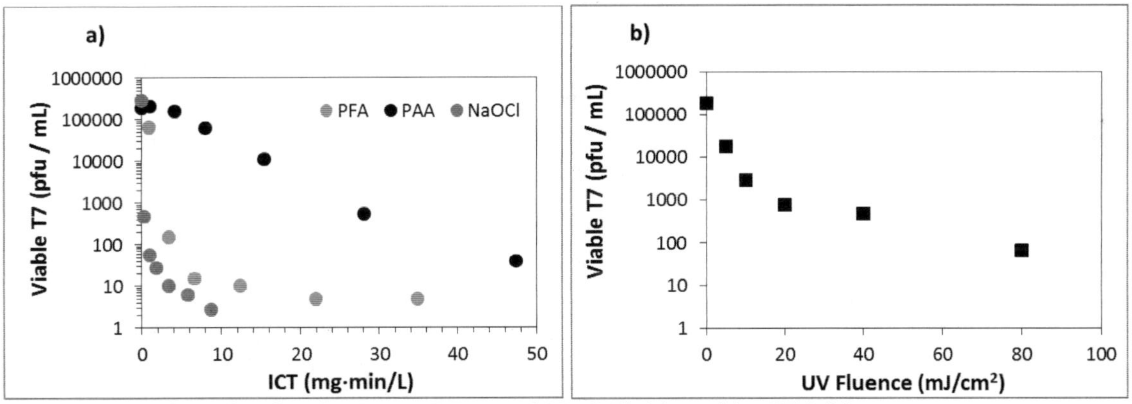

Figure 3: T7 inactivation by PAA, PFA and NaOCl (a) and by UV (b).

Concentration profiles

Based on the fecal bacteria inactivation dose response data, design calculations were performed for the secondary effluent to determine the required initial concentrations for PAA and PFA, and the resulting final effluent concentrations, for three HRTs (5, 10, and 25 min). These calculations allowed us to build "concentration profiles" for each disinfectant, which show the required initial concentrations at a defined HRT, and the residual final concentrations to achieve different log reductions for the target organisms: *E. coli* (EC), fecal coliforms (FC), total coliforms (TC) and enterococci (Entero).

Figures 4 and 5 show concentration profiles of PAA and PFA as a function of I-CT to achieve different log reductions of the target organisms investigated, respectively at 10-minute and 25-minute HRTs. The circles in the figure indicate the required I-CT ranges to achieve the log reduction for each organism indicated in the corresponding text (e.g. in Figure 4, the blue circle indicates the region of I-CT and the relative initial and final concentrations to achieve up to 3 log reduction of EC, up to 3 log reduction of FC, and so on). As shown in the figures, the required initial concentration of PAA and the final residual is always higher than that required using PFA to achieve several log reductions for each microbial target. Also, for the longer HRT, the PFA residual is low (less than 1 mg/L).

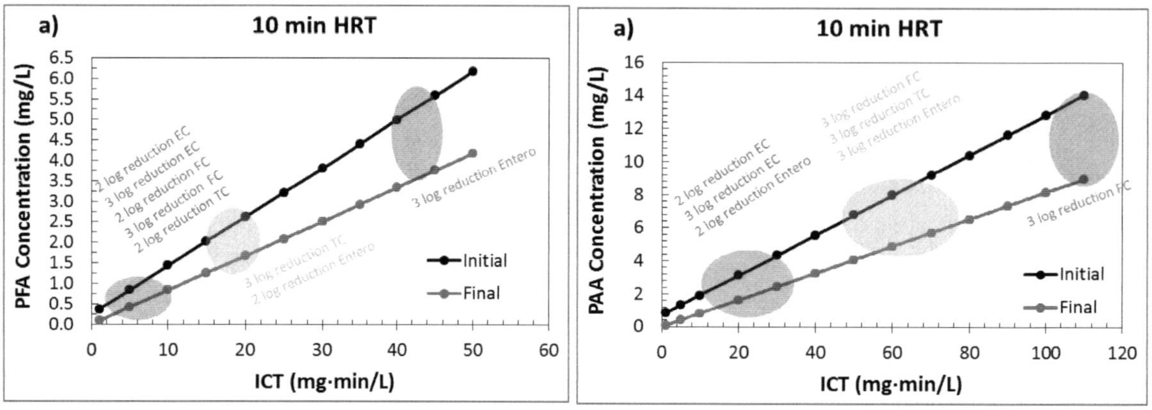

Figure 4– PFA (a) and PAA (b) concentration profiles as a function of I-CT for 10 min HRT for the secondary effluent.

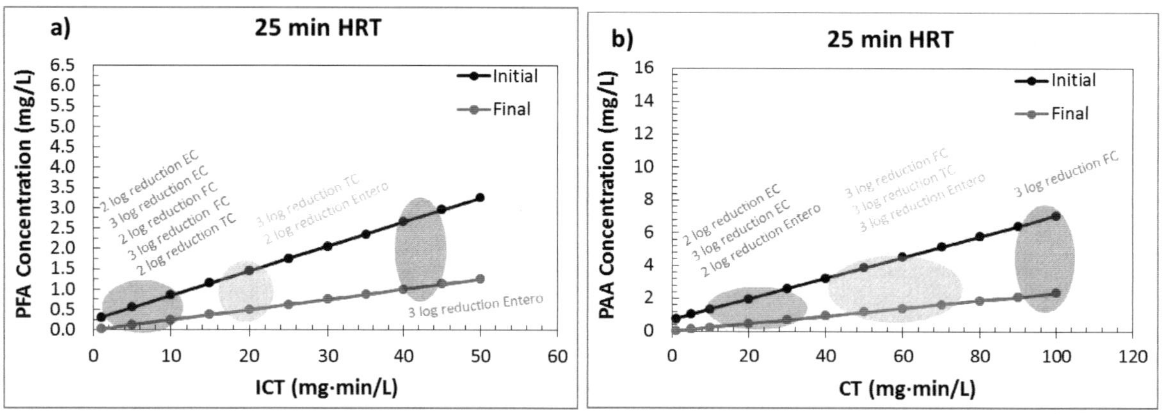

Figure 5– PFA (a) and PAA (b) concentration profiles as a function of I-CT for 25 min HRT for the secondary effluent

CONCLUSIONS

This study showed that for the primary and secondary effluents which were tested, PFA required a lower I-CT dose than PAA, and comparable to that of hypochlorite, to achieve high levels of inactivation of indicator bacteria. Due to the high initial concentrations of target indicators in primary effluent, PFA could achieve up to 3 log reductions at lower I-CT than PAA and NaOCl, whereas the effect of tailing in the secondary effluent at 3 log reduction resulted in higher I-CT than that required in the primary effluent for the same log removal.

MS2, T1, and T7 were all very susceptible to inactivation by UV. For MS2, peracids required high I-CT doses (e.g. for PAA and PFA, I-CT was greater than 30 mg·min/L) to achieve 1 log removal of MS2. PAA could not achieve more than 1 log reduction of T1 at I-CT greater than 50 mg· min/L, whereas 2 log reduction of T1 was achieved by PFA at I-CT of 30 mg·min/L. For T7, 3 log reductions were achieved by PFA at I-CT less than 10 mg·min/L, whereas PAA required up to I-CT of 50 mg·min/L for the same log reduction.

Based on the different scenarios investigated at varied HRTs, the initial concentration of PFA was lower than that of PAA. Also, in most cases, the PFA residual was low, possibly negating the need for quenching. These results suggested that PFA can be used as a fast-acting disinfectant, providing flexibility in applications such as short HRT applications as well as combined sewer overflows (CSOs).

REFERENCES

APHA, 2017. Standard methods for the examination of water and wastewater. American Public Health Association, Washington D.C.

Beck S. E., Wright H. B., Hargy T. M., Larason T. C., Linden K. G., (2015). Action spectra for validation of pathogen disinfection in medium-pressure ultraviolet (UV) systems. Water Research. Volume 70, 2015, Pages 27-37, ISSN 0043-1354

Bell K., Block P., Amador J., (2015). Peracetic Acid (PAA) Disinfection - Upgrading WWTPs to reduce ecosystem toxicity. Proceedings of the Water Environment Federation, WEFTEC 2015: Session 600 through Session 608, pp. 318-327(10).Water Environment Federation. https://doi.org/10.2175/193864715819556219

Chhetri R.K., Thornberg D., Berner J., Gramstad R., Öjstedt U., Sharma A.K., Andersen H.R., (2014). Chemical disinfection of combined sewer overflow waters using performic acid or peracetic acids. Sci. Total Environ., 490 (2014), pp. 1065-1072, 10.1016/j.scitotenv.2014.05.079

Fallon, K.S., Hargy, T.M., Mackey, E.D., Wright, H.B., Clancy, J.L., (2007). Development and characterization of nonpathogenic surrogates for UV reactor validation. J. Amer. Water Works Assoc. 99 (3), 73e82.

Gehr R., Chen D., Moreau M., (2009). Performic acid (PFA): tests on an advanced primary effluent show promising disinfection performance. Water Science & Technology—WST, 59.1. IWA Publishing 2009.

Grenier M., Nutt S.G., Delaura T.J., Seggewiss G., (2015). State of the industry update on the use of peracetic acid for municipal wastewater disinfection. Proceedings of the Water Environment Federation, WEFTEC 2015: Session 600 through Session 608, pp. 2112-2134(23). Water Environment Federation. https://doi.org/10.2175/193864715819555977

Luukkonen, T., Heyninck, T., Rämö, J., & Lassi, U., (2015). Comparison of Organic Peracids in Wastewater Treatment: Disinfection, Oxidation and Corrosion. Water Research, 85, 275-285. http://dx.doi.org/10.1016/j.watres.2015.08.037

LuukkonenT. & Pehkonen S.O., (2017). Peracids in water treatment: A critical review, Critical Reviews in Environmental Science and Technology, 47:1, 1-39, DOI: 10.1080/10643389.2016.1272343

Ragazzo, P., Chiucchini, N., Piccolo, V., & Ostoich, M., (2013). A new disinfection system for wastewater treatment: performic acid full-scale trial evaluations. Water Science & Technology, 67(11), 2476. http://dx.doi.org/10.2166/wst.2013.137

Ragazzo, P., Feretti, D., Monarca, S., Dominici, L., Ceretti, E., & Viola, G. et al. (2017). Evaluation of cytotoxicity, genotoxicity, and apoptosis of wastewater before and after disinfection with performic acid. Water Research, 116, 44-52. http://dx.doi.org/10.1016/j.watres.2017.03.016

2018 WEF DISINFECTION and REUSE SYMPOSIUM
Portland, Oregon

Abstract: Water, Sanitation and Hygiene (WASH)

26 February 2018 pgary@Tulane.edu

MANAGING CHLORINATED WATER FOR EBOLA DISINFECTION *

The epic Ebola epidemic in West Africa # stands as the most devastating outbreak of the Ebola virus in recent history. To stop the spread of the virus and to treat infected individuals, the response required round-the-clock execution of extreme infection prevention controls (IPC), which depended on the availability of ultra-ready chlorinated water, provided by WASH Teams.

To better understand the disinfection challenges impacting IPC, vital WASH functions (Water, Sanitation & Hygiene Functions) at two (2) Ebola Treatment Centers in Sierra Leone during the Ebola Outbreak are presented for discussion. Focus of the presentation is the all critical role of clean, safe chlorinated water – the essential and critical agent for achieving absolute disinfection to halt, overcome, and ultimately eliminate Ebola transmission.

Production and management of chlorinated water, principal role of a WASH Team, are critical for disinfection with patient care, staff protection, and cleaning protective gear (PPE).

Two different concentrations of chlorinated water are prepared by the WASH Team to support IPC Protocols for mitigating the Ebola Virus: 0.5% and 0.05% sodium hypochlorite by weight.

Using data collected at two sister Ebola Treatment Centers, located at Makeni and Kambia in Sierra Leone, the critical roles of High Chlorinated (0.5%) and Low Chlorinated (0.05%) Water are presented and reveal two operational relationships during the 2015 Ebola Outbreak:

> Use of High Chlorinated Water with Daily Patient Population and Staff Entry/Exit through high risk zones (HRZ)in the Treatment Centers, as shown in attached Figures 1 and 2.

> Use between Low Chlorinated Water with Daily Patient Population and Staff Entry/Exit through high risk zones (HRZ) in the Treatment Centers, as shown in attached Figures 3 and 4.

Also presented are unique lessons learned with disinfection challenges in extreme environments, plus overcoming limited access to reliable water sources (ground/surface) in remote situations.

Field data analyses was conducted with STATA 13 (STATA Corp, Texas, USA), spanning both univariable and multivariable linear regression analysis, specifically to examine differences in the impact of High and Low Chlorinated Water for disinfection in two Ebola Treatment Centers.

---

* *Authors:* *Lee P. Gary, Jr. / Strategic Management Services – USA.*
Kari F. Brisolara / LSU School of Public Health.
Robert S. Reimers / Tulane University – Emeritus.

West Africa: *Guinea, Liberia and Sierra Leone – WHO 2017.*
Confirmed Cases 28,616 / Confirmed Deaths 11,308.

Increasing In-Home Water Availability in Remote Alaska's Homes with Water Reuse

Aaron D. Dotson
University of Alaska Anchorage
Email: addotson@alaska.edu

ABSTRACT

KEYWORDS
On-site Water Reuse, Hygiene, Remote, Rural, Water Quantity, Health

INTRODUCTION

The 2010 United States (US) Census data revealed that the State of Alaska has the largest percentage of homes without piped water and sewer (or comparable services, e.g. well and septic). The majority of these homes are located in remote communities. Unlike rural communities found in the continental US, these remote communities are not connected by roads and can only be reached via air or water (e.g. summer river/ocean or winter ice road/snowmachine travel). The cold temperatures, Arctic features (e.g. permafrost), and lack of access to these communities makes them particularly challenging. Further, as these communities are predominantly Alaskan Native and are likely to have limited economic resources (e.g. jobs and/or industry). Further, there is potential for the existence of multiple governing bodies (e.g. municipal, tribal and regional corporations) that may increase complexity of community wide improvements.

A recent study found that gastrointestinal, skin, and respiratory diseases occurrence rates were greater in unserved communities (those without in-home running water) (Hennessy et al., 2008). While considered unserved, these communities all maintain a watering point or washeteria (i.e. a watering point, showers, toilets, and laundry facilities) that provides water that meets primary drinking water standards. Through this provision of water and use of traditional water sources, people living in these communities only utilize about 6 liters (1.5 gallons) of water per person per day (Thomas et al., 2016). At this utilization rate, it is likely that greater disease occurrence likely associated with low water quantity, not water quality. Thus disease transmission is proposed to be water wash related and not waterborne. The four village study found that interventions that increased in-home water use reduced disease occurrence rate (Thomas et al., 2016). The interventions in this study were installation of piped water and sewer service as well as education to improve water use habits within the home through additional water use for hygiene and household cleaning.

While there are relatively few communities remaining, about 30, that are considered unserved, there are many homes within served communities that are unserved. These unserved homes are a particular challenge with capital costs in excess of an estimated $200,000 per home to establish new service (DEC, 2018). Further, many of these unserved homes are geographically located in areas that either in the floodplain or have been severely affected by climate change. These communities in particular will not meet the funding requirements for construction of permanent water and sewer infrastructure. The task of serving the unserved in served communities, unserved communities, and/or communities preparing to move requires a different approach that conventional methods (e.g. piped services, on-site well/septic).

Two such projects are underway to improve the health and wellbeing of those living in unserved homes. These projects are the Alaska Water Sewer Challenge (AWSC) and the Alaska Native Tribal Health Consortium (ANTHC) Portable Alternative Sanitation System (PASS). These projects take different approaches to the challenge. The University of Alaska Anchorage (UAA) team participating in the AWSC has taken the approach of increasing water availability through on-site water reuse for non-potable direct human contact, a urine diverting dry toilet (UDDT), and disinfection of self-hauled drinking water (i.e. washeteria or traditional). The PASS has taken the approach of improved in-home/on-site water and wastewater management of self-hauled water. The PASS includes a UDDT and a low-flow handwashing sink where urine and greywater are discharged on-site to the subsurface. The source of water for these PASS fixtures is a 100 gallon in-home water tank where during filling the water is filtered and the disinfected by the user. This manuscript will describe the current status of these two alternatives.

DISCUSSION

Detailed information regarding study methodology and specific outcomes of the described projects can be found elsewhere and additional publications are in preparation as of June 2018.

Urine Diverting Dry Toilet

The use of a UDDT is common to both the UAA AWSC prototype and PASS system. UAA has leaned on the lessons learned from PASS when considering the UDDT. Originally, the PASS systems were installed with a commercially available toilet manufactured by Separette (ANTHC, 2016). While functional, this toilet has a variety of features (e.g. a rotating solids collection containers) that many were concerned would fail with daily use as well as the plastic base flexed under the weight of larger users. Further, in the case of toilet failure (i.e. frozen urine discharge line), it was observed that users would revert to use of a honeybucket (5-gallon bucket with toilet lid) which is notably less sanitary. To this end, ANTHC commissioned Lifewater Engineering to construct a simpler toilet with more robust materials with a fallback operation that would allow for continued use even when urine discharge failed. Lifewater achieved this by constructing the toilet from HDPE used in their on-site wastewater treatment system, a hinged urine diverting toilet seat, a 5-gallon bucket for solids collection, and an attachable small (1-gallon) urine collection container when subsurface discharge was unavailable. Similar to the Separette, the Lifewater toilet used an always-on ventilation fan that creates a negative pressure on the solids collection receptacle for odor control and drying of the solids. This toilet adaptation physically felt much more sturdy to the user and was installed during Fall of 2017 in new installations and subsequently replaced the Separette installations in prior installations. In addition to the UDDT, the PASS system uses a custom built small dry urinal as through end-user feedback it was identified that men desired to urinate standing up which was not easily achieved using the urine diverting toilet seat. During UAA AWSC prototype testing, the UDDT was not tested due to on-going PASS work but it is noted that operation of the UAA AWSC prototype will require a small amount of urine to operate most effectively which a homeowner or operating company may choose to collect from the UDDT.

University of Alaska Anchorage's Alaska Water Sewer Challenge Prototype
The UAA AWSC prototype utilized three systems; UDDT (described in the prior section), a drinking water system and a water reuse system. The system was designed such that self-haul practices of 6 liters (1.5. gallons) per person per day could be maintained while water availability was increased.

The drinking water system was designed to only address microbial contamination that may be present in the traditional source or acquired during self-haul of washeteria water. The system included a 5-gallon water storage vessel built into a counter top, water was drawn from the vessel with a 1 gallon per minute demand pump, pumped through a 1 micron nominal cartridge filter, an ultraviolet disinfection reactor (> 120 mJ/cm^2) and delivered to two drinking water spigots (kitchen sink and bathroom sink). The drinking water system was to be used for drinking, cooking and activities such as tooth brushing where water was to be orally consumed. While drinking water usage is highly variable it was assumed that 2 liters (0.5 gallon) per person per day would be used.

The water reuse system is quite complex, located in a separate treatment building (10-foot long shitting container) and at this current junction, not suitable for installation within a home. This prototype system was operated as if installed in a four person home where 15 gallons of water per person per day for non-potable direct contact uses was available. The water reuse system collected water from the kitchen sink (2.5 gallons per person per day), the bathroom sink (hand washing after toilet use and other hygiene practices), the shower (one 6 minute shower per person per day at 1 gallon per minute), and a combo clothes washer and condensing dryer (1 load per day, about 12 gallons of water). For prototype testing, greywater was synthetically produced according to NSF-350 (NSF, 2014) and a team derived kitchen sink water inclusive of items common to remote Alaska communities (i.e. berries, fish, meat, oil, rice). Additionally, an equivalent of urine was added to the system as if each person urinated in the shower daily during use. While the system produces a water that meets direct potable reuse guidelines (Water Reuse, 2015) it was not intended for consumption. Fixture collected greywater is pumped to an equalization tank (located in the home). The equalization tank allows fixtures to be operated on demand while allowing the downstream settling tank to be filled in a pulse and allowed time for settleable material to settle. Greywater is allowed to settle for at least three hours prior to being decanted into the next process. Under prototype operation, sludge from the settling tank accumulates at a rate of about 1 liter per month. The settled greywater is pumped into a sequencing biological reactor (SBR). The SBR was operated with about 1 g/L of aerobic biomass. To avoid excess foam production, the SBR was intermittently aerated for 4 hours prior to 1 hour of constant aeration. Subsequently, the biomass was allowed to settle in the SBR for about 3 hours prior to decanting through a 1 micron nominal filter into the ultrafilter feed tank. About 1 gallon of sludge was removed from the SBR weekly. This SBR operational procedure was selected to allow an successful operation without daily operation control requirements. Once in the ultrafilter feed tank, the water would be treated through a PW4040F membrane where the filtrate was directed to the reverse osmosis feed tank and the retentate would be directed back to the ultrafilter feed tank. The ultrafilter would operate until its feed tank was reached a minimum set-point and at that time would return this water to the equalization tank. After returning its water, the ultrafilter was rinsed with fully treated water. The reverse osmosis membrane was operated similar to the ultrafilter, sending treated water to the wash water storage tank through high dose ultraviolet disinfection and returning retentate to the reverse osmosis feed tank. Unlike the ultrafilter, once the water in the feed tank was concentrated it was not returned. This concentrate was maintained in the tank for seven days until it was wasted. To maintain wash water quality, the wash water tank was intermittently ozonated using an ultraviolet light based ozone generator. The wash water tank was delivered to the household fixtures using a demand pump.

The described configuration requires that homeowner withdraw 10 gallons per week of concentrated greywater from the system weekly. Further, 10 gallons of a source water of the homeowners choosing needs to be added to the system through a household fixture. Additionally, daily inclusion of a small amount of urine is needed to ensure a highly active biomass. Under these operating conditions, the prototype creates water that has less than 1 mg-C/L, turbidity less than 0.5 NTU, and is well disinfected (12-log virus, >10-log bacteria). Further, this system meets the goals of the project at large with projected costs of less than $160,000 capital cost and $135/month operational.

Portable Alternative Sanitation System

The PASS system was initially implemented eight homes in Kivalina (ANTHC, 2016) in 2015 and subsequently in a number of communities in interior Alaska during the Fall of 2017. The system included a 100 gallon tank where water of the homeowners choosing (i.e. rainwater or washeteria water) is pumped into the tank through screen, a sediment filter, and filter rated to remove 3-log protozoan cysts. Once in the tank, homeowners are advised to disinfect the tank using liquid bleach per a protocol outlined in the owner's manual. Water from the tank is fed by gravity to a low flow wash basin for handwashing. The system also includes a UDDT and urinal. Greywater and urine are disposed of into the subsurface near the home in a very small soak away pit. The monthly operating cost of the system was about $30 per month. Since the report in 2017, ANTHC has learned and adapted their system to meet homeowner use patterns and address system failures. The installations of Fall 2017 and revision to original installation are currently being studied and findings will be reported on soon. With increasing number of installations, ANTHC is looking to further study actual health improvements in homes with a PASS installed in the homes.

CONCLUSION

AWSC and PASS have made notable strides to improve the lives of those in remote Alaska that are currently without access to adequate and well managed water and wastewater. While the current approach of the two projects are notably different, they are both striving for the same underlying goals. The UAA AWSC prototype has a long way to go prior to in-home implementation. To bring the UAA prototype into a household, a full detailed design will need to be performed, advanced programming and operational logic developed, and real-world operational validation performed in with and without willing test subjects (preferably in a location easily accessible by designers). For the PASS system, there was never any notable laboratory testing due to process simplicity. The approach that ANTHC has taken has enabling rapid field adaptation based on operational feedback by users, maintainers, and installers. Through this approach ANTHC has gained significant support of the users, communities and tribal organizations and requests new installations grows greater by the day. ANTHC's next focus will be to capture insights into measurable health benefit of their system versus commonly practiced self-haul techniques.

ACKNOWLEDGEMENTS
The author would like to thank the Alaska Department of Environmental Conservation and the United States Environmental Protection Agency for funding associated with the Alaska Water Sewer Challenge. Further, UAA's progress on the Challenge would not have been possible without the significant student support, most notably Gregory Michaelson, Cara Lucas and Kaitlin Mattos. And the support of its subcontractors Karl Linden of University of Colorado, Greg Jones of GV Jones & Associates and Kurt Vause of StreamlineAM. And finally, the author thanks ANTHC for their willingness to include him in their PASS work during his 2016/2017 sabbatical.

REFERENCES
Alaska Native Tribal Health Consortium (ANTHC). 2016. "Portable Alternative Sanitation System: Final Report | Kivalina, Alaska." Accessed 2018/07/02 at https://anthc.org/wp-content/uploads/2016/01/Kivalina-Report-E-Version.pdf

Department of Environmental Conservation (DEC). 2018. "Alaska Water Sewer Challenge" Accessed 2018/07/02 at https://dec.alaska.gov/water/water-sewer-challenge/

Hennessy, Thomas W. et al. 2008. "The Relationship Between In-Home Water Service and the Risk of Respiratory Tract, Skin and Gastrointestinal Tract Infections Among Rural Alaska Natives." American Journal of Public Health. 98 (11): 2072-78.

National Sanitation Foundation (NSF). 2014. "NSF/ANSI 350 – 2014: On-site Residential and Commercial Water Reuse Treatment Systems.

Thomas T. K., T Ritter, D. Bruden, M. Bruce, K. Byrd, R. Goldberger, J. Dobson, et al. 2016. "Impact of providing in-home water service on the rates of infectious diseases: results from four communities in Western Alaska." Journal of Water and Health 14(1): 132-141.

Water Reuse Foundation. 2015. "Framework for Direct Potable Reuse." Project Number 14-20. ISBN: 978-1-941242-30-8

DRS 2018 Session 09 Supporting Abstracts

AGENDA

Session 09: Residuals, Sanitation and Hygiene: Concerns, Design and Application for Low Technology Processing

Tuesday, July 31
1:30 PM – 4:45 PM

Moderators: Leonard Casson, University of Pittsburgh; Krishna Krishnapurs, Parkson, Inc.; Robert Reimers, Tulane School of Public Health and Tropical Medicine

1:30 PM	**Sampling and Disinfection of Wastewater relative to Parasites and the Role of associated Surrogates, Indicators and Non-Vital Models** Dwight Bowman, Cornell University Dept Of Micro & Imm
2:00 PM	**Effective Treatment of Biosolids and Manures – Multiple Additives/Factors for Disinfection, Stabilization and Product Quality** Robert Reimers, Tulane University/Consultant
2:30 PM	**Low-Cost, Low-Tech Biosolids Treatment via Combined Long-Tern Storage (Lagoon) and Air Drying: A Comparison of Two Pilot-Scale Studies** Jennifer Becker, Michigan Technological University
3:00 PM	**Networking Break**
3:45 PM	**Effectiveness of Tropical Lagoon Treatment for Disinfection of Domestic Septage** Joel Nkiama Konde, Kinshasa School of Public Health, University of Kinshasa
4:15 PM	**Usage of Low-Cost Bio Electro Technology to Convert Class B Biosolids Processes (Aerobic and Anaerobic Digestion) to Class A Disinfected and Stabilized Biosolids** Adam Faschan, Ardurra Group
4:45 PM	**Symposium Adjourns**

Sampling and Disinfection of Wastewater relative to Parasites and the role of associated Surrogates, Indicators, and Non-vital Models.

Angenent LT[1], Bowman DD[2], Butkus M[3], Harroff LA[1], Islam N[1], Labare M[4], Liotta JL[2], and Marcano V[2].
1 Biological an Environmental Engineering, Cornell University
2 Microbiology and Immunology, Cornell University
3 Geography and Environmental Engineering, United States Military Academy
4 Chemistry and Life Sciences, United States Military Academy

Our work over the years has focused on work with *Ascaris suum* of pigs as a surrogate that can be added to wastewater systems at levels that allow the testing of disinfection processes. It has the advantage of being of only minimal risk to humans and having the same physical-chemical properties as the human *Ascaris*, *Ascaris lumbricoides*. Similar work has also focused on using murine Giardia, *Giardia muris*, as a surrogate for human giardiasis , and then using the actual wild-type form of *Cryptosporidium parvum* from calves as an organism to test disinfection processes with various methods for killing cysts and oocysts. It is also hoped tha someday indicator organisms may be identified in sewage sludges that can be

Recently work between our group and the faculty of Biological and Environmental Engineering at Cornell and the Departments of Chemistry and Life Science and Geography and Environmental Engineering at West Point has focused on the treatment of *Ascaris suum* eggs in biosolids with fatty acids. This extends from work Published in 1953 by Dr. Takeyama in the Osaka university Medical School Journal suggesting that short-chain fatty acids have the capability to penetrate and kill parasite eggs in sewage. Using this process we have continued to work towards looking at the levels of fatty acids that can kill eggs in acid digesters and in dry toilets. This work is beginning to produce fruition, and we recently published on methods utilizing the carboxylate platform and chain elongation to produce n-butyric acid and n-caproic acid-via the anaerobic fermentation of human fecal material. *Ascaris* eggs were then deactivated through exposure to these carboxylic acids. Using batch experiments with human fecal material as substrate, we accumulated n-butyric acid and n-caproic acid at total concentrations (uncharged acid plus conjugate base) of 257 and 27.1 mM, respectively. We showed that carboxylic acids at these concentrations inactivated Ascaris eggs when the pH was below the pK(a) for the acids, causing them to exist primarily in the uncharged forms. These results indicated that in biological production of carboxylic acids in human fecal material provides a promising method of pathogen inactivation and may lead to new developments in sanitation technology and treatment of fecal sludge. Results of other past and ongoing studies will also be reported.

Effective Treatment of Biosolids and Manures – Multiple Additive/Factors for Disinfection, Stabilization and Product Quality

Reimers, R.S.[1], Brisolara, B.F.[2], Krishnapurs, K.[3], Smith, J.E.[4], and Rubin, A. R.[5]

[1]Texas A&M University, Houston, Texas
[2] Louisiana State University School of Public Health, New Orleans, Louisiana
[3]Parkson, Inc., Fort Lauderdale, Florida
[4]USEPA (Retired), Cincinnati, Ohio
[5]North Carolina State University, Raleigh, North Carolina

Introduction

Chemicals assist in disinfection by raising/lowering the pH, enhancing exothermal production of energy, oxidation or reduction reactions with the residuals and the application of non-charged disinfectants. The biological processing assists disinfection by autothermic biological activity (increasing the temperature), reduction of the degradable organics (reducing the shielding of pathogens and available nutrients) and production of biocidal agents in the degradation process (fermentation or anaerobic processing). Most disinfection processes combine the physical, chemical and biological factors to optimize residual disinfection. Chemicals are now assisting stabilization processes (ammonia, ferric chloride and chlorine dioxide) along with those addressed in the 503 regulations.

Disinfection

The main physical parameter is the increase in the temperature. Except in heat drying, temperature always acts in concert with other stressors. Processes such thermophilic digestion and advanced alkaline stabilization use either biological activity or quicklime hydrolyzation to produce the elevated temperatures. Studies at Tulane University have been able to isolate the temperature effects through their work with *Ascaris* egg inactivationi[1]. If temperature is the primary stressor, temperatures of greater than 70°C are required or pasteurization, but with the utilization of abiotic chemicals coupled with temperature, inactivation temperature can drop to 52 to 55°C as noted in anaerobic digestion and alkaline treatment. It should be emphasized that closed systems are much more effective. In a closed system the volatile inactivating agent can attain higher concentration as a result of the Henry Gas Law[2,3]. In addition, the usage of electrolytic current has been noted to catalyzed exothermic chemical reactions assisting the biosolids disinfection[4].

Various chemicals have been utilized to assist in the disinfection of biosolids for over fifty years. These chemicals are classified based on the mode of disinfection and stabilization. The following chemicals are utilized in various disinfecting processes as listed in Table 1.

Table 1: Chemicals Used for Disinfection of Biosolids

Alkaline Agents	Acid Trimming Agents	Oxidation Reduction Control Agents	Non-Charged Disinfectants
Lime Cement Kiln Dust Portland Cement Alkaline Fly Ash Silicates Spent Bauxite Hydroxide anions	Sulfuric Acid * Nitric Acid Phosphoric Acid Sulfamic Acid* Sodium Bisulfate *Makes the sludge cake thixotropic.	Ozone Peroxide Ferrate Chlorine Dioxide Chlorite Chlorate Hypochlorite Hypochlorous Acid Peracetic Acid	-Ammonia (Alkaline Treatment) -Amines (Alkaline Treatment/Composting) -Organic Acids, Aldehydes, and Ketones (Anaerobic Digestion and Composting) -Nitrous Acid (Acidic Treatment)

The usage of chemicals is generally a part of most residual disinfection processes. These processes can occur in acidic range pHs less than 3, in the neutral range at pHs between 5 to 9 and in the alkaline range at pHs greater than 10. In most cases, they will be conjunction with physical and biological processes.

Biological processing has been very effective in digestion, composting and storage. In these processes, there is mechanical or autothermal heating. In addition, biocidal inactivation has been observed during lagoon storage. The anaerobic treatment takes approximately 40% less than compared to aerobic processes until temperature reaches around 50 to 55°C due to the biologically produced biocidal byproducts.

In addition, if the solids content of anaerobic biosolids increases, the inactivation rates tended, to increase. . For example, parasites and bacteria tended to die-off 5 times faster with an increase in solids from 4% to 24% while the viruses die-off at 2.5 times greater rates with high solids for the lagoon stored anaerobic digested biosolids from Chicago. Soils tend to inhibit the die-off of parasites and viruses by 3 to 5 times in non-tilled or lagoon-stored biosolids as shown in Table 3[5]. The rationale for these factors is the combination of biological activity, temperature and produced biocidal agents. These agents are probably ammines, ammonia and other non-charged species, such as organic acids, aldehydes and alcohols, probably effect process inactivation, but data, has not appropriately been developed, though, Japanese work has suggested potential inactivation. [6,7] .

Almost all processes developed to inactivate pathogens utilize a combination of physical, chemical and biological stressors. For example, Table 2 illustrates the combination of these agents. This Table notes the importance of the multistressors in the processing of municipal sludges and residuals. There is more investigations related to integrating these stressor in the process design of disinfections. Most of these processes have 3 to 4 stressors involved in affecting the pathogens.

Table 2: Parameters of Influence in Pathogen Inactivation in Biosolids

Biosolids' Disinfection Process (Number of Stressors)	Irradiation	Temperature	Solids Content	NH_3	Organic By-Products	Desiccants
Composting (3)	-	+	-	±	+	-
Anaerobic Digestion (3)	-	+	+	-	+	-
Aerobic Digestion (2)	-	+	+	-	-	-
Lagoon Storage (3)	-	+	+	-	+	-
Air Drying (4)	+	+	+	-	-	+
Alkaline Stabilization (4)	-	+	+	+	-	+
Gamma/Beta (1)	+	-	-	-	-	-

In summary, disinfection process utilize many physical, chemical and biological stressor works in concert to disinfect pathogens in biosolids and manures. By optimizing these factors, the cost of these processes can be reduced substantially. The anaerobic digestion, composting, air drying and alkaline stabilization processes are effected by at least three to four of these stressors. Currently, there is a great deal of process development in this area, but until we understand how to integrate these factors and explain how the mechanisms work, there will be a lot of skepticism by both the public and regulatory agencies.

Vector Attraction Reduction and Stabilization

In the 503 regulations (40 CFR Parts 257 and 503), the stabilization of municipal sludge is ascertained for the long term by either the Saturated Oxygen Uptake Rate (SOUR) test or the volatile suspended solids (VSS) reduction. The short-term stabilization of alkaline processes is based on high alkaline dosages or solids content. In the future, the content of ferric chloride, chlorine dioxide or CO_2 release will be incorporated[9,10]. When looking at carbon dioxide respiration and oxygen uptake rates, we will most likely base our procedure on the current carbon dioxide respiration regarding compost stability [11,12]. Our results may be evaluated using the Compost Stability Index, which is based on the level of microbial activity in a sample by monitoring respiration. This index will be similar for the SOUR test and will require a similar acclimatization step. See Table 3 below:

Table 3: Compost Stability Index [11,12]

RESPIRATION RATE (mg CO2-C/g compost carbon day)	**RATING**	**CHARACTERISTICS**
<2	**Very stable**	▪ Well cured compost ▪ No odors ▪ No continued decomposition
2 – 5	Stable	▪ Cured compost ▪ Limited odor potential ▪ Minimal impact on soil carbon and nitrogen dynamics
5 – 10	Moderately Unstable	▪ Uncured compost ▪ Minimal odor production ▪ Addition to soil may result in nitrogen immobilization ▪ High phytotoxicity potential ▪ Not recommended for growing plants from seed
10 – 20	Unstable compost	▪ Very immature compost ▪ High odor and phytotoxicity potential ▪ Not recommended for growing plants from seed
>20	Unstabilized material	▪ Extremely unstable material ▪ Very high odor and phytotoxicity potential ▪ Not recommended for use

In the last two years, Water Environment Research Foundation Study has looked at assessment of acceptable exceptional quality biosolids and has developed a stability model to assess potential odor. The model looked at the variables of total volatile solids, ammonia, total iron, total aluminum, strontium, methionine and SOUR. This model could predict the stability of a sludge cake[13].

Product Development

An area that requires more site/region specific research with fertilizer/soil amender usage of biosolids is the nature of crop yields, crop qualities, and nutrient values. Significant data has been gathered over the past decade on this topic including more recent work] found that biosolids application improved grain yield and protein content, and also resulted in lower bulk density, greater total carbon, mineralizable nitrogen, and extractable phosphorus[14]. Soil organic matter has also been analyzed after long-term biosolids application (1972 – 2004) and was shown to resist decomposition more than with traditional fertilizers indicating an increase in stability which contributes to soil C sequestration[15]. They attribute this increase in stability to the increase in amorphous Fe and Al (2 – 7 fold increase with biosolids application[15]. Preliminary data indicates that usage of biosolids could have as much as impact as biotechnology on agriculture production.

The aim of this paper is to explore areas where products can generate enough profit to absorb the cost to municipalities or farms and still yield a fair amount of income [50% profit or more]. As shown in Table 2, market values of soil and soil amender products range from 3 to 70 dollars per dry ton. On the other hand, cost of sludge treatment has fluctuated over the past 20 years. Therefore, processing cost can be 10 to 100 times that of the final products value. This type of rationale is not viable for small and intermediate sized municipalities or farms. As a result of this concern, there are 3 potential markets of value-added fertilizers:

1. Engineered soils [immobilize heavy metals such as lead, cadmium nickel, etc.],
2. Organically enhanced agricultural and/or ornamental horticultural fertilizers and
3. Turf grass

Of particular importance to this process is the market analysis and assessment of market variety. The market assessment goes well beyond agricultural land application and includes use on specialty crop, utilization in horticulture and high quality turf, horticultural uses in nursery and landscape operations including manufactured soil, use in storm water BMP operations such as bio-retention cells, phytoremediation projects, utilization in solid waste facilities as daily, interim and final cover for low temperature anaerobic reactors, and utilization as structural fill for buildings and roads and many more potential uses. For products used in traditional agricultural, silvicultural and horticultural activities consulting the local USDA Census of Agriculture and local USDA NRCS soil surveys are critical resources for market identification. The development of a commercial fertilizer denotes high percentages of nitrogen, phosphorous, potassium, and other elements, and must be added to an existing process such ammonia sulfate fertilizer process or alkaline process. the percentages are shown in Table 4.

Local users are key to a beneficial use plan and assessment of their needs is critical to a resource recovery effort. Ideally markets are local, say within 50 to 100 miles of a resource recovery facility. So what might professionals consider initially in development of a beneficial use plan?

1. What are local agricultural, silvicultural, and horticultural markets for a biosolids derived product?
2. What are the product qualities required by those various markets and can the biosolids material (and potentially other local materials) be commingled to produce those value added materials?
3. Does the generator have samples of product that will be produced or have access to product from potential sister operations, are senses involved? Can potential users see, smell, touch and evaluate nutrient levels in product
4. Will the product be supplied in bulk loads, 0.5 ton to 1 ton super sacks, or bags? What does the user community desire?
5. Has the generator developed GIS or GOOGLE maps showing potential users? Are these general agriculture or specialty crop operations with special product quality needs? [15]

Table 4: Mineral Nutrients (Percent Dry Weight) related to Biosolids

Parameter	Percent Dry Weight	Total Fertilizer Value
Total Nitrogen as N	3.5 to 6.4	8.0
Organic Nitrogen as N	2.0 to 4.5	8.0
Phosphorus as P	0.8 to 3.9	3.5
Phosphorus as P_2O_3	1.8 to 8,7	8.0
Potassium	0.2 to 0.7	5.6
Potassium as K_2O	2.2 to 8.8	8.0

The reason that the application of biosolids has been due to the following reasons. The biosolids contains micronutrients vitamins and complex organic carbon. It enhances the mineral availability and uptake by plants. The organics in the biosolids adsorbs water thereby reducing water requirement and improves soil drainage properties. The nitrogen and phosphorus are organically bound, which causes slow release of nitrogen and phosphorus. The treated biosolids contain plant disease biocidal organics and byproducts after treatment. The cost of biosolids products are elucidate in Table 5[16].

Table 5: Potential Value of Fertilizers containing Biosolids

Agriculture Applications	Cost per Dry Ton ($)	Comments
Fill Cover	3 to 15	Used in Landfills/Construction Applications
Top Soil	10 to 30	Biosolids Blended with Inorganic Ingredients
Soil Amenders	20 to 70	Utilizing Disinfected Biosolids
Heat Dried Biosolids (HDB)	40 to 120	The HDB produces a Low Level Fertilizer
Milorganite	100 to 150	HDB refined over the last 50 years
Enhanced Organic Fertilizers	400 to 600	Utilizes Biosolids with Inorganic Fertilizer Production

References

1. Mbela, K.K., "Evaluation of Temperature Effects on Inactivation of *Ascaris* eggs in both Aerobic and Anaerobic Digestion Processes," Master's Thesis, Tulane University, New Orleans, Louisiana (September 1989).
2. Fitzmorris, K.B., Reimers, R.S., Oleszkiewicz, J.A., and M.D. Little, 2007. "Decrease of time for pathogen inactivation in alkaline disinfection systems using pressure." Vol. 79, Water Environment Research, Alexandria, Virginia, pp. 388-395.

3. Reimers, R.S., Fitzmorris, K.B., Smith, J.E., Boyd, G.R., and Bowman, D. D., 2004. "State of Art in Treatment and Survival of Pathogens in Biosolids," Journal of Residuals Science and Technology, Volume 1, Number 2 , DEStech Publications Inc., Lancaster, Pennsylvania, pp. 93-104.
4. Elektorowicz, M., Safael, E., Reimers, R. and Oleszkiewicz. 2014. "Controlled Pasteurization, Disinfection and Sterilization of S by BioElectro Process," Water Supply Quality, PZITS, Torun, Poland.
5. Pillai, S.D.and Reimers, R.S., 2010. "Disinfecting and Stabilizing Biosolids using E-bean and Chemical Oxidants, WERF Report U4R06, Water Environment Federation, Alexandria, Virginia, 303pp.
6. Reimers, R.S., Bowman, D.D., Schafer, P.L., Tata, P., Leftwich, B.D. and Atique, M.M., 2001."Factors Affecting Lagoon Storage Disinfection of Biosolids," Proceedings of Joint WEF/AWWA/CWEA Specialty Conference entitled "Biosolids 2001", Water Environmental Federation, Alexandria, VA, CD-ROM
7. Yoshida, S., "Resistance of Ascaris Eggs," 1920.J. Parasitology, Vol. 6, pp. 132-139.
8. Takeyama, O., 1951. "Ovicidal Influence of Several Substances Found in Feces and Urine, Especially of Those Contained in Night-Soil," (in Japanese) Osaka Daigaku Igaku Zasshi, Vol. 2 (1):27-37 (Cited by Morishita, K. 1972), "Studies on Epidemiological Aspects of Ascariasis in Japan
9. Xu, Y., Reimers, R.S. and Englande, A.J.,2013. "Development of Heat Dried Biosolids Product for Application in the Wetlands," WEFTEC Chicago, Alexandria, Virginia CD-ROM641
10. Reimers, R.S., Pratt-Ward, L.S., Bowman, D.D. and Oleszkiewicz, J.A., 2007. "Usage of Nitrous Acid and Chlorine Dioxide in Residuals' Disinfection and Stabilization," International Water Association Specialty Conference Proceedings Moving Forward Wastewater Biosolids Sustainability: Technical, Management and Public Synergy, International Water Association, Greater Moncton Sewerage Commission, Moncton, New Brunswick, Canada. pp. 263-268
11. Epstein, E. and Kobnke, H. 1957. "Soil Aeration as affected by Organic Matter Application." Soil Sci. Soc. Amer. Proc. 21(6): 585-588.
12. Barha, R. and Pramer, 1965. "Features of a Flask Method for Mearsuringthe Persistence and Biological Effects of Pesticides in the Soil." Soil Sci.(100):68-70
13. Taylor, M. 2017. "Quantification and Predtive Modeling of Biosolids Odors." WEFTEC Proceedings, Alexcandria VA, CDROM
14. Cogger, C.; Bary, A.; Kennedy, A.; Fortuna, A., 2013." Long–Term Crop and Soil Response to Biosolids Applications in Dryland Wheat". J. Environ. Qual., 42, 1872–1880.
15. Rubin, A.R. and Thomas Cross, P.E.; Developing a Sustainable Biosolids Management Plan for Mt. Dora, Fl. Presentation to NC AWWA WEA Meeting, 2012
16. Fitzmorris-Brisolara, K.B., Reimers, R.S., Smith, J.E., Rubin, A.R. , Parry, D.L., Donovan, J.F., Faschan, A., Wanstrom, C. and Zappi, P. (2016). "Management Protocols for Development of Value-Added Products," WEF Residuals and Biosolids Specialty Conference, Milwaukee, Wisconsin.

Low-cost, low-tech biosolids treatment via combined long-term storage and air drying: A comparison of two pilot-scale studies

Jennifer G. Becker[1], Karina Eyre[1], Tanner Keyzers[1], Eric A. Seagren[1]

1. Michigan Technological University, Houghton, MI, U.S.A.

Introduction: Land applied biosolids in the U.S. are classified as Class B biosolids, which still have detectable levels of pathogens and indicator organisms (PIOs) present, and can only be applied to land that will not be used for production of food, or Class A biosolids, which must essentially contain below detectable concentrations of PIOs and can be used for all land applications. Water resource recovery facilities (WRRFs) producing Class B biosolids face growing challenges including increased public scrutiny, lack of resources, and increasing energy costs. Accordingly, many WRRFs across the U.S. are seeking to upgrade to Class A biosolids production, which opens additional avenues for land application and distribution of biosolids. Unfortunately, although some WRRFs are able to implement the generally high-cost, high-tech treatment systems currently accepted for producing Class A biosolids, the vast majority of U.S. WRRFs serve small communities that lack the resources to implement these systems. As a result, there is a growing need for low-cost, low-tech (LCLT) treatment options for Class A biosolids production. An example of a LCLT treatment option that has been successfully implemented is the combined high-solids long-term storage and air-drying process used at the Metropolitan Water Reclamation District of Greater Chicago. However, widespread adoption and regulatory approval of such LCLT methods has been stymied by a lack of information on the impacts of environmental factors and biosolids properties on PIO inactivation during LCLT treatment.

Objectives: The overall goal of this multi-phase project is to develop a rational and universal approach for the design of LCLT Class A biosolids treatment processes. As part of the effort to achieve this goal, pilot-scale studies have been performed at two collaborating WRRFs, the Portage Lake Water and Sewage Authority (PLWSA) in Houghton, MI, and the Gogebic-Iron Wastewater Authority (GIWA) in Ironwood, MI. During these studies, fundamental information on the impact of key process parameters on the inactivation of PIOs is being collected during high-solids, long-term storage (Stage 1) followed by air drying (Stage 2). This presentation focuses on comparing the key process parameters and PIO inactivation during the two treatment stages at the two plants. Collection of such data is important because if LCLT processes are to achieve wide spread application, the results must be reproducible and predictable.

Methods: PLWSA and GIWA are similarly sized WRRFs (design flow ~3-3.5 MGD) that treat their solid residuals using anaerobic digestion followed by polymer addition and dewatering via belt-filter press. GIWA and PLWSA store their biosolids in uninsulated storage sheds during winter months. To examine the effect of indoor versus outdoor storage, Stage 1 of this pilot scale study incorporates two treatments performed in triplicate at each site, with pilot test beds located inside an uninsulated storage shed in Treatment 1, and outside in Treatment 2. For each treatment, two sets of triplicate test beds were constructed (4 ft. high, 8 ft. long, and 4 ft. wide), with biosolids initially placed in 3 ft layers. Fecal coliforms (FC), enteric virus, and helminth ova abundance data are needed to establish that the biosolids can be characterized as Class A.

Specifically, to establish PFRP equivalency, a 3-log reduction in enteric viruses, and a 2-log reduction in helminth ova must be demonstrated. Therefore, the boxes were equipped with sampling stacks that held sentinel chamber devices containing biosolids inoculated with *Ascaris* ova and human poliovirus. *Ascaris* ova sentinel chambers were also buried in suet feeders in each box. The sentinel chambers allow for exchange of gases and liquids, but prevent the release of the pathogens into the environment. Ambient weather data and in-situ measurements of biosolids temperature were collected continuously. The biosolids were also sampled at regular intervals to determine moisture content (as total solids (TS)), organic matter (as volatile solids (VS)), NH_3 content, volatile fatty acids (VFAs), alkalinity and pH. Sampling was halted over the winter when the biosolids froze, and was resumed when they thawed in spring. In addition, molecular and culture based methods were used to quantify FC, coliphage, enteric virus, and helminth ova abundance. After approximately one year of storage in the test beds at PLWSA, the biosolids were moved to an outdoor air-drying operation. The indoor and outdoor samples were composited respectively, and were placed on filter press belts, which allowed water to drain into the soil below. Temperature sensors and the *Ascaris* ova and virus containers were all transferred to the air-drying operation. The biosolids were shaped into windrows approximately 18-20 inches high and were turned twice a week. The air-drying period at PLWSA lasted for a total of 9 weeks from late August through the end of October. At GIWA, the transition to the air-drying stage will occur after approximately two years of high-solids long-term storage, to examine the impact of one to two years storage time on PIO reduction during high solids, long-term storage.

Results: Importantly, the overall data trends from the pilot-test for the two facilities are very similar, indicating the potential for good reproducibility of results. Nevertheless, there were some interesting differences as well. For example, during Stage 1, biosolids stored in the test beds at both plants initially exhibited a decrease in pH over time through the fall. However, while the pH decrease at PLWSA continued slowly and plateaued at around pH 7 after winter, the rate of decrease accelerated at GIWA over the spring, with the outside boxes dropping below pH 5, and the inside boxes dropping below 6. Similarly, biosolids stored at both plants experienced an increase in TS over time. However, while the TS at PLWSA reached 25-30%TS and seemed to plateau there after a year, at GIWA the TS continues to slowly trend upward, with all boxes at >40%TS. Interestingly, at both WRRFs, VS continued to trend slowly down over the course of a year, decreasing roughly 5 to 10% since the beginning of the study. During the fall, ammonia levels in Treatments 1 and 2 tended to stay roughly the same, or increase, likely due to ammonification. However, during the spring and summer, the ammonia levels in the outside boxes at PLWSA and GIWA decreased, while increasing in the inside boxes. Decreasing ammonia outside could be tied to nitrification; for the inside boxes, the increasing ammonia could be associated with continued ammonification. Trends in PIOs were similar between the two WRRFs, with a reduction of FCs to below Class A levels (< 1000 MPN/g TS) after overwinter storage. Nevertheless, the environmental conditions after one year of storage were not adequate to cause the change in spiked viable *Ascaris* ova required for Class A biosolids. The Stage 2 results will be compared when the GIWA data are available in early summer 2018.

Conclusions: Parallel pilot-scale studies simulating LCLT biosolids treatment via combined long-term storage and air drying are being performed over one to two years at two similar WRRFs. Trends to-date in key environmental conditions and physical/chemical parameters, as well as PIOs, are generally very similar between the two test locations. Importantly, these data demonstrate the potential for obtaining reproducible and predictable outcomes using LCLT treatment processes in different locations.

Effectiveness of Tropical Lagoon Treatment for the Disinfection of Domestic Septage

Konde NN[1], Brisolara KF[2], Kiyombo MK[1], Reimers RS[3], England AJ[3], and Bowman DD[4]

1. Department of Environmental Health Sciences, University of Kinshasa School of Public Health.
2. Department Environmental and Occupational Health Sciences, LSU School of Public Health
3. Department of Global Environmental Health Sciences, Tulane University SPHTM
4. Department of Microbiology and Immunology, Cornell University

In this work conducted under wet tropical climate conditions, septage from septic tanks was stored into lagoons to pursue pathogens inactivation of to the safe reuse of the resulting biosolids without threat to the public health and the environment.

Four lagoons with square section of 4m and 2m, respectively at the top and the bottom, and with 3m depth each, were designed in pairs corresponding to total solids contents, less than 5% versus 8% to 12%, and their exposition or not to rainfall. They were located at 4.4180°S and 15.3065°E in a wet tropical climate. The total annual rainfall on the site of this experiment is 1.77 m with daily average temperature of 26.2°C (range: 22°C - 30°C) and an average relative humidity of the air of 74.8%. Lagoons were filled from June 2013 to August 2014 by vacuum trucks emptying septic tanks. Three to five cycles of filling, settling of the solids, discarding the upper liquid fraction of the septage, and refilling until the desired solids content level was reached. Within each pair of solids content level, one lagoon was provided with a plastic sheet as protection in case of rain event. Along with local climate characteristics, abiotic parameters of the stored fecal sludge were also measured. The die-off of pathogens was monitored using coliforms bacteria and *Ascaris* eggs as indicator organisms. Sampling events were conducted from September 2014 to September 2015, with one sampling event per month. At each sampling event, two sentinel chambers containing *Ascaris* eggs were withdrawn from each lagoon as well a composite sample of the stored sludge. Pathogen survival was monitored through total and fecal coliforms densities as well as *Ascaris* ova counts and viability.

Statistical analyses did not reveal any difference between the four lagoons in terms of measured abiotic parameters except for the total solids and volatile solids contents ($p=0.000$). Within pairs of lagoons, there was not any difference for all abiotic parameters within "solids contents" level but a trend of difference was observed for the volatile solids for the lagoons with solids contents ranging from 8% to 12% ($p=0.090$).

The two lagoons in the high solids content level (8% to 12%) had more than 3 log reduction for the density of *Ascaris* eggs while in those with a low solids content (less than 5%), the inactivation observed was between 1and 2 log reduction, at the end of this experimental study. As for total coliforms organisms, there was a 5 log reduction in the high solids content lagoons after 8.7 months while in those with low solids content, 4 log reductions occurred near the end of the study. Four log reductions of fecal coliforms organisms were observed after 5.3 and 9.3 months of fecal sludge storage in high and low solids content lagoons, respectively.

The results of this study suggest that tropical lagoon storage with at least ten percent solids content, can be an effective option for the disinfection of fecal sludge, provided the availability of the land. However, when the solids content of the stored fecal sludge is less than this value, lagoon treatment must be combined with a further treatment such as drying beds or co-composting with organic domestic solid wastes for effective control of pathogens.

Key words: Tropical wet climate, lagoon storage, septage fecal sludge disinfection, solids content, inactivation of *Ascaris* eggs.

Table 1: Summary statistics of abiotic parameters of the stored septage by lagoon

Parameter	Descriptive Statistics	Lagoons with high solids content		Lagoons with low solids content	
		Protected from rain (L101) n = 14	Unprotected from rain (L102) n = 14	Protected from rain (L201) n = 14	Unprotected from rain (L202) n = 14
Temperature (°C)	Maximum	30.7	31.5	31.6	31.5
	Median	27.2	28.1	28.7	28.7
	Minimum	22.8	23.7	23.7	24.5
pH	Maximum	7.6	7.4	7.5	7.4
	Median	6.8	6.9	6.8	6.9
	Minimum	6.6	6.7	6.6	6.7
Conductivity (µmS/m)	Maximum	3265.0	3320.0	2965.2	2970.0
	Median	2063.0	1831.0	1896.3	1792.5
	Minimum	918.5	958.0	951.5	850.5
ORP (mV)	Maximum	-35.0	-60.0	-39.0	-59.5
	Median	-182.5	-171.0	-183.0	-161.8
	Minimum	-310.0	-303.0	-305.4	-313.0
Total solids (% of dry weight)	Maximum	10.2	9.3	4.2	4.2
	Median	8.3	7.9	3.2	3.1
	Minimum	7.6	7.0	2.5	2.1
Volatile solids (% of dry weight)	Maximum	8.6	7.7	3.6	3.9
	Median	7.0	6.6	2.7	3.0
	Minimum	5.6	4.9	1.4	1.6
NH_4^+ (mg/L)	Maximum	0.36	0.30	0.25	0.23
	Median	0.10	0.09	0.08	0.08
	Minimum	0.03	0.05	0.04	0.03

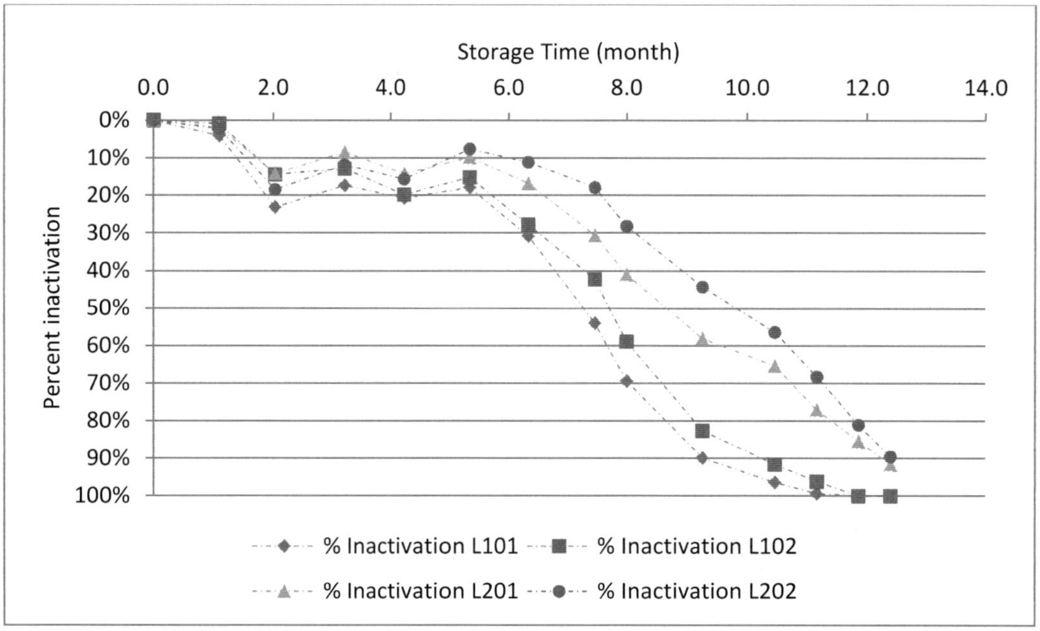

Figure 1: Inactivation of *Ascaris* eggs over the storage time in month by lagoon

Usage of Low-Cost Bio Electro Technology to Convert Class B Biosolids Processes (aerobic and anaerobic digestion) to Class A Disinfected and Stabilized Biosolids

Maria Elektorowicz[1], Adam Faschan[2], , Robert S. Reimers[3], Paul Zappi[4] and Jan A. Oleszkiewicz[5]

[1]Department of Building, Civil and Environmental Engineering, Concordia University, Montreal, Canada
[1]Ardurra Group - Engineering and Consulting, Metairie, Louisiana
[3]Department of Global Environmental Health Science, Tulane University, New Orleans, Louisiana
[4]Wastewater Operations Division, Department of Public Works and Engineering, Houston, Texas
[5]Department of Civil Engineering, University of Manitoba, Winnipeg, MB, R3T 2N2, Canada

ABSTRACT

Novel BioElectro process uses enhanced electrokinetic phenomena to produce exothermic reactions for pathogens inactivation. The technology was demonstrated to achieve various levels of disinfection of biosolids. Pasteurization and even sterilization of anaerobically digested biosolids were demonstrated. The research showed reduction of *Clostridium perfringens* spores by 8 log within less than two hours, while 3 log reduction was reached within less an hour. Simultaneously, properties of the biosolids were changed leading to improved dewatering. The new process is ready for full scale application. One application can be for pre-treatment before ambient temperature aerobic digestion which disinfects the solids and facilitates thermophilic digestion, thus generating a Class A finished product.

The cost of ensuring that aerobically-digested solids meet Class A standards is often prohibitive—particularly at small to mid-size facilities where there is a limited resale market. Subsequently, many municipalities are searching for alternative technologies that are efficient and cost-effective. The BioElectro process provides a potential solution to more energy intensive options, since it generates exothermic reactions in biosolids containing ammonia due to combination of a low voltage DC field with additives. Preliminary studies have shown that temperature could rise to 50 °C and 70 °C within less than half an hour to an hour, respectively, using BioxyS™ and a voltage gradient below 5 V/cm. The heat generated in the BioElectro process facilitates subsequent thermophilic digestion—decreasing the time needed for biosolids stabilization and ensuring biosolids disinfection.

BioElectro provides the potential to retrofit existing aerobic digesters to achieve a Class A biosolids through disinfection and thermophilic digestion, without increasing the existing digestion volume. Bench-scale studies have shown that the process would require small batch or continuous flow tank (due to short retention time) prior to existing digesters, with a system of electrodes supplied with a low DC voltage gradient. The existing WAS feed can be diverted prior to digestion or a portion of existing digesters can be utilized. The process can be cost effective for smaller utilities because of the low construction costs, the potential to reduce existing digester volumes, and changes to sludge properties that improve dewatering and reduce operational costs. This technology was reviewed through the LIFT Technology Scan program and received favorable comments.

Accounting for Future Potable Reuse when Implementing Disinfection Solutions Today

June Leng
Jacobs, Sacramento, CA
Email: june.leng@jacobs.com

INTRODUCTION

To address increasing water shortages in California and a few other southern states in the United States, the current trend in water reuse is moving from non-potable reuse towards direct or indirect potable reuse to maximize the potential of all available water resources. Significant efforts have been made among water agencies and the water industry to define the guidelines and criteria needed for potable reuse in order to address advanced treatment of both microbiological and chemical contaminants. Reduction of chemical contaminants of concern, such as N-Nitrosodimethylamine (NDMA) and 1, 4–Dioxane, often requires the treatment train to include an advanced oxidation process (AOP). One of the AOP technologies has demonstrated its effectiveness involving UV in combination with addition of hydrogen peroxide, ozone, or more recently proposed hypochlorite, to destroy persistent chemical contaminants. A number of UV-AOP processes are currently in operation in Southern California for NDMA or 1,4-Dioxane reduction.

UV technologies have been widely applied in disinfection in compliance with non-potable reuse criteria established by water authorities, e.g., the California Water Recycling Criteria (California Code of Regulations, Title 22). With the increasing interest in potable reuse, it is a natural reaction for most water recycling plants using UV systems to explore the option to implement UV-AOP as part of their advanced treatment process for potable reuse. However, the typical UV dose requirements for non-potable reuse disinfections are far less than what are required in a UV-AOP process, i.e., less than 100 mJ/cm^2 versus hundreds or often over 1,000 mJ/cm^2 for UV-AOP. This poses significant challenges to the conversion of a UV facility from UV-disinfection to UV-AOP. To address these challenges, a study was performed to identify issues such as regulatory compliance and technology applicability. The study also provides a potential implementation approach including facility needs, costs and a sensitivity analysis.

OBJECTIVES AND APPROACH

The purpose of the study is to provide current and future UV owners a potential pathway of UV implementation for potable reuse. In particular, the study focuses on future owner of UV systems that are being designed for UV-disinfection for non-potable reuse to achieve the treatment goal for UV-AOP for potable reuse with robustness, reliability and operational flexibility.

The objective of the study is threefold: (1) Identify the current regulatory requirements of non-potable reuse vs. potable reuse; and define the potential impact on UV system conversion. 2) Identify UV technologies for implementation and future conversion which are applicable, cost effective, and sustainable. 3) Present estimated costs in a range of UV system capacities comparing UV-disinfection for non-potable reuse and UV-AOP for potable reuse; provide implementation costs (per mgd) for UV disinfection system in four selected disinfection capacities or recycling flows that are typical reuse flow projections.

The initial step of the study was to collect and summarize the available and anticipated treatment requirements for UV in non-potable and potable reuse, including indirect and direct potable reuse. In addition, a UV technology that was most applicable for UV-disinfection to UV-AOP conversion was selected for sensitivity analysis. Findings from the initial steps of the study are described in the following sections.

NON-POTABLE REUSE vs. POTABLE REUSE

UV technology, when employed as part of the treatment processes for water recycling, has different treatment objective or treatment target in non-potable reuse and potable reuse applications.

In non-potable reuse application, UV is the disinfection process for reduction of microbial containments, or pathogen inactivation. UV light prevents microorganisms from reproducing and infecting by photochemically damaging its nucleic acids. UV disinfection typically follows tertiary filtration or membrane processes. Delineated in California Title 22 non-restricted reuse criteria, the water reuse requirements pertaining to the disinfection process include bacteria and virus counts limitations: *"Disinfected by a disinfection process that has been demonstrated to inactivate and/or remove 99.999 percent of the plaque forming units of F-specific bacteriophage MS2, or polio virus in the wastewater."* Disinfection must meet the following criteria: *"the median concentration of total coliform bacteria measured in the disinfected effluent does not exceed an MPN of 2.2 per 100 milliliters utilizing the bacteriological results of the last seven days for which analyses have been completed."* UV design doses are well defined for achieving the regulatory requirements for non-potable reuse. The UV design guidelines set forth by National Water Research Institute (NWRI) are widely adopted for UV disinfection system design. Pilot demonstration is not required. The NWRI Guideline also provides testing protocol for performance validation of installed UV systems.

In potable reuse application, UV is the advanced oxidation process (AOP) for reduction of chemical containments, or micro-pollutants. The UV-AOP mechanisms are photolysis and hydroxyl radical generation, which involves chemicals as source of hydroxyl radicals in addition to high energy input. UV-AOP is one of the three typical treatment technologies of Full Advanced Treatment (FAT) following microfiltration and reverse osmosis (RO). In addition to microbial contaminant criteria, California Title 22 indirect potable reuse criteria include specific requirements for the oxidation process in FAT. An oxidation process such as UV-AOP has to demonstrate consistent destruction of a bundle of selected indicator chemicals or *"provide no less than 0.5-log reduction of 1,4-dioxane."* Site-specific bench or pilot studies are required to determine UV doses to achieve the half log 1,4-dioxane reduction. UV doses of UV-AOP to achieve the 0.5-log 1,4-dioxane reduction have been shown in the realm of UV photolysis, which is approximately an order of magnitude higher than what is required for UV-disinfection. Under UV photolysis, pathogen reduction becomes a side benefit. California Title 22 limits the pathogen reduction benefit from UV-AOP. For each regulated pathogen (*enteric virus, Giardia cyst*, or *Cryptosporidium oocyst*), FAT is credited for maximum 6-log reduction credit.

Comparison of UV-disinfection for non-potable reuse and UV-AOP for potable reuse in terms of treatment objectives, testing requirements and energy requirements is summarized in Table 1.

Table 1. UV-Disinfection vs. UV-AOP: Treatment Objectives, Energy and Testing Requirements

	UV-DISINFECTION	UV-AOP
Treatment Target	Pathogens	Chemical contaminants
UV Dose	Usually defined for log reduction credits or numerical target	No set dose target, but relies on reaction kinetics. Site-specific bench or pilot studies to determine dose
Off-site Testing	3rd party validations	Not required and not a common practice
On-site Testing	Spot-check bioassay if non-potable reuse (only in California)	Pilot demonstration and full-scale performance usually required
Energy Requirement	Relatively less ambiguous due to defined target dose	Variable due to target contaminant, oxidant concentration, UV lamps (low pressure or medium pressure), and water quality (competitor scavengers, e.g., Nitrate)

UV TECHNOLOGY APPLICABILITY

Current UV installations for non-potable reuse applications are mainly based on two major UV configurations: an open-channel UV system and a closed-vessel UV system. Open channel UV systems are common in non-potable reuse applications, particularly with upstream tertiary treatment processes using media filtration. Closed-vessel UV systems are commonly installed downstream of membrane filtration, such as a membrane bioreactor in non-potable reuse applications. Both systems can achieve the disinfection goal for the current design as well as have the potential to be expanded to provide the higher UV doses needed for UV-AOP. However, membrane technologies that have been employed in indirect or direct potable reuse treatment trains, such as microfiltration and reverse osmosis, typically produce pressurized effluent. As such, using closed-vessel UV systems in the UV-AOP process downstream is more appropriate to handle the pressurized effluent.

The closed-vessel UV system is equipped with either low pressure high output (LPHO) or median pressure (MP) lamps. The closed-vessel UV system has the capability of being upgraded and expanded from regular disinfection to a UV-AOP reactor. UV manufacturers offer the closed-vessel UV system for both disinfection and UV-AOP applications. Though modifications may be needed, the UV systems to be installed in the current non-potable reuse design can be employed within its lifetime in the UV-AOP treatment process for future potable reuse treatment applications.

Table 2 summarizes the comparison of UV technology application in UV-disinfection and UV-AOP in terms of its status of application in water recycling, lamp technologies and reactor configurations.

Table 2. UV-Disinfection vs. UV-AOP: Technology Applicability

	UV-DISINFECTION	UV-AOP
Low Pressure High Output Lamps	Common	Common
Medium Pressure Lamps	Common, smaller footprint but relatively higher energy demand	Common, smaller footprint and used for seasonal contaminant treatment
Open-Channel	Common, but mostly for non-portable reuse or secondary effluent	Less common to accommodate pressurized upstream flow
Closed-Vessel	Common for drinking water and non-portable reuse applications	Dominating

Closed-vessel UV system equipped with low pressure high output UV lamps is used as the UV technology in conversion sensitivity analysis.

SENSITIVITY ANALYSIS

As a part of the study, a sensitivity analysis was performed to evaluate the impact of the UV-disinfection to UV-AOP conversion in terms of facility requirements and costs. The sensitivity analysis included four selected capacity or flow scenarios that are in the capacity range of typical water recycling facilities.

UV systems with various flow capacities in the evaluation are developed based on the treatment target and design criteria. For UV-disinfection, the UV system receives membrane filtration effluent and provides disinfection prior to the designated non-potable uses. For UV-AOP, the UV system receives RO permeate and provides oxidation of chemical contaminants prior to the designated indirect or direct potable uses.

In both non-potable and potable reuse applications, major components of the UV system include lamps and reactors. All UV facilities in the evaluation are sized based on a closed-vessel UV system equipped with low pressure high output amalgam lamps.

Disinfection with UV is achieved with two functional parameters: UV dose and UVT, which represents recycling flow water quality. The required dose value for non-potable reuse applications is well defined via bioassay testing. In this evaluation, a UV dose of 80 mJ/cm^2 and a minimum UVT of 65% are designated for membrane filtration effluent based on NWRI UV Disinfection Guidelines for Water Reuse (2012).

Electrical Energy per Order of contaminant removed (EEO) is used in the sizing of and control of UV-AOP systems. EEO is an empirical function representing various design conditions that potentially impact the oxidation process, such as lamp output, flow rate, and contaminant log reduction kinetics. In this evaluation, an EEO of 0.3 kWh/1,000gal/order is used for UV system sizing. This EEO number is within the range that have been proven effective and reliable for 0.5-log 1,4-dioxane reduction in a number of pilot and full-scale operations. In addition, minimum UVT of 95%, which is typical for microfiltration and RO permeate, is used for UV system sizing.

Evaluation criteria used as the basis for UV facility sizing and cost estimates are summarized in Table 3. These design criteria served as the basis for UV equipment sizing, facility layouts, and O&M cost estimates.

Table 3. UV-Disinfection vs. UV-AOP: Evaluation Criteria for Sensitivity Analysis

	UV-DISINFECTION	**UV-AOP**
UV SYSTEM CAPACITY	Assumed for the purpose of the sensitivity analysis - 2 MGD; 4 MGD; 8 MGD;10 MGD	
TREATMENT TARGET	For non-potable reuse: - Total coliform 2.2 MPN/100mL - 5-Log polio reduction	For potable reuse: - 1, 4 – Dioxane 0.5 Log reduction
SYSTEM COMPONENTS	- Closed-vessel UV system with low pressure high output lamps	- Closed-vessel UV system with low pressure high output lamps; - H_2O_2 as oxidant or source of hydroxyl radical - Sodium hypochlorite for hydrogen peroxide residual quenching
DESIGN UV TRANSMITTANCE (UVT)	Membrane permeate 65%	Microfiltration and RO permeate 95%
DESIGN DOSE	NWRI guideline dose criteria: - 80 mJ/cm^2	EEO (Electrical Energy per Order of contaminant removed) based dose criteria: - 0.3 kWh/1,000gal/order - 5 mg/L H_2O_2

The UV-AOP process requires a combination of UV with chemical oxidants, such as hydrogen peroxide, ozone, or more recently proposed hypochlorite, to enhance the generation of hydroxyl radicals to more effectively destroy the persistent chemical contaminants. Therefore, to implement the UV-AOP process, a chemical injection and storage or ozone generation system may need to be added. In the sensitivity analysis, hydrogen peroxide is used as the oxidant for hydroxyl radical generation, and sodium hypochlorite is used as quenching chemical for residual hydrogen peroxide.

UV facilities are sized based on the criteria in Table 3. The UV equipment used for sensitivity analysis is selected from the UV technologies that have been through the validation process with sizing algorithm established and approved for use in water recycling applications. The UV system selected for evaluation is a closed-vessel system equipped with 250-watt UV lamps. Redundancy is typically required for reuse facilities. In closed-vessel UV system, redundancy is provided with an extra parallel train. UV-Disinfection and UV-AOP system components and their sizing are summarized in Table 4.

Table 4. UV-Disinfection vs. UV-AOP: System Sizing Under Selected Treatment Capacities

	2 MGD	**4 MGD**	**8 MGD**	**10 MGD**
UV-Disinfection				
Number of reactor trains	3	4	6	8
Number of duty reactors	2	3	5	7
number of duty lamps	144	216	360	504
Maximum power draw, kW	36	54	90	126
UV-AOP				
Number of reactor trains	3	5	8	10
Number of duty reactors	2	4	7	9
number of duty lamps	144	288	504	648
Maximum power draw, kW	36	72	126	162

The monetary analysis of the UV systems was performed for the equipment cost and O&M cost of each UV system. The equipment cost of the closed vessel UV system selected for this analysis was obtained from an equipment manufacturer that provides the UV equipment that meets all the evaluation criteria in Table 3.

Annual O&M cost of a UV system mainly consist of power cost, lamp and ballast and other consumable parts replacement cost. O&M cost of UV-AOP systems also includes the cost of chemicals that are used as oxidants or source of hydroxyl radicals. For annual O&M cost estimate, assumptions were made for unit cost of power and chemical purchase. Major replacement costs for UV system are obtained from the UV equipment manufacturer. Annual replacement cost of each major part is prorated based on its warranted life time. The O&M cost in this evaluation only includes major items that would differentiate the UV systems in the evaluation: power, chemical cost, and consumable replacement. Common O&M costs, such as labor and compliance testing costs, are not included.

Unit flow costs, or $/MGD, were obtained by normalizing the equipment cost, power cost and annual O&M cost with the system treatment capacity. Normalized cost curves are developed with $/MGD over the range of evaluated system capacity scenarios and presented in Figure 1 through Figure 3.

- Figure 1 – UV equipment cost per MGD system capacity
- Figure 2 – Annual power consumption per MGD system capacity
- Figure 3 – Annual O&M cost per MGD system capacity

Figure 1 through Figure 3 shows the cost impact of expansion from UV-Disinfection to UV-AOP in terms of energy cost and annual O&M cost, as well as additional UV equipment investment. These figures can be used when estimating conversion installation cost, O&M cost and power consumption of a UV system with UV system treatment capacity close to the treatment capacity in the evaluation (i.e. 2 MGD to 10 MGD).

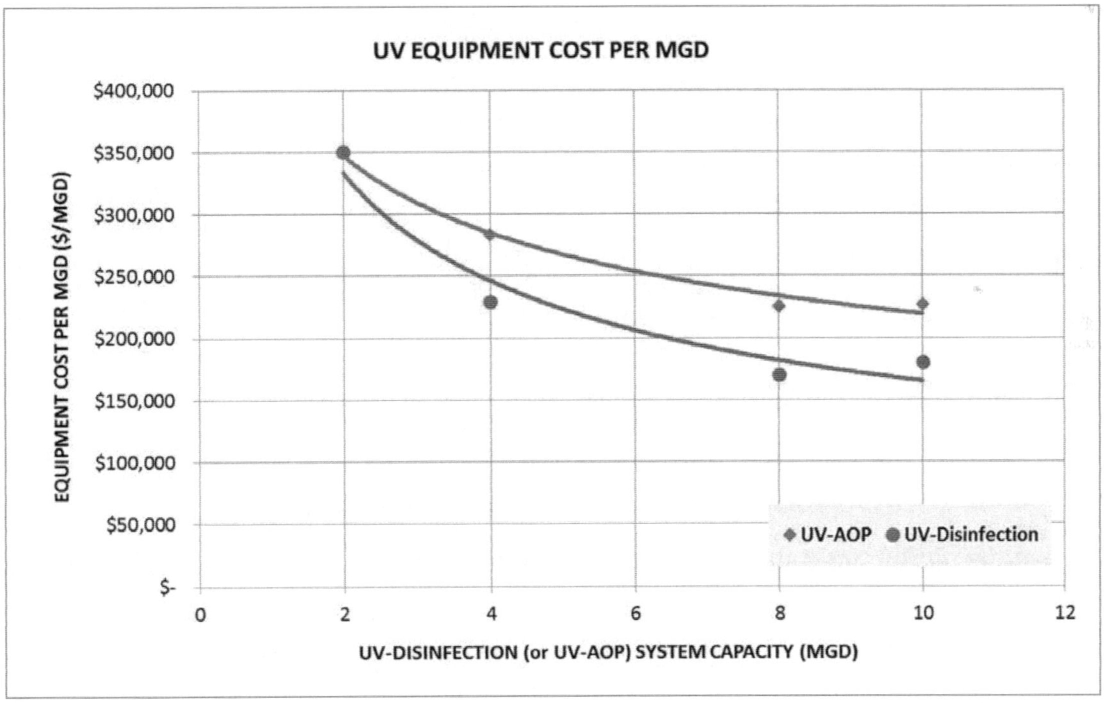

Figure 1. UV equipment cost per MGD system capacity

The unit equipment cost curve demonstrates a common Economy of Scale in which the UV equipment cost per MGD decreases with increasing system capacities. The blue line is always above the red line showing that, in general, additional UV equipment is required when converting UV-disinfection to UV-AOP.

However, the cost increase is not quite in proportion to the required dose increase. As discussed previously, the typical UV dose requirements for non-potable reuse disinfection are far less than what are required in a UV-AOP process. The dose for UV-AOP is typically ten times of the dose for UV-disinfection. The small increment equipment cost is also shown in Table 4, where one or two extra reactors are required for conversion in the capacity range of the evaluation.

The other system design parameter, UVT, plays an important role in equipment expansion for conversion from UV-disinfection to UV-AOP. From non-potable reuse to potable reuse, more advanced or rigorous treatment, such as FAT including ultrafiltration and RO, is employed upstream of UV-AOP in order to meet stringent water quality criteria for potable reuse. FAT significantly improves effluent UVT, from 65% to greater than 90%. The general increase in the UVT could significantly compensate the increase in dose requirement, therefore reduce the amount of incremental equipment.

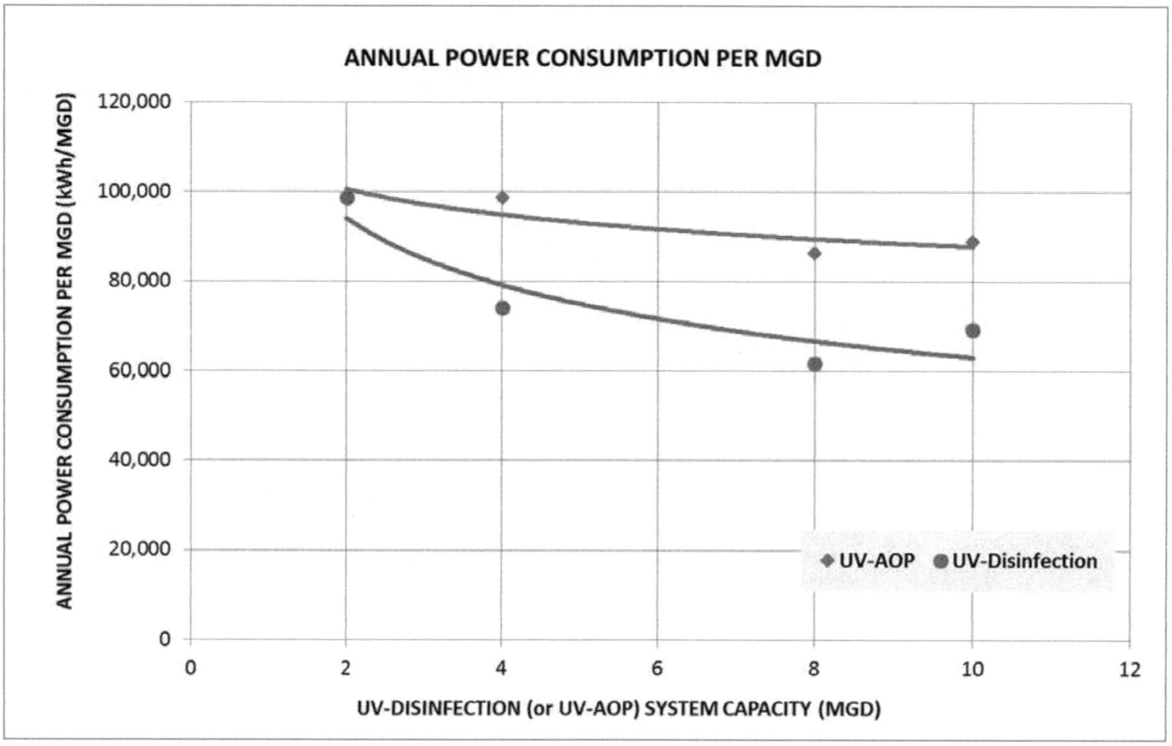

Figure 2. Annual power consumption per MGD system capacity

Power consumption increases with increasing amount of duty equipment. Therefore, the increase of power consumption for conversion follows similar trend of equipment cost increase (Figure 1).

The power consumption for a UV-disinfection system is in a range of 60,000 kWh to 90,000 kWh per year per MGD treatment capacity of the system. The power consumption for a UV-AOP system is in a range of 90,000 kWh to 100,000 kWh per year per MGD treatment capacity of the system. The power consumption increases 20% to 30% per MGD treatment capacity for conversion. The smaller the system, the higher the unit power consumption is due to the fact that UV system is modulating by design. Smaller systems typically consist of limited number of reactors which results in less flexibility in operation for power saving.

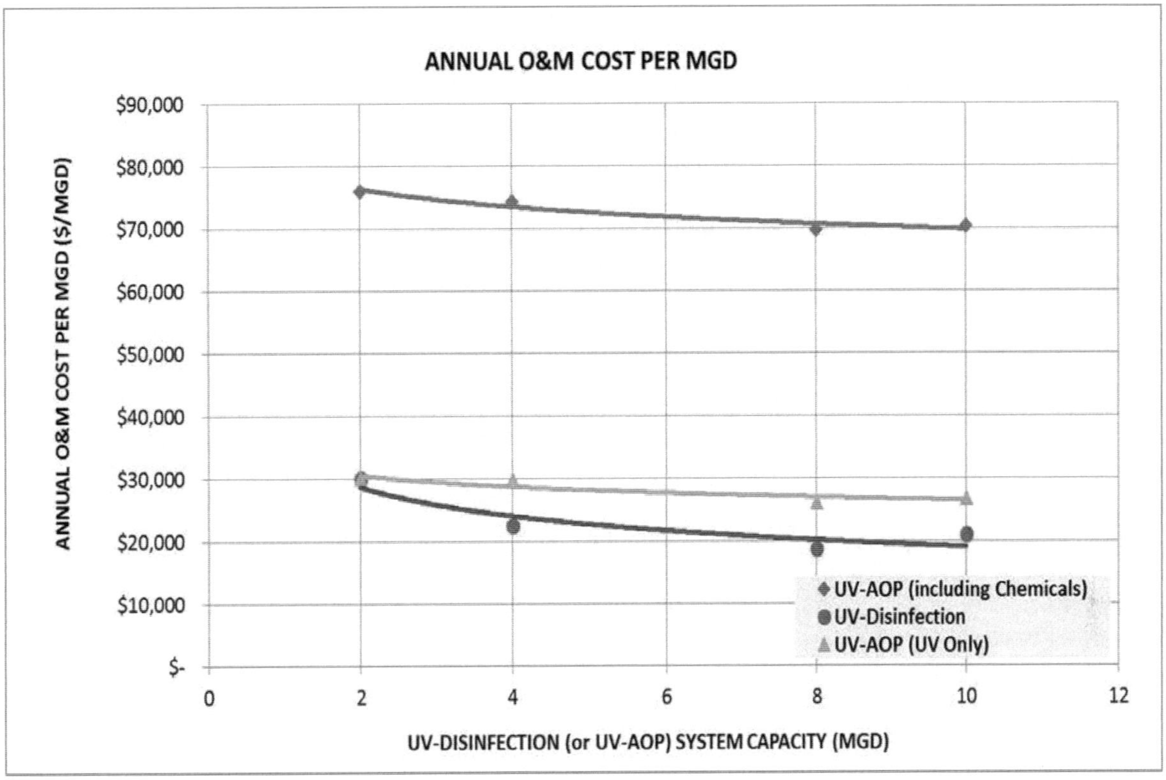

Figure 3. Annual O&M cost per MGD system capacity

Two major O&M cost items for a UV system is power consumption and annual equipment consumable parts replacement. These two major cost items almost share half and half of the total annual O&M cost of the systems in the sensitivity analysis. Both increases with increasing amount of duty equipment for the conversion from UV-disinfection to UV-AOP. Annual O&M cost increase for UV system only shows approximately 30% to 40% for the scale of treatment capacities in the evaluation.

The significant increase of annual O&M cost is from the use of chemicals in the UV-AOP process. Figure 3 shows, when cost of chemicals is included, the annual O&M cost per MGD treatment capacity of a UV-AOP system is approximately 3 to 4 times higher than that of a UV-disinfection system typically with no chemical involved in operation. Chemicals not only serve as the oxidant or source of hydroxyl radicals in the AOP process, but also are needed to quench the residual oxidant post AOP process. A typical UV-Peroxide AOP process would require a hydrogen peroxide concentration from 2 mg/L up to greater than 10 mg/L depending on water quality and the level of treatment is needed. In this sensitivity analysis, hydrogen peroxide is used as oxidant and sodium hypochlorite is used as the residual hydrogen peroxide quenching agent. Hydrogen peroxide dose of 5 mg/L was used in the AOP calculation. A 10% hydrogen peroxide reduction across the UV reactor was assumed, with 90% left for residual quenching by chlorine. Quenching chlorine is provided with 12.5% sodium hypochlorite solution.

CONVERTING UV-DISINFECTION TO UV-AOP

Converting a UV facility from UV-disinfection to UV-AOP is feasible. System conversion or facility planning needs to consider the significant difference of UV implementation in terms of treatment target and design criteria for non-potable vs. potable reuse. The sensitivity analysis in this study has shown, for any existing UV facilities, upgrading from UV-disinfection to a UV-AOP is a viable and cost-effective approach to keep the facility in compliance with the stringent water quality criteria for reuse applications. For treatment facilities that are implementing a non-potable reuse UV-disinfection today while bearing the potential capability of future upgrade for potable reuse using UV-AOP, the sensitivity analysis herein provides key considerations for facility planning and a glimpse of cost impact.

UV Technology Selection

One key consideration is technology applicability. With particular treatment requirements for both non-potable reuse UV-disinfection and potable reuse UV-AOP, there are a limited number of UV technologies that possess the effectiveness and capability for both. Using closed-vessel UV systems in the UV-AOP process downstream might be more appropriate; the selected UV system may need to have the capability of being upgraded and expanded from UV-disinfection to a UV-AOP reactor.

Additional Capital Investment

Additional capital investment is expected for conversion. The higher level of UV doses can be achieved by either adding more UV equipment, or by lowering the treated flow rate significantly as well as taking advantage of the better water quality from advanced treatment processes upstream of the UV systems. Higher UVT could help increase the delivered dose for UV-AOP application. However, if the current flow capacity is desired to be maintained the same for the potable reuse applications, it is necessary to consider a capacity expansion to accommodate the UV-AOP dose requirements.

Facility Requirements

From the sensitivity analysis presented herein, the equipment expansion is expected to be 20 to 30 percent. The capital cost of the expansion is site specific, and is not only from the equipment addition but the expansion and upgrade of facilities that associated with the UV system operation. UV facility conversion could include, but not limited to the followings:

- More UV equipment: UV system expansion can be achieved by addition of parallel trains or addition of reactor in series, or swap lamps with higher wattage lamps, etc.
- Chemical storage and feed facilities.
- Additional process monitoring and control equipment.
- Update programming to accommodate the higher dose needs.
- Integrate UV systems with oxidant dosing equipment.
- Shelter or enclosure for UV equipment, power panel, as well as control equipment.

Additional O&M Cost

Additional O&M cost is expected for conversion. Conversion from UV-disinfection to UV-AOP requires higher UV doses therefore additional UV equipment. Annual O&M cost is expected to increase with increasing energy consumption and cost for parts replacement from the additional UV equipment. The sensitivity analysis has shown that the significant O&M increase is due to the addition of chemicals in UV-AOP. UV photolysis alone can destruct some chemical contaminants (e.g., NDMA), but mostly the UV-AOP process requires a combination of UV with oxidants such as hydrogen peroxide, ozone, or more recently proposed hypochlorite to enhance the generation of hydroxyl radicals to more effectively destruct the persistent chemical contaminants. It is common that multiple chemicals are used in UV-AOP operation, as oxidant for hydroxyl radicals, residual oxidant quenching, and in some cases required as residual for post-AOP disinfection. O&M cost for chemical use is mainly from annual chemical purchases.

AOP Validation and Demonstration

The UV-disinfection system for non-potable reuse is typically selected and designed with UV technology that is being validated by independent third-party entities following strict validation protocols. The design conditions UV-disinfection systems are typically well defined and verified with many full-scale installations currently in operation. However, the UV systems used in AOP processes are usually not validated due to the lack of appropriate testing surrogates at the dose levels required in AOP processes.

Because of that, pre-installation bench scale testing and piloting are necessary to establish robust design criteria. Bench-scale treatability study is typically conducted to confirm effectiveness and efficiency of UV-AOP for designated treatment target, such as 0.5 log reduction of 1,4-dioxane, or 1.2-log reduction of NDMA, etc. After bench-scale treatability study, onsite pilot study with site-specific water and quality is recommended. The major elements and key considerations of onsite pilot study include:

- Seasonable variation in water quality to be accounted for if applicable
- Optimize UV energy requirement and chemical dosage
- Design engineer or independent 3rd-party to be involved along with technology vendor
- Regulators onboard to facilitate the approval process

Post-installation field validation or demonstration is required for UV-AOP to confirm the effectiveness and efficiency on the treatment performance.

ACKNOWLEDGEMENT

The authors gratefully acknowledge the City of San Louis Obispo Water Resources Recovery Facility and City of San Mateo Clean Water Program imposed the challenge and provided the motivation and opportunity for this study. The acknowledgement also extends to staff from Trojan Technologies and Xylem Wedeco, for providing UV technology and cost information in support of the analysis.

REFERENCE

Ultraviolet Disinfection Guidelines for Drinking Water and Water Reuse (2012) 3rd Edition, Published by National Water Research Institute in collaboration with Water Research Foundation

Title 22 Code of Regulations, Regulations Related to Recycled Water, June 18, 2014 (Revisions effective on 6/18/14) by California Department of Public Health

Expert Panel Final Report: Evaluation of the Feasibility of Developing Uniform Water Recycling Criteria for Direct Potable Reuse, August 2016, prepared for California State Water Resources Control Board

From Collection Systems to Tap – Resilience of Treatment Processes for Potable Reuse

Justin Mattingly[1*], Sharon Waller[2], Patti Craddock[3], Stuart Khan[4]

[1] Water Research Foundation
[2] Sustainable Systems LLC – Consulting
[3] Short Elliott Hendrickson Inc.
[4] University of New South Wales, Australia

[*] Email: jmattingly@waterrf.org

ABSTRACT

Design guidelines for resilience in direct potable reuse (DPR) have been developed with consideration of three discrete stages including wastewater source control, wastewater treatment, and advanced water treatment. Additionally, the distribution system for advanced treated water is also considered. Risk assessment provides a framework where the hazard analysis and critical control point (HACCP) methodology can be integrated to enhance vulnerability analysis, and asset characterization can also be supplemented by considering resilience principles. Resilience, defined as rapid recovery after failure, was found to depend significantly on process robustness and redundancy, storage, and integration of operational considerations during design including monitoring, process controls, communication protocols, emergency response plans, well defined operations and maintenance plans, and operator training to reduce response time.

KEYWORDS: Potable Reuse, Water Reuse, Advanced Treatment, Source Control, Resilience, Risk Management, Utility Management

INTRODUCTION

Water shortages, the impacts of climate change, and a greater understanding of available technologies are motivating water agencies to expand and secure their water portfolios. Included in the water supply options being considered are indirect potable reuse (IPR) and direct potable reuse (DPR). To implement potable reuse responsibly, there is a need to better define resiliency and develop methods to enhance the resiliency of a potable reuse system. This work combines risk assessment (RA) (AWWA, 2010; AWWARF, 2002; ISO, 2009) with 1) hazard analysis and critical control point (HACCP) systems, and 2) "4R" principles for measuring reliability through redundancy, robustness, and resilience.

Reliability is dependent on redundancy, robustness and resilience, which are collectively referred to as the 4Rs (Haas and Trussell, 1998; Pecson et al., 2015). Reliability is the ability to consistently meet water quality objectives. Redundancy is provided via parallel process trains, alternate assets including source, replacement parts, or above minimum treatment measure by metrics such as log reduction value (LRV) credits (Tchobanoglous et al., 2015). Robustness is the resistance to failure through 1) diversity of contaminants removed via an individual unit treatment process, 2) multiple treatment barriers, or 3) resistance to failure. Finally, resilience implies rapid recovery from failure by detection and response. These 4R principles can be applied to asset characterization and identification of risk mitigation strategies. HACCP can be applied to the vulnerability analysis step of risk assessment to identify points where operation may be monitored and controlled (Fig. 1).

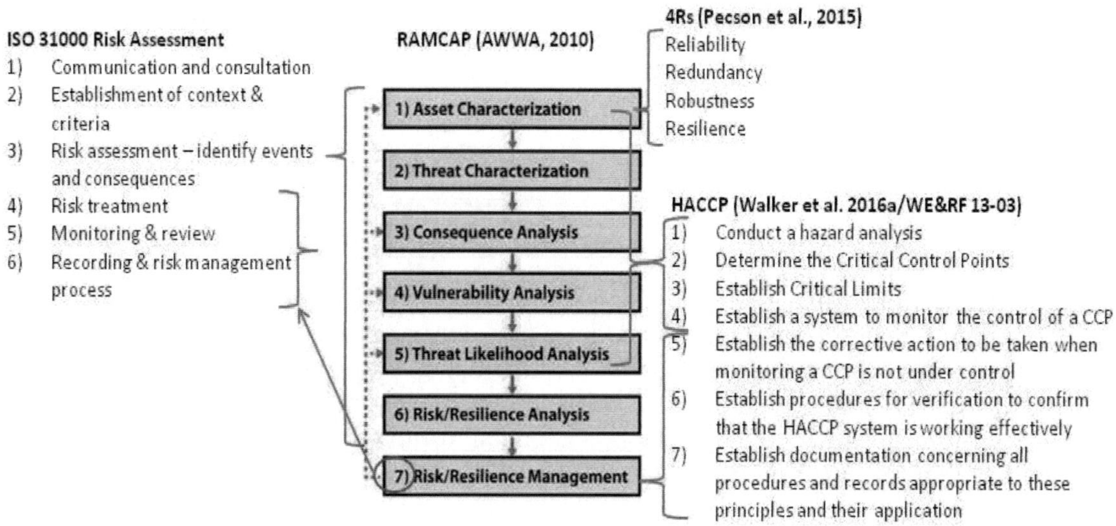

Figure 1. Resilience Assessment Combines 4R Risk Assessment and HACCP Methods (Waller et al., 2018)

The research presented here developed guidelines for three separate stages of the DPR water cycle: wastewater source control (including industrial pretreatment and source control programs), wastewater treatment (WWT), and advanced water treatment (AWT). Ideally, these three stages would be owned and operated by the same entity for DPR applications. However, wastewater source control and WWT in the U.S. are typically owned and operated by one entity while AWT and/or drinking water treatment are owned and operated by a separate entity. These entities (industry and general public in partnership with wastewater agency, the water reuse agency, and the drinking water agency respectively) should demonstrate high levels of communication, cooperation, and engagement in producing a high quality final product from municipality to municipality. Based on existing infrastructure, ownership, and management typical of today's water and wastewater industries, consideration of these three separate stages is useful for developing strategies aimed at producing a reliable direct potable reuse supply. However, continually seeking opportunities for more aggregated management as "one water" will likely aid long-term product improvement.

The results of this research are further detailed in the Water Research Foundation report, *From Collection Systems to Tap: Resilience of Treatment Processes for Direct Potable Reuse (Reuse-14-13)*, published in June 2018. Guidelines for designing resilient DPR process trains from wastewater source through wastewater treatment and advanced water treatment are presented with the principles of reliability, redundancy, robustness, and resilience integrated into existing risk methodology. Risk of treatment failure can never be completely eliminated. Therefore, process monitoring is critical since response depends on detection. Hazardous events including process malfunction, wastewater source variability, natural disasters, and malevolent attack, and dependency on third party providers such as power supply, can lead to consequences that affect utility objectives associated with both water quality and quantity.

METHODOLOGY

Three example DPR process trains were identified to represent common processes used for DPR applications with variations for applications where brine disposal or storage may be limited (Fig. 2). A gap analysis was performed to identify information needed for resilience guidelines focusing on the treatment processes commonly considered for DPR treatment trains. Technology providers were contacted to develop design checklists related to resilience features, and utilities were surveyed for significant failure modes and associated process monitoring strategies for unit processes commonly considered for DPR applications. Survey findings and a literature review were then compiled to identify hazardous events that could lead to water quality failure (Fig. 3), and what detection and response actions could be applied to avoid failure or mitigate consequences. HACCP analysis used by some utilities and in previous studies (Halliwell et al., 2014; Salveson et al., 2014; Walker et al., 2016a) provided information on critical control points and where monitoring and control may be applied. Survey findings were then compiled into a hazardous analysis catalog supplemented with information from published literature and manufacturers for all unit treatment process units considered.

Example DPR Process Trains	Rational and Key Points to Consider
DPR 1: Membrane Train → WWS → WWT → MF → RO → UV+AOP → CR → ESB →	Typical IPR or DPR process train including ESB used in areas where brine discharge is feasible.
DPR 2: Non-Brine Producing Train → WWS → WWT → O_3+BAC → GAC → UV+AOP → CR → ESB →	Non-brine producing for areas where opportunities for brine discharge is limited.
DPR 3: Limited Storage Train → WWS → WWT → O_3+BAC → MF → RO → UV+AOP → CR →	O_3 and BAC is added to membrane-based AWT process train.

Figure 2. Example DPR Process Trains (Waller et al., 2018)

Key: CR = chlorine disinfection and residual for distribution; DPR = direct potable reuse; ESB = engineered storage buffer; GAC = granular activated carbon; IPR = indirect potable reuse; MF = microfiltration; O_3+BAC = ozone & biological activated carbon; RO = reverse osmosis; UV+AOP = ultraviolet irradiation & advanced oxidation; WWS = wastewater source control; WWT = wastewater treatment

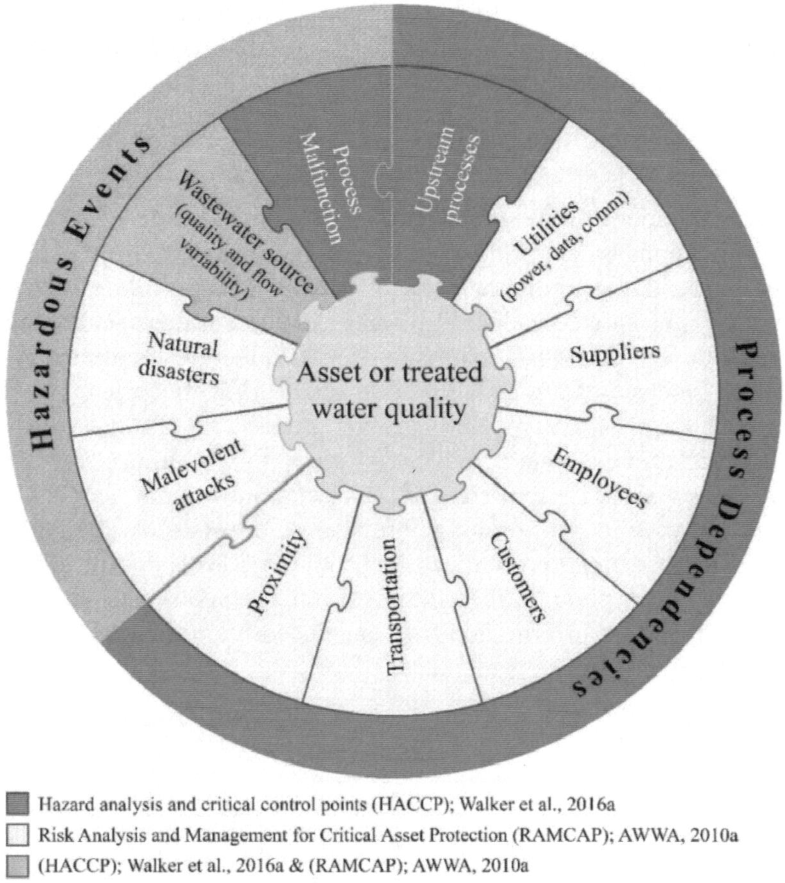

Figure 3. Types of Hazardous Events (Waller et al., 2018)

The basic steps of risk assessment include assessment of events and their consequences and integration of risk-based decisions into daily operations and management (ISO, 2009). Risk can be represented as a range (high/low) or a calculated value (risk = C*V*T, [AWWA, 2010; AWWARF, 2002]) based on consequences (C) and the likelihood of failure, which can be further parsed as likelihood of event occurrence also referred to as the threat (T) multiplied by the likelihood the event will result in a failure also referred to as the vulnerability (V). The hazard analysis catalog developed in this work provides a valuable resource for hazardous event characterization and vulnerability analysis, which is additionally supported by HACCP and 4R principles to aid identification of risk mitigation measures. Further detail on consequence analysis is provided by others (AWWA, 2010; Matthews, et al., 2014; Panico, et al., 2015).

RESULTS

The guidelines produced from this work cover background on how risk assessment may be applied to three example DPR process trains with respect to the 4R principles, a hazard analysis catalog for unit treatment processes commonly considered for DPR, discussion of resilience by DPR stage (wastewater source control, wastewater treatment, and advanced water treatment), and example applications of the combined RA/4R/HACCP approach.

Asset Characterization
Assets including wastewater source and common DPR treatment processes have been characterized based on the 4R concept (reliability, redundancy, robustness, and resilience) to identify relative resilience features and opportunities for risk abatement. Assets were characterized for log reduction value (LRV) credited by regulatory agencies, chemical removal efficiency, availability of on-line process monitoring, and speed of recovery features such as modularity of design. The distribution system should also be considered as an asset during risk assessment as water quality may degrade after treatment from the time it enter the distribution system to delivery to the consumer. Potable water distribution systems may be a significant challenge to monitor and maintain water quality due to hydraulic dead ends and biofilm growth prone to select water chemistry and warmer climates (Waller, 2014).

Hazardous Event Characterization
Hazardous events and dependencies applicable to unit treatment processes commonly considered for DPR are presented in Tables 1-3. Hazardous events include equipment failure events similar to those considered in a HACCP analysis (Walker et al., 2016a) while dependency events such as a power outage may be largely out of a utility's control. While many of these upstream dependencies are intuitive, others may be a bit less obvious. For example, the presence of large organic molecules in membrane treatment can contribute to RO membrane fouling due to the adsorption of organic molecules to the membrane surface. In addition, biological growth in a BAC filter can be limited by the availability of organic carbon and nutrients, in particular orthophosphate, resulting in a dependency. However, nutrients are not necessary for GAC. The HACCP process provides greater detail on process malfunction, and therefore may be considered a micro-level analysis either imbedded in a risk assessment framework or performed in parallel to risk assessment.

Table 1. Hazardous Events for Membrane Unit Treatment Processes (Waller et al., 2018)

Hazardous Event	**MBR**	**MF/UF**	**RO**
Loss of filter integrity (membrane, O-rings, piping)	X	X	X
Mechanical failure of auxiliary process components	X	X	X
Online monitor failure	X	X	X
Membrane exposure to oxidizers	X		X
Incorrect chemical addition (scaling or precipitation)	X	X	X
Particulate fouling (SDI)			X
Chemical damage to membrane			X
Non-conforming brine and CIP, prevents disposal			X
Upstream Dependencies	**MBR**	**MF/UF**	**RO**
Chemical Addition	X	X	X
Silt density index			X
Oxidizers			X

Large organic molecules (>10,000 daltons)			X
Suspended solids		X	X
Biological treatment process	X		
Chemical fouling			X
Downstream Consequences	**MF/UF**	**RO**	**MBR**
Brine disposability		X	
Permeate stability		X	
Solids Handling	X		X
Dissolved solids		X	
Suspended solids	X		X

Table 2. Hazardous Events and Dependencies for Advanced Oxidation Processes (Waller et al., 2018)

Hazardous Event	**UV/H_2O_2**	**O_3/pH**
Analyzer failure	X	X
Low UV transmittance of influent[1]	X	
Low UV lamp ballast power	X	
Insufficient CT for LRV	X	X
Air pocket in reactor (in-vessel type UV reactor)	X	
Increase in influent bromine, nitrate, nitrite and/or NOM	X	X
Insufficient residual control of oxidant chemical	X	X
Upstream Dependencies	**UV/H_2O_2**	**O_3/pH**
UV absorbing constituents	X	
Scavenging constituents	X	X
pH		X
DBP precursors	X	X
Organic carbon	X	X
Downstream Consequences	**UV/H_2O_2**	**O_3/pH**
Oxidizers	X	
DBP precursors	X	X

[1] Either UV transmittance or UV absorbance may be monitored. UV absorbance and transmittance can be related by the Beer-Lambert law where: Absorbance = 2- $\log_{10}$(transmittance, %).

Table 3. Hazardous Events and Dependencies of GAC/BAC Unit Treatment Processes (Waller et al., 2018)

Hazardous Event	GAC	BAC
Breakthrough of a contaminant	X	
Change in TOC composition and/or loading	X	
Analyzer failure	X	X
Water quality characteristic changes that affect biological activity		X
Upstream Dependencies	**GAC**	**BAC**
Dissolved Oxygen		X
Turbidity and macromolecular organics	X	X
pH		X
Temperature	X	X
Nutrients (in particular orthophosphate)		X
Downstream Consequences	**GAC**	**BAC**
UV Transmittance	X	X

Vulnerability Analysis

A vulnerability analysis of hazardous event-asset pairs determines the likelihood of failure for a given hazardous event. The analysis is based on the ability to detect, delay, and respond to potential failures within an allowable failure response time specific to the combined DPR treatment train and storage residence time. HACCP analysis provided significant additional guidance to determine a point where monitoring may best be applied to detect process malfunction and apply control for correction.

Risk and Resilience

Resiliency for DPR systems involves planning, design, operations, maintenance, and administration activities that all need to be considered at project concept. Resilience may be enhanced by process redundancy, source redundancy (which provides opportunities for blending), storage, real-time monitoring, and automated response to decrease response time to system upsets. Diversity of supplies via blending or standby alternate sources enhance resilience when there are doubts about the quality of any one source. The integration of operational planning including monitoring and control actions in the design phase is critical to achieve the level of resilience required for DPR. In cases where real-time monitoring methods may not be available, increased redundancy and robustness combined with on-line monitoring of a surrogate parameter may be recommended to achieve desired resilience. Fail-safe design strategies such as automated diversion of off-spec water may avert a water quality failure despite unit process failure. In cases where storage may be limited, on-line monitoring may be necessary to provide sufficient resilience. Other operational considerations during the design phase include operations and maintenance standard practices, operator training (Walker et al., 2016b), communications procedures to coordinate between DPR stages (typically comprising different agencies and departments within an agency), and efficacy of wastewater treatment and source control programs (Rimer et al., 2016).

DISCUSSION AND CONCLUSIONS

Resilience, defined as rapid recovery after failure, was found to depend significantly on process robustness and redundancy. Storage may be needed to increase allowable failure response time, and integration of operational planning in the design phase (real-time monitoring and automated response) may reduce overall response time. Other resilience considerations include dependencies on upstream processes and third party providers (e.g., power and communication); process robustness for removal of key contaminants such as compounds of emerging concern; establishment and monitoring of utility specific intermediate treatment goals for each DPR water cycle stage; and minimizing failure response time including process monitoring and response. Future research needs include risk assessment for wastewater treatment and wastewater source control stages and specific processes (GAC, BAC, and MBR) to identity any potential LRV, CCP, and monitoring recommendations.

Significant contributions of this work include: 1) the hazardous analysis catalog of actual and anticipated hazardous events (failure modes) for DPR unit treatment processes and potential associated responses, and 2) an integrated framework of risk assessment methodology developed to compare risk and resilience for the water industry with HACCP analysis and resilience principles.

REFERENCES

ASTM. (2014). Standard Test Method for Silt Density Index (SDI) of Water (ASTM D4189).

AWWA. *AWWA J100-10 (R13) Risk and Resilience Management of Water and Wastewater Systems (RAMCAP)*; American Water Works Association: Denver, CO, 2010.

AWWARF. *Risk Assessment Methodology for Water Utilities (RAM-W), Second Edition*; AWWA Research Foundation: Denver, CO, 2002.

Haas, C. N., and Trussell, R. R. Frameworks for Assessing Reliability of Multiple, Independent Barriers in Potable Water Reuse. *Water Science and Technology* **1998**, *38*(6), 1-8.

Halliwell, D., Burris, D., Deere, D., Leslie, G., Rose, J., and Blackbeard, J. *Utilization of Hazard Analysis and Critical Control Points Approach for Evaluating Integrity of Treatment Barriers for Reuse (WRRF-09-03)*; WaterReuse Research Foundation: Alexandria, VA, 2014.

ISO. *ISO 31000:2009(E) Risk management — Principles and guidelines. 5th Edition*; International Organization for Standardization; Switzerland, 2009.

Matthews, J. C., Piratla, K., and Matthews, E. *Disaster resilience of drinking water infrastructure systems to multiple hazards.* Paper presented at the Structures Congress 2014, April 3, 2014 - April 5, 2014, Boston, MA, 2014.

Panico, A., Iasimone, F., Fabbrocino, G., and Pirozzi, F. Vulnerability assessment of drinking water treatment plants. *Chemical Engineering Transactions* **2015**, *43*, 2317-2322.

Pecson, B., Trussell, R. S., Pisarenko, A. N., and Trussell, R. R. Achieving Reliability in Potable Reuse: The Four Rs. *American Water Works Association Journal* **2015**, *107*(3), 48-58.

Rimer, A., DeCarolis, J., and Sathyamoorthy, S. *Evaluation of Source Water Control Options and the Impact of Selected Strategies on DPR (Reuse-13-12)*; Water Environment & Reuse Foundation, Alexandria, VA, 2016.

Salveson, A., Salveson, M., Mackey, E., and Flynn, M. *Application of Risk Reduction Principles to Direct Potable Reuse (WRRF-11-10*); WateReuse Research Foundation, Alexandria, VA, 2014.

Walker, T., Stanford, B. D., Burris, D., and Vickers, J. *Development of Operation and Maintenance Plan and Training and Certification Framework for Direct Potable Reuse (DPR) Systems (Reuse-13-13)*; Water Environment & Reuse Foundation, Alexandria, VA, 2016b.

Walker, T., Stanford, B. D., Khan, S.J., Robillot, C., Snyder, S., Valerdi, R., Dwivedi, S., & Vickers, J. *Critical Control Point Assessment to Quantify Robustness and Reliability of Multiple Treatment Barriers of a DPR Scheme (Reuse-13-03)*; Water Environment & Reuse Foundation, Alexandria, VA, 2016a.

Waller, S. 2014. *Real-Time Monitoring Tools to Characterize Microbial Contaminants in Reclaimed Water: State-of-the-Science Assessment (WRRF 11-06)*; WateReuse Research Foundation, Alexandria, VA, 2014.

Waller, S., Craddock, P., and Khan, S., *From Collection Systems to Tap:Resilience of Treatment Processes for Direct Potable Reuse (Reuse-14-13)*; The Water Research Foundation, Alexandria, VA, 2018.

Tchobanoglous, G., Cotruvo, J., Crook, J., McDonald, E., Olivieri, A., Salveson, A., and Trussell, R.S. *Framework for Direct Potable Reuse*; American Water Works Association, National Water Research Institute, Water Environment Foundation, WateReuse Research Foundation, Alexandria, VA, 2015.

Design of a Managed Aquifer Recharge and Advanced Treatment Facility

Andrew Newbold[1*], Lauren Zuravnsky[2], Charles Bott[2], Germano Salazar-Benites[2], Chris Wilson[2], Dwayne Amos[1], David Briley[1]

[1]Hazen and Sawyer, Virginia Beach, Virginia
[2]Hampton Roads Sanitation District, Virginia
*anewbold@hazenandsawyer.com

ABSTRACT

The Sustainable Water Initiative for Tomorrow (SWIFT) Research Center represents a critical step toward meeting Hampton Roads Sanitation District's (HRSD) future regulatory requirements by advancing the conceptual development of SWIFT program. The SWIFT Research Center includes a 1 mgd advanced water treatment facility that provides a platform for research, public outreach, and operator training. The operation of the facility will inform full scale implementation of the program.

The Research Center includes operational flexibility beyond what is typical for full scale water treatment facilities. The facility is designed to allow for future changes to the treatment train and the process can operate in a variety of treatment configurations to allow HRSD to determine the most effective and efficient mode of operation to meet their water quality goals. The approach taken at the Research Center reflects HRSD's adaptive management approach to defining regulatory requirements for full-scale SWIFT facilities.

KEYWORDS

Advanced treatment, groundwater, managed aquifer recharge, water reuse, design-build procurement

INTRODUCTION AND BACKGROUND

Hampton Roads Sanitation District (HRSD) has the capacity to treat approximately 249 million gallons per day (mgd) of wastewater and serves nearly 1.7 million people in the southeastern region of Virginia. The Hampton Roads region is faced with several water related challenges including the restoration of the Chesapeake Bay, unsustainable withdrawal of water from the Potomac Aquifer System (PAS) resulting in depletion of groundwater resources, intrusion of saltwater into the PAS, and relative sea level rise. HRSD has recognized that an innovative solution will be needed to manage regulatory requirements while simultaneously addressing regional water challenges. In response to these challenges, they embarked on a multi-phase managed aquifer recharge program; the Sustainable Water Initiative for Tomorrow (SWIFT).

The drinking water supply for Hampton Roads consists of a combination of groundwater and surface water sources. Currently, wastewater is treated at one of 13 HRSD treatment plants and discharged into the Chesapeake Bay – which releases into the ocean or is discharged directly into the ocean – with no beneficial use. With SWIFT in operation, the highly treated effluent from

HRSD's treatment plants will be treated to drinking water quality using an Advanced Water Treatment (AWT) process and used to recharge the aquifer. At build out, SWIFT will enable HRSD to recharge the PAS with over 100 mgd of highly treated water.

HRSD conceived a multi-phase implementation plan for the SWIFT program as follows:

Phase 1 – Concept Feasibility: This phase consisted of a high-level feasibility study to determine if additional investment into the concept was warranted. The results from this phase showed that the benefits of the SWIFT program could outweigh the costs and that further investigation was needed.

Phase 2 – Concept Development & Pilot Testing: HRSD performed room-scale piloting of two alternative treatment trains using treatment plant effluent from their York River Treatment Plant facility. Full Advanced Treatment (FAT) consisting of ultrafiltration, reverse osmosis, and ultraviolet disinfection/advanced oxidation was piloted alongside a carbon-based treatment train consisting of flocculation, sedimentation, ozone, biofiltration, granular activated carbon adsorption, and ultraviolet disinfection. The results from the side-by-side piloting showed that the carbon-based train was equally effective at treating the water as the FAT train, and the carbon-based train produced water that was more compatible with the aquifer at a lower overall cost to the program.

Phase 3 – Concept Demonstration: Design and construction of the SWIFT Research Center, a one million gallon per day (mgd) advanced water treatment facility intended to demonstrate that the advanced water treatment facility will produce water that meets the water quality goals and is compatible with the receiving aquifer. The other goals for the facility include collecting 18 months of AWT performance data, aquifer hydraulic response data, Soil Aquifer Treatment (SAT) data, as well as providing a platform for staff and operator training and public outreach. Finally, the facility will help HRSD define regulatory requirements for full-scale implementation.

Phase 4 – Facility Plan Development: Master planning of advanced water treatment facilities including any required upgrades to existing treatment plants (TPs) and identifying opportunities for consolidation of AWT facilities.

Phase 5 – Implementation Plan: Detailed planning and design of AWT facilities and TP upgrades.

Phase 6 – Full Scale Facility Implementation: Construction, commissioning, and startup of SWIFT AWT facilities.

HRSD selected design-build project delivery for the SWIFT Research Center. The advantages of this approach included single-point responsibility for project delivery, allowing for overlap of the design and construction phases of the project, and the high degree of collaboration between the owner, the designer, and builder that is typical of design-build project delivery.

IMPORTANCE OF TREATMENT FLEXIBILITY

The SWIFT Research Center is designed to bridge the gap between the room-scale pilot demonstration and full-scale implementation of the SWIFT advanced water treatment process. Accordingly, HRSD included flexibility as a key goal of the facility to allow for testing design and operational concepts at a meaningful scale before full scale implementation. Another objective for the facility is to define regulatory requirements for the program. HRSD can adjust the multibarrier approach to treating the water and protecting human health to demonstrate the performance and reliability of the treatment process. Below is a summary of these flexible treatment elements.

Process Area Configuration

The SWIFT Research Center process area is designed so that the treatment process can be modified in the future. Process, chemical, and sample piping is exposed and installed in a trench running the length of the process area instead of being buried or encased in the slab. This will facilitate future modifications to the treatment train should they become necessary.

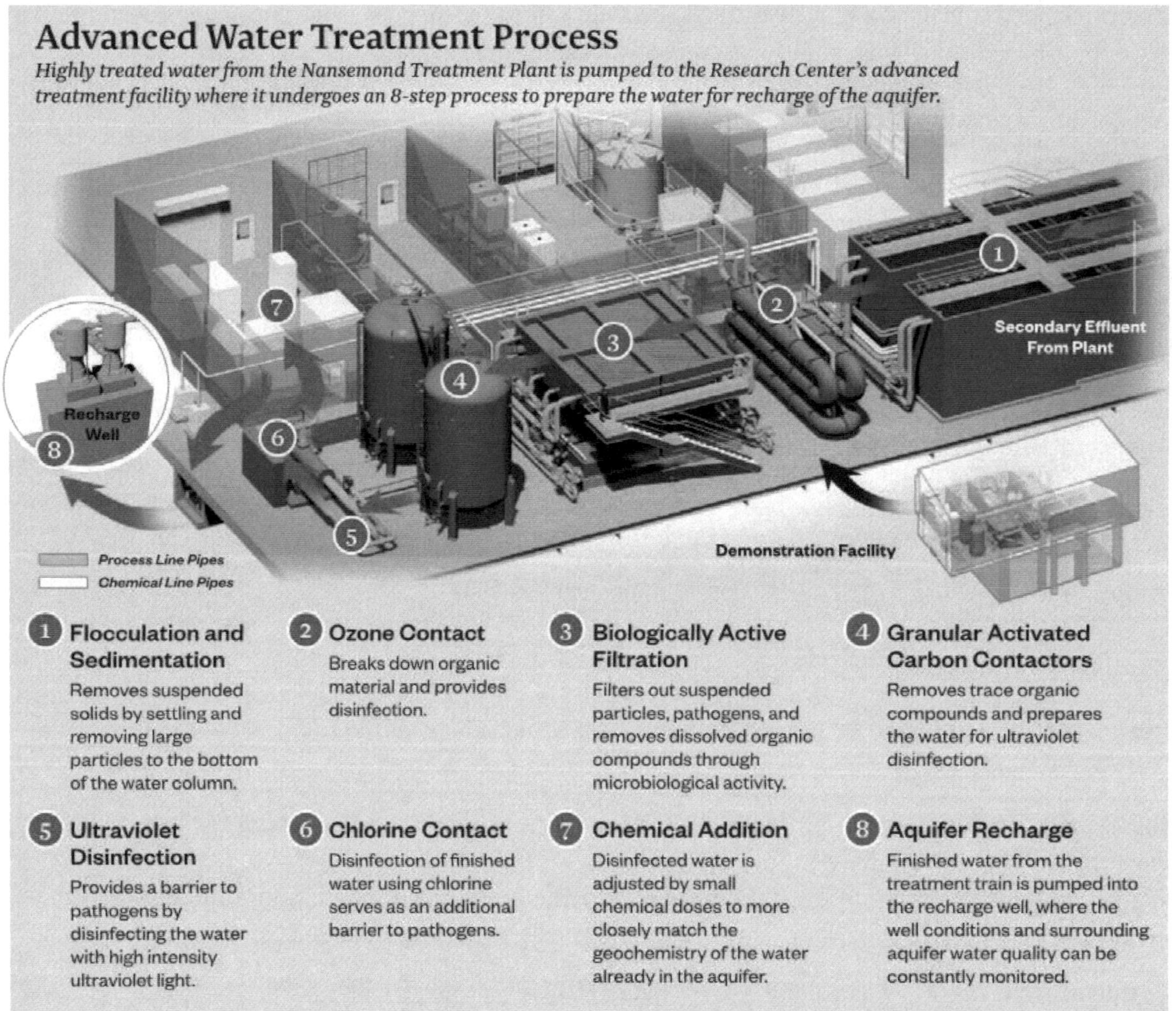

Figure 1 – SWIFT Research Center Process Layout

Flocculation and Sedimentation

The SWIFT Research Center process flow is split between two integrated flocculation and sedimentation tanks. The flow to each train can be controlled independently to allow the operators to adjust the flocculation tank detention times and the surface overflow rate of the sedimentation tank inclined plate system.

Settled solids can be removed from the bottom of the sedimentation tank and removed from the process or can be recirculated back to the first flocculation tank to improve process performance. During the operation of the room-scale process, HRSD found that recirculating solids significantly improved settled water turbidity.

Ozone

The SWIFT Research Center supports multiple ozone application methodologies depending on the treatment goals for ozone and for the facility in general. If the treatment goal for ozone is to oxidize organics and create assimilable organic carbon (AOC) for the biofiltration process then ozone can be added using residual control, or ultraviolet light absorbance (UV-254) reduction control. If the treatment goal for ozone is to serve as a barrier to viruses or protozoan cysts then ozone can be dosed based on maintaining a set log reduction value (LRV) that automatically adjusts the ozone dose based on the LRV setpoint, the water temperature, the location of the residual monitoring point along the pipeline reactor, and the process flow rate.

The serpentine pipeline ozone contactor includes six ozone residual monitoring ports along its length. HRSD can relocate the ozone residual monitoring probes to any of these ports, program the change into the control system, and optimize ozone dose, pathogen inactivation performance, and their disinfection byproduct control strategy. Finally, the ozone system is designed for a wide range of dosages (4 – 20 mg/L) to meet expected organics demand and the potential for intermittent nitrite demand for ozone from the advanced biological nutrient removal process upstream of the SWIFT Research Center at the Nansemond Treatment Plant.

Figure 2 – SWIFT RC Ozone Contactor with Multiple Residual Monitoring Points

Biologically Active Filtration

The SWIFT Research Center utilizes four Biologically Active Filters (BAFs) to provide removal of assimilable organics and to serve as a barrier to pathogens. Three of the filters currently have a conventional modular underdrain system with a gravel layer for retaining media. The fourth filter has a modular underdrain system that includes a slotted media retainer and no gravel layer. With four available filters, HRSD can experiment with different empty bed contact times and hydraulic loading rates to optimize the approach for full-scale facilities.

Pathogen inactivation

Though pathogen inactivation is not anticipated to be a regulated parameter for the operation of full-scale SWIFT facilities, the proposed treatment goals are consistent with conservative approaches: at least 12-log removal of viruses and 10-log removal of *Cryptosporidium* and *Giardia.* One of the goals for the SWIFT Research Center is to determine the performance of SAT as a barrier to pathogens. As such, the Research Center process can be operated to achieve pathogen inactivation by claiming a combination of available inactivation credits. See Table 1 below for the range of pathogen inactivation credits that can be claimed from each unit process at the SWIFT Research Center to meet the 12-10-10 inactivation goal.

Table 1 – SWIFT Research Center Pathogen Inactivation Credits

Parameter	Floc/Sed & BAF	Ozone	UV	Cl2	SAT	Total
Enteric Viruses	2	0 – 3	4	4	0 – 6	12 – 19
Cryptosporidium	4	0	4	0	0 – 6	10 – 14
Giardia	2.5	0 – 1.5	4	0	0 – 6	10 – 14

Conclusions

HRSD's SWIFT managed aquifer recharge program has the potential to provide multiple benefits to the Hampton Roads region. A critical part of initializing the program is determining the regulatory and non-regulatory treatment goals for the program. The SWIFT Research Center was conceived to advance the understanding of these treatment goals and help define the regulatory and treatment framework for the program. Incorporating sufficient operational flexibility to support future changes to the program was a key aspect to the Research Center. The facility includes several features that achieve the level of flexibility that HRSD needs to allow the program to develop through technological innovation, regulatory acceptance, and operational improvements.

Planning for a Potable Reuse Future While Maximizing Recycled Water Today

Jasmine Diaz, PE[1]
Mike Falk, PhD, PE[2]

[1] Water Systems Consulting, Inc.
[2] HDR, Inc.

ABSTRACT
The City of San Luis Obispo's (City) Water Resource Recovery Facility (WRRF) has the capacity to treat 5.1 million gallons per day (mgd) and has been distributing tertiary recycled water via a City-wide purple pipe system since 2007. In 2014, the WRRF Project began to upgrade the facility to comply with new National Pollutant Discharge Elimination System (NPDES) limits that will go into effect Fall of 2019. As the City navigated the planning phase for the WRRF Project, maximizing recycled water emerged as a paramount goal for the project and for the City leading to a series of studies that identified potable reuse as the most beneficial long-term plan. The initial study identified existing facility improvements and future facility requirements necessary to support a potable reuse plan. After the City quantified their recycled water production capacity, a subsequent analysis was conducted to determine the most beneficial use of the resource by analyzing five (5) alternatives from a triple bottom line approach. Two (2) alternatives were selected that would allow the City to maximize beneficial reuse in the short-term and in the long-term. The preferred short-term approach was to prioritize customer deliveries where recycled water would be used during months when demand was lowest and supply was greatest (September to May). The preferred long-term approach was to position the City for future production of a potable reuse supply. To align with long-term potable reuse planning, several elements were introduced to the WRRF Project, such as upgrading the secondary treatment process to a membrane reactor to produce high quality water for potable reuse that also allows the City to more reliably meet upcoming NPDES permit requirements. To realize the short-term goal, the City is in the process of identifying existing demand and understanding the influent flow characteristics of the WRRF.

KEYWORDS: Recycled water, maximizing production, Triple Bottom Line, potable reuse planning, Program Charter

INTRODUCTION

The City of San Luis Obispo's (City) Water Resource Recovery Facility (SLO WRRF), located in California, has the capacity to treat 5.1 million gallons per day (mgd) and has been distributing tertiary recycled water via a City-wide purple pipe system since 2007. The City's goal is to maximize recycled water production and target delivery to the highest demand and best use of the drought-resistant water supply. This goal will be primarily addressed during the City's on-going WRRF Project, which includes upgrading the entire facility to meet stricter water quality requirements for continued discharge into San Luis Obispo Creek.

The SLO WRRF was originally constructed in the early 1900s and has since been upgraded and expanded numerous times. Since the last major upgrade in 1994, the National Oceanic and Atmospheric Agency (NOAA), has required the City to release a minimum flow of approximately 1.6 mgd to the San Luis Obispo Creek. This discharge requirement, along with the City's desire to create a robust recycled water program, has led to conflicting priorities that the City has had to navigate during the planning phase of the WRRF Project.

The City made the decision early on in the WRRF Project to define their existing recycled water production capabilities, and their short- and long-term customer demands, through a collaborative workflow that involves various stakeholders and Triple Bottom Line (social, economic and environmental) evaluations in workshop settings.. The proactive approach has led to a series of studies that have informed and supported decision making, thereby enhancing the City's WRRF Project and recycled water program.

METHODOLOGY

At the start of the WRRF Project, the City decided to go through a chartering process to establish the foundation for decisions to be made for the WRRF Project. The chartering process, facilitated by HDR, included a collaborative discussion to define the following:

- Mission
- Vision
- Objectives and Performance Measures
- Guiding Principles

The participants in the chartering process consisted of members of the Project Team that included Water Systems Consulting Inc. (WSC), HDR Inc., RRM Design, and the City (Utilities Department). Once the Program Chartering process was complete, a charter signing took place to document the commitment to the Program Charter.

Figure 1 illustrates the culmination of the collaborative effort and provides the foundation for the WRRF Project Program to ensure that the Project Team "delivers a water resource recovery facility in partnership with stakeholders that provides economic, social and environmental value to our community" (Figure 1).

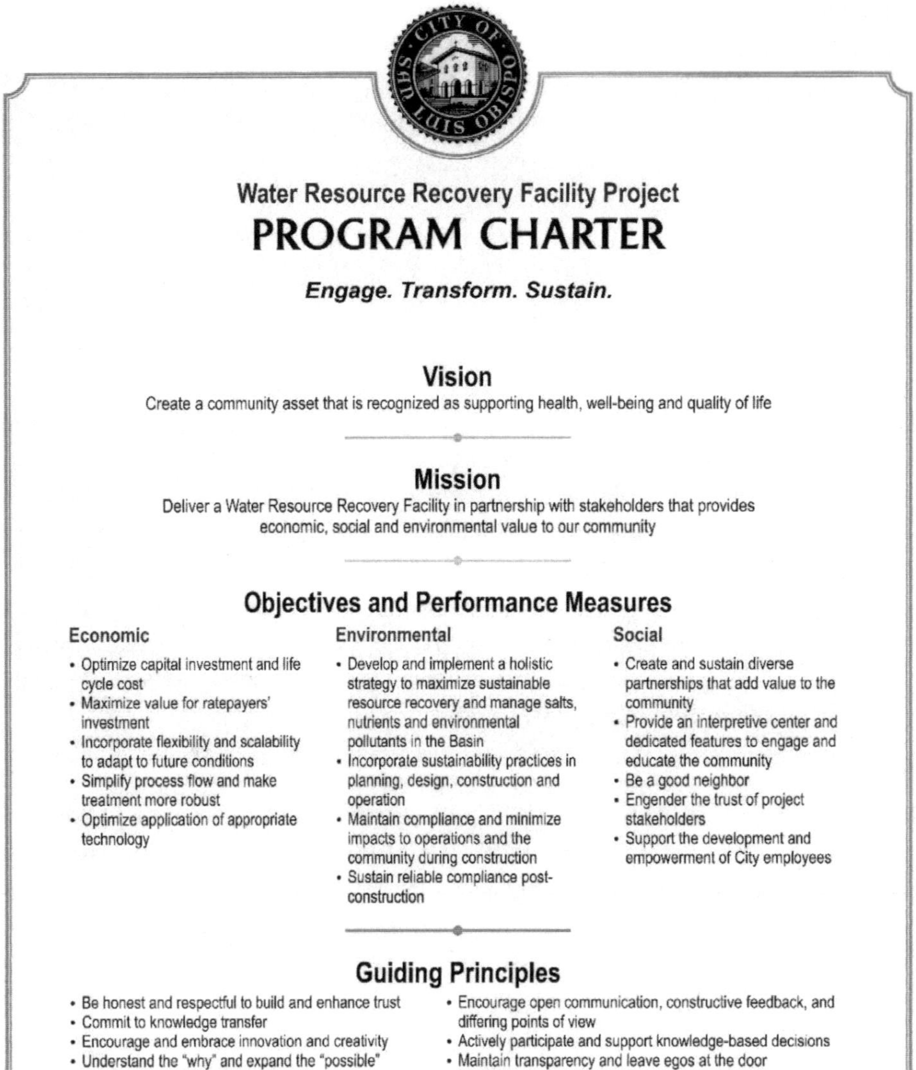

Water Resource Recovery Facility Project

PROGRAM CHARTER

Engage. Transform. Sustain.

Vision

Create a community asset that is recognized as supporting health, well-being and quality of life

Mission

Deliver a Water Resource Recovery Facility in partnership with stakeholders that provides economic, social and environmental value to our community

Objectives and Performance Measures

Economic

- Optimize capital investment and life cycle cost
- Maximize value for ratepayers' investment
- Incorporate flexibility and scalability to adapt to future conditions
- Simplify process flow and make treatment more robust
- Optimize application of appropriate technology

Environmental

- Develop and implement a holistic strategy to maximize sustainable resource recovery and manage salts, nutrients and environmental pollutants in the Basin
- Incorporate sustainability practices in planning, design, construction and operation
- Maintain compliance and minimize impacts to operations and the community during construction
- Sustain reliable compliance post-construction

Social

- Create and sustain diverse partnerships that add value to the community
- Provide an interpretive center and dedicated features to engage and educate the community
- Be a good neighbor
- Engender the trust of project stakeholders
- Support the development and empowerment of City employees

Guiding Principles

- Be honest and respectful to build and enhance trust
- Commit to knowledge transfer
- Encourage and embrace innovation and creativity
- Understand the "why" and expand the "possible"
- Maintain a high standard of performance
- Be productive while holding ourselves and others accountable
- Encourage open communication, constructive feedback, and differing points of view
- Actively participate and support knowledge-based decisions
- Maintain transparency and leave egos at the door
- Celebrate our successes while being candid about and learning from our mistakes
- Have fun!

PREP FOR TOMORROW

Figure 1. WRRF Program Charter (2014) Developed in Partnership with the City of San Luis Obispo, HDR, WSC, and RRM Design and then revised in 2016 to include Rincon, Jacobs (formerly CH2M) and newly elected City Council members.

The Project Team developed an evaluation process to identify criteria and assign scores to each decision under consideration based on the Program Charter. The Project Team first defined a set of criteria that match the City's key goals and objectives. Weighting factors were then assigned to each criterion to reflect the relative importance of the different criteria. The Project Team then assigned scores for each decision under consideration based on how well it met the criteria. Scores and weights were combined to yield an aggregate score for each alternative. This system provides a fairly objective outcome from a group of individual's subjective inputs.

Evaluation Criteria
The Project Team identified several criteria that could be used to measure how well a conceptual alternative meets the City's objectives. These criteria were developed in part from the Program Charter and divided into three categories that represent the three contributors to a Triple Bottom Line: economic, social, and environmental. A total of 13 Triple Bottom Line criteria were identified, as described in Table 1.

Table 1. Evaluation Criteria based on the Program Charter (Facilities Plan, HDR and WSC, 2016)

Category	Criterion	Description
Economic	Value for ratepayers' investment	This criterion is based on how the value received by the City relates to the required investment.
	Flexibility	Short term flexibility is the ability to adapt to changes in regulations or unexpected conditions. Long-term flexibility is the ability to adapt to long-term changes such as demographic changes or changes in water use patterns.
	Robustness	The reliability of the equipment or processes that will be needed to implement a strategy, and how well the equipment and processes can be expected to stand the test of time, including reliable compliance.
	Simplicity	A higher degree of simplicity indicates less potential for process failure or operator error due to the complexity of an element or strategy.
	Scalability	This criterion represents the degree to which a strategy that works well on a limited or trial basis can be scaled up to address City-wide issues and accommodate future growth.
Social	Creates and sustains partnerships	The alternative is conducive to fostering partnerships with other project stakeholders.
	Positive community engagement	This criterion gauges how an alternative will affect the City's ability to engage community members and build broad-based community support for the WRRF and its mission.
	Good neighbor	The alternative allows the City and the WRRF to build goodwill with citizens and businesses near the WRRF.
Environmental	Maximize sustainable resource recovery	The alternative helps the City meet its goal of recovering available resources from wastewater.
	Manage salts and pollutants in the basin	The management of salts and nutrients is an increasingly important issue for wastewater dischargers and recycled water producers; this criterion gauges how an alternative affects the City's ability to effectively manage these constituents.
	Construction impacts to the community	The short-term impacts related to construction of improvements (e.g., noise, dust, vibration).
	Operations impacts to the community	The long-term impacts related to new operations (e.g., increased truck traffic for hauling organic waste).
	Reliable compliance	The alternative's impact on the City's ability to consistently meet environmental regulations.

Weighting of Criteria
The Project Team would evaluate the relevance and relative importance of each criteria and depending on the decision under consideration, weighting of 1, 2, or 3 was applied to each criteria in a collaborative group setting before the scores were applied to each element. Table 2 shows an example of how the weighting criteria could be applied before each element was evaluated.

Table 2. Example of Weighting Applied to Decision Criteria

Category	Criterion	Weight
Economic	Value for ratepayers' investment	3
	Flexibility	2
	Simplicity	3
	Scalability	1
	Robustness	2
Social	Creates and sustains partnerships	2
	Positive community engagement	2
	Good neighbor	3
Environmental	Maximize sustainable resource recovery	2
	Manage salts and pollutants in the basin	2
	Construction impacts to the community	1
	Operations impacts to the community	2
	Reliable compliance	3

Assignment of Scores
The Project Team developed a scoring guide to facilitate the assignment of scores for each element within the decision under consideration. For each criterion, an element was given a score of 1, 3, or 5. These scores were assigned in a workshop setting and entered in a spreadsheet for calculation of an aggregate score. The guidelines that were developed to help the scoring assignment process are shown in Table 3.

Table 3. Example of Guidelines for Assignment of Scores (WRRF Facilities Plan (2016), (HDR & WSC)

Category	Criterion	Score 1	Score 3	Score 5
Economic	Value for ratepayers' investment	Low relative value for overall investment	Moderate value for overall investment	High value for overall investment
	Flexibility	Less flexible	Typical	More flexible
	Robustness	Less robust; higher risk	Typical	More robust; less uncertainty
	Simplicity	More complex	Typical	Less complex
	Scalability	Not scalable	Neutral	Scalable
Social	Creates and sustains partnerships	Impedes partnerships	Allows partnerships	Fosters partnerships
	Positive community engagement	Discourages positive community engagement	Permits positive community engagement	Encourages positive community engagement
	Good neighbor	Expected to harm relations	Neutral	Expected to improve relations
Environmental	Maximize sustainable resource recovery	Reduces recovery of resources	Neutral	Maximizes resource recovery
	Manage salts and pollutants in the basin	Reduces ability to manage	Neutral	Enhances ability to manage
	Construction impacts to the community	Significant negative impacts	Typical	Minimal impacts or positive enhancements
	Operations impacts to the community	Significant negative impacts	Typical	Minimal impacts or positive enhancements
	Reliable compliance	Uncertainty in compliance	Typical	Increased confidence in compliance

APPROACH/RESULTS

As the WRRF Project progressed through the planning and design phase, the City was faced with difficult decisions on the future of the recycled water system. As the City navigated these decisions, three (3) different recycled water (RW) program studies were sequentially completed to inform decision makers by answering key questions and defining prior unknowns. These studies provided valuable information for the planning and design of the WRRF Project, which is well integrated with the recycled water program. Figure 2 and Figure 3 demonstrates when each study that was initiated and where the findings were incorporated. A question goes into each study, and although we get a good result and great information, there are questions that come out of each study that require further investigation. The City's ability to rapidly and concurrently proceed through these studies with common approaches is what made the planning successful. This was more advantageous than one large, ultra-comprehensive planning study that takes would normally take 2 years to develop. With each study, the City's confidence grew and more informed decisions could be made.

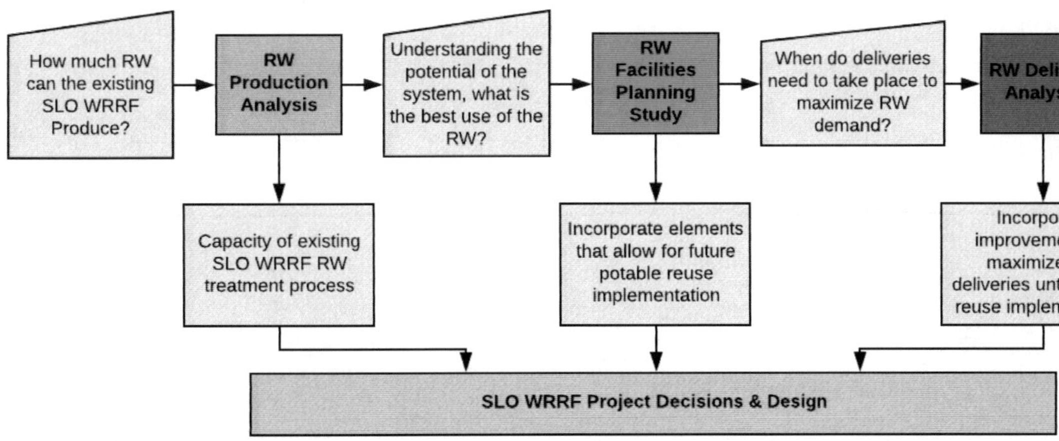

Figure 2. Connections Between the Studies and the Larger WRRF Project

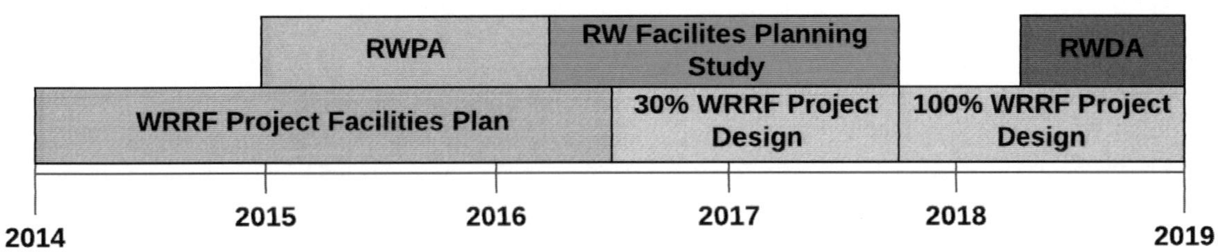

Figure 3. WRRF Project Planning, Design, and RWPA, RWFPS, and RWDA Timelines

Recycled Water Production Analysis (RWPA) (WSC, 2015)

To understand how to maximize resource recovery, the City needed to understand the current capacity and limitations of the SLO WRRF recycled water treatment process. The RWPA focused on the tertiary treatment system's infrastructure capacity and the secondary effluent supply to the tertiary treatment system to estimate how much Title 22 compliant disinfected tertiary treated water could be produced at the WRRF before the WRRF Project was completed. The RWPA also considered the 1.6 mgd downstream flow requirement that was assumed to be in place indefinitely for the purposes of the analysis.

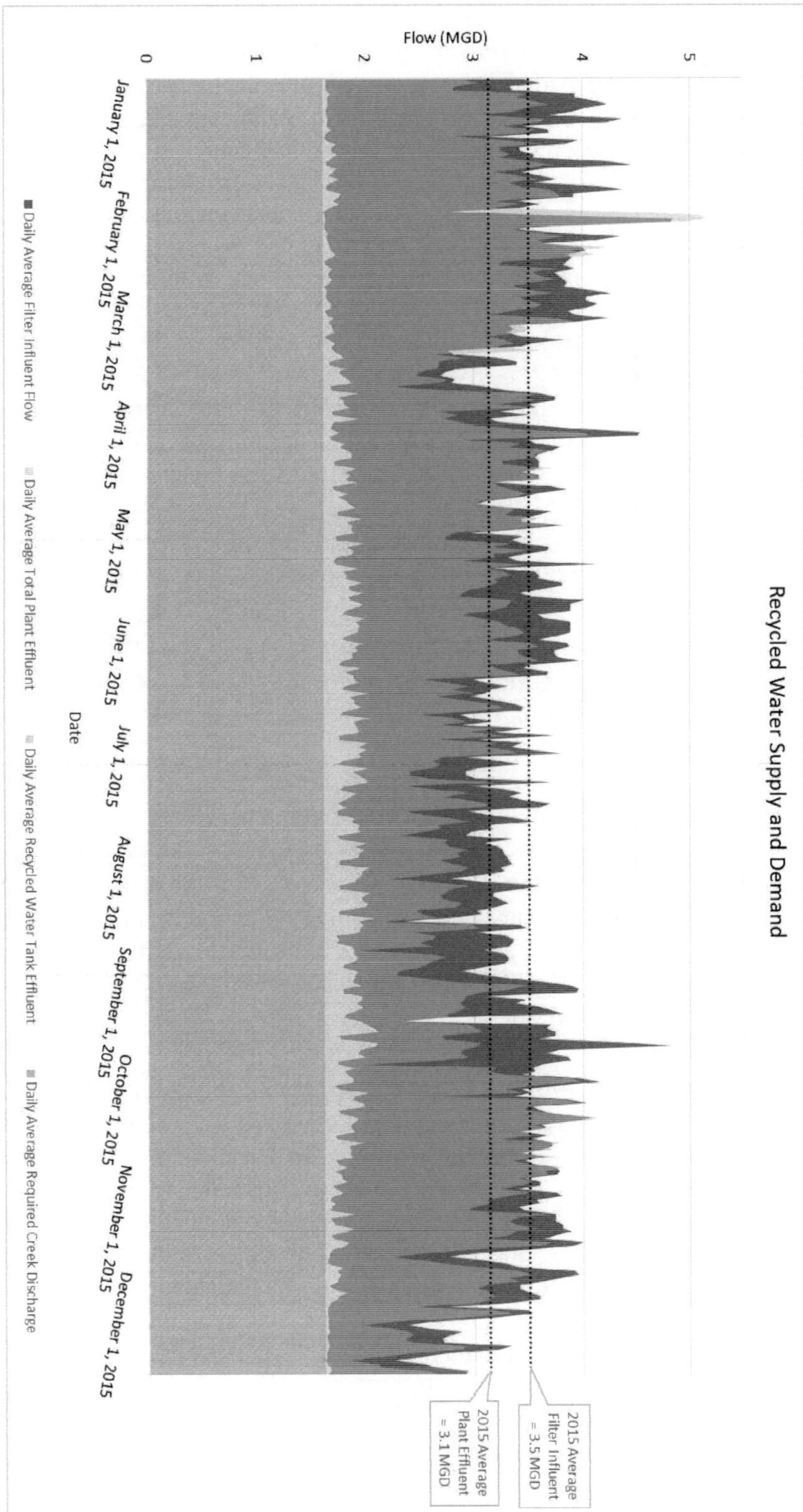

Figure 4. 2015 City of San Luis Obispo WRRF Recycled Water Supply and Demand, 2015 (RWPA, WSC 2015)

In 2015, WSC estimated that the WRRF's infrastructure can produce and deliver up to 3.5 mgd at any given time, while the average daily secondary effluent supply available for recycled water ranges between 1.5 and 1.9 mgd as demonstrated in Figure 5. The largest limitation of the WRRF's recycled water system is limited by the flows received by the WRRF. Table 3 summarizes the recommendations that resulted from the RWPA.

Table 4. Recycle Water Production Analysis Infrastructure Improvement Recommendations for the City of San Luis Obispo Water Resource Recovery Facility (WSC, 2015)

Recycled Water Vault Valve Automation	repair the existing actuator on the butterfly valve to achieve automation or replace the butterfly valve to more finely control the flow that goes to the creek versus recycled water. This is key in reducing the amount of treated effluent discharged to the creek to the minimum requirement, to the extent that recycled water demand is available. One option for a replacement valve would be a Cla-Val flow control valve.
Chlorine Dosing Automation	automate the recycled water chlorine dosing system to increase adaptability of the system to changing flow through the channels, to ensure only the chlorine that is needed to meet Title 22 standards is used, and to streamline the process for the operators.
Filter Feed Complex Pump Automation	automate filter feed pumping to reduce large jumps in flow into the filters, increase stability in the rest of tertiary system, and improve operational efficiency.
Backwashing Schedule	adjust the backwashing schedule so that backwashes occur during times of low recycled water demand. Consider peak electricity demand charges when selecting backwashing schedule.
Delivery Schedule	deliver some recycled water to customers during the day to increase recycled water deliveries and increase the working volume of the recycled water storage tank. Additional analysis could be performed to optimize the schedule to align with operator practice and contracts with the City for recycled water.

Recycled Water Facilities Planning Study (RWFPS, HDR & WSC, 2017)
The RWPA was able to quantify the amount of recycled water the WRRF could produce at a given time based on the infrastructure in place (before WRRF Project) which was estimated to be between 1,500 and 1,900 acre-ft per year (AFY). This information was used to provide the basis for the development of the RWFPS, which was developed by WSC and HDR and evaluated the following alternatives using the methodology previously described. At the time the RWFPS was initiated the five (5) alternatives below were considered most viable.

Alternative 1: Deliver tertiary-treated recycled water from the WRRF to agricultural customers in the Edna Valley irrigation area;

Alternative 2: Deliver tertiary-treated recycled water from the WRRF to Cal Poly;

Alternative 3: Provide groundwater recharge within the San Luis Obispo Valley Groundwater Basin for indirect potable reuse (IPR);

Alternative 4: Provide direct potable reuse (DPR); and

Alternative 5: Construct a new satellite WRRF to serve Cal Poly and the surrounding areas.

Table 4 displays the results from the screening process for evaluating each alternative. Based on the results of the screening process, a more detailed evaluation was conducted of the top 3 scoring alternatives to provide the City with a more detailed implementation plan for implementing the chosen alternatives.

Table 5. Recycled Water Facilities Planning Alternative Screening Results (WSC & HDR, 2017)

Alternative	Non-Weighted Score	Weighted Score
Alternative 1: Deliver Recycled Water to The Edna Valley irrigation area Agricultural Interests	28	**86**
Alternative 2: Deliver Recycled Water to Cal Poly	24	**74**
Alternative 3: Indirect Potable Reuse	24	**80**
Alternative 4: Direct Potable Reuse	22	**76**
Alternative 5: Construct a New Satellite WRRF to Serve Cal Poly and the Surrounding Areas	14	**40**

The chosen alternatives were to 1) deliver recycled to the Edna Valley irrigation area and 2) plan to implement potable reuse in the future (combination of alternative 3 and 4). Alternative 3 and 4 were combined to leave room for flexibility in the future based on the acceptable technologies at the time.

Recycle Water Delivery Analysis (RWDA) (WSC, 2019)
As the RWPA identified, the most significant limiting factor of the WRRF's recycled water system is not receiving sufficient influent flow quantities to the WRRF. WSC in partnership with the City are in the process of developing a tool that will allow them to develop a more detailed yearly delivery schedule to accommodate increasing demand for recycled water and to schedule deliveries to Edna Valley. The RWPA is scheduled to be completed in 2019.

WRRF Project
The City initiated the WRRF Project initially to meet new effluent water quality requirements mandated by the Central Coast Regional Water Quality Control Board (CCRWQ). The City could have easily chosen the more traditional route with a municipal wastewater treatment upgrade, but the City saw the WRRF Project as an opportunity to accomplish more. The City didn't plan to initiate the three (3) studies previously mentioned, rather, the studies were initiated as a result of a need to work through constraints and opportunities available for maximizing recycled water production and allowing the City to expand their water portfolio with a renewable, drought-resistant water supply.

Considering that the WRRF Project is currently on schedule to completed at least three (3) years after 100% design is complete, the RWPA provided the City with a way to maximize recycled water production in the meantime.

The RWFPS provided the City with better options that would allow them to continuously maximize the beneficial reuse. When the City made the decision to plan for a potable reuse future, the WRRF Project was in the 30% design stage. During this stage the WRRF Project decided to re-evaluate the options that were previously explored for the secondary treatment process. The result from the evaluation created a significant change by choosing to move forward with a membrane bioreactor system instead of a traditional activated sludge process.

The RWDA will provide the City with information needed to maximize the recycled water deliveries before and after the WRRF Project is constructed. The information derived from the study may be incorporated in the WRRF Project during construction to repurpose assets.

Through the WRRF Project journey the City was able to develop short-, medium-, and long-term goals to realize social, environmental, and economical benefits for the City and the region. Figure 4 provides a summary of goals the City plans to achieve at the end of their water supply evolution.

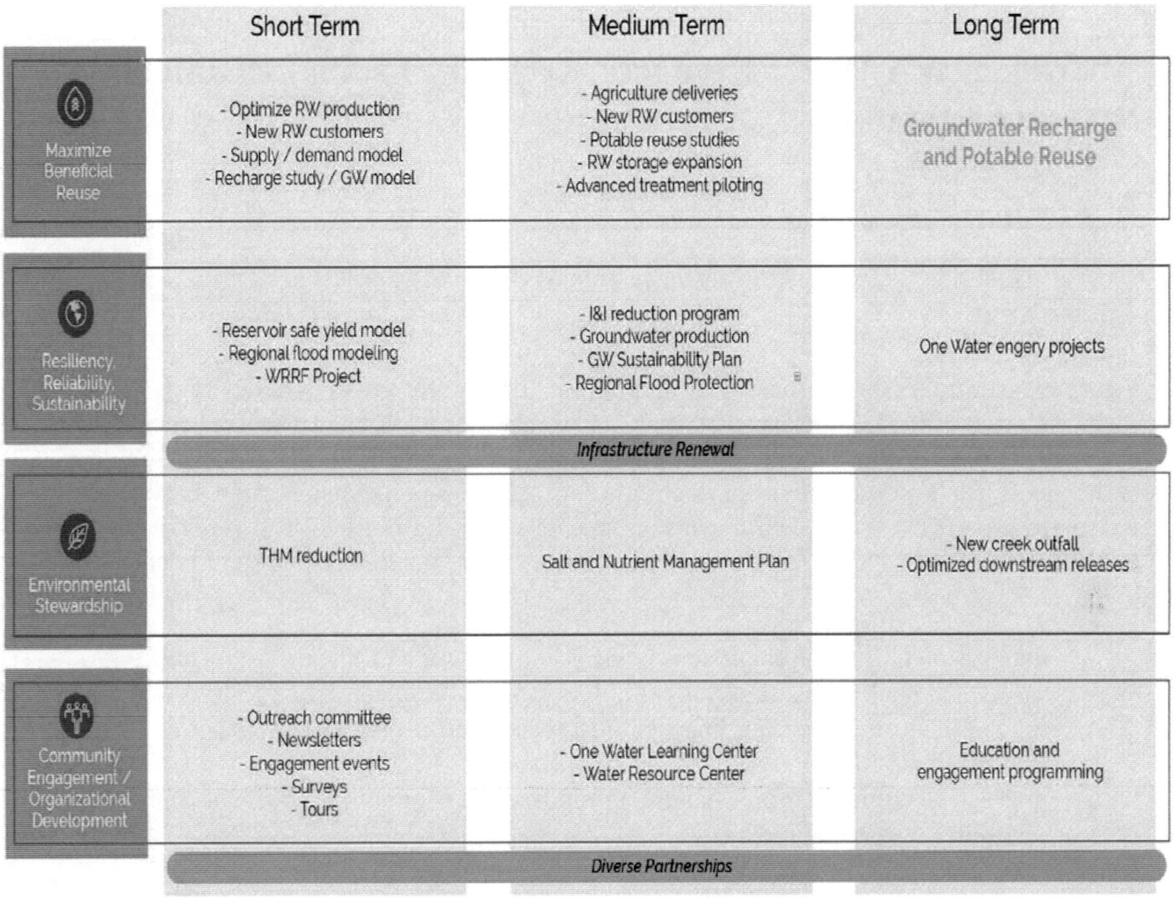

Figure 5. City of San Luis Obispo Water Resiliency Planning Summary

CONCLUSION

The City has taken a proactive approach to fully understanding their existing recycled water production capabilities and customer demands, including developing anticipated short-term and long-term future demands. The proactive approach has led to a series of studies that fed back into the larger WRRF Project and enhanced the project at during different stages within the project. The incremental steps taken to realize the City's goal to maximize recycled water production has led the City to make a significant change to the WRRF Project that will allow them more flexibility to implement potable reuse in the future, while maximizing recycled water production today.

Aberdeen, WA Water Resource Recovery Facility Disinfection System Evaluation

Jay Swift, P.E., Gray and Osborne, Seattle, WA

ABSTRACT

This paper describes an alternatives evaluation for upgrading the system used for disinfection of effluent discharged by the City of Aberdeen's Wastewater Treatment Plant (WWTP). The City currently uses gas chlorination for disinfection of WWTP effluent and a gas sulfur dioxide system for dechlorination. Although the disinfection system has proven to be safe and effective in meeting the City's NPDES permit limits, the City is evaluating options for upgrading or replacing the existing disinfection system because of:

- the age of equipment, approaching the end of its useful life,
- safety and health considerations for plant staff and the public associated with transport, storing and handling of chlorine and sulfur dioxide gases, both of which are extremely toxic
- more stringent regulations and corresponding costs of compliance, and decreasing reliability of gas shipments.

Nine alternatives were identified and screened, and a preferred alternative identified.

KEYWORDS

Disinfection, Chlorination, Safety, Hypochlorite

BACKGROUND

The City of Aberdeen, WA currently uses gas chlorination for disinfection of WWTP effluent and a gas sulfur dioxide system for dechlorination. Although the disinfection system has proven to be effective in meeting the City's NPDES permit limits, the City wishes to evaluate the possibility of upgrading or replacing the existing disinfection system due to the following issues:

1. Gaseous chlorine and sulfur dioxide are highly toxic. Although the City has not had any major accidents or releases of gaseous chlorine or sulfur dioxide, the current system does present a safety risk to the WWTP staff and the public.

2. Shipments of gaseous chlorine and sulfur dioxide (in particular) have become less reliable. There are fewer manufacturers and suppliers of

chlorine and sulfur dioxide as a result of more stringent regulations. Jones Chemical is the only chlorine gas vendor in Washington.

3. In order to stay below the threshold quantity limits defined by the Environmental Protection Agency Risk Management Plan (RMP) Rule (40 CFR 68), federal Occupational Safety and Health Administration (OSHA) Process Safety Management Regulations (29 CFR 1910.119), Washington State Department of Labor and Industries Division of Occupational Safety and Health Process Safety Management (PSM) Regulations (WAC 296-67), the City is required to minimize the amount of these gases onsite. Combined with the unreliability of deliveries, there is a risk of permit violations.

4. Although portions of the system have been upgraded recently, some of the equipment and piping is approaching the end of its useful life.

Aberdeen WWTP Liquid Stream Treatment

The Aberdeen WWTP receives wastewater from the City of Aberdeen commercial residential and industrial users, the City of Cosmopolis, the Stafford Creek Correctional Center, a small amount of leachate from the closed county landfill operated by LeMay Solid Waste, and septage from septic tanks trucked into the facility. The WWTP discharges to Grays Harbor from an outfall located in the northern Grays Harbor channel at the mouth of the Chehalis River.

The WWTP uses a conventional activated sludge process with primary clarifiers, diffused air aeration basins with selectors, secondary clarifiers, and chlorine disinfection. The wastewater from the collection system enters the influent pump station before being pumped up to gravity flow through the facility. Septic tank waste hauled to the facility by truck is dumped into the system downstream of the influent sampling station. Therefore, the septic wastes are not accounted for in the influent loading. The landfill leachate is pumped through a force main within the City's collection system, and is thus passed through the headworks and accounted for in the influent loading.

The grit removal system in the headworks has not been used for many years since it proved to be inadequate, so grit removal is accomplished on the primary clarifier sludge using a hydrocyclone. Some rocks and coarse grit settle and accumulate in the influent pump station wet well and in the influent channel prior to the step screens.

After screening, the wastewater flow is split to the two existing primary clarifiers. A third primary clarifier is planned for the future. From the primary clarifiers the flow is joined and again split to feed two aeration basins. Each aeration basin has two selectors (one anoxic and one anaerobic) selectors and an aeration basin utilizing fine bubble aeration. There are provisions for a third aeration basin for future flows.

The flow from the aeration basins is combined and split to three secondary clarifiers. The two original clarifiers have an integral outer ring which functions as the chlorine contact chamber. The outfall was replaced in 2014 and now consists of a 36-inch HDPE SDR 11 pipe that discharges approximately 135 feet from the shore at MLLW in 18 feet of water. The outfall is in the northern channel of the Chehalis River, but the water is considered a part of Grays Harbor Estuary and a marine water.

Existing Disinfection System

The existing disinfection system provides chlorination using gaseous chlorine and dechlorination using gaseous sulfur dioxide. Chlorine contact is provided in channels forming an annular ring around Secondary Clarifiers 1 and 2. (The effluent from the third clarifier must be sent back and mixed into the existing chlorine contact chambers.)

Chlorine flash mixers are installed in chambers immediately upstream of the contact channels. Sulfur dioxide is added at the effluent Parshall flume.

The chlorination and dechlorination process systems are continuously controlled and monitored by three oxidation-reduction potential (ORP) meters: one in each of the chlorine contact channels about 15 feet downstream of the rapid mix basin and one in the common discharge chamber downstream of the final effluent Parshall flume. Total residual chlorine is monitored by collection of a single grab sample in the common discharge chamber downstream of the Parshall flume for permit compliance. Total chlorine residual is also measured from grab samples collected at the locations of the ORP probes in the east and west chlorine contact channel. The target ORP setting generally corresponds to total chlorine dose residual of 0.5 to 0.75 mg/L.

Chlorine dose is controlled to maintain an operator adjustable ORP set point. Similarly, the sulfur dioxide dose is controlled to maintain an ORP set point corresponding to a chlorine residual of less than 0.08 mg/L.

Much of the chlorination equipment was installed during the 1977 to 1981 expansion, including the majority of the piping. Dechlorination equipment was installed in a subsequent upgrade. The chlorinators and sulfonators including the injectors were replaced in 2000 and the new units rebuilt in 2008 to 2009. The ORP meters are about 15 years old, and will need to be replaced within the next 3 years. All of the remaining chlorination/dechlorination equipment will need to be replaced within the 20-year planning period, if the City retains gas chlorination /dechlorination as the means of disinfection.

In the Gas Storage room, each chlorine and sulfur dioxide cylinder is connected to an automatic switchover vacuum regulator and is sitting on a scale. Each set of cylinders feeds into a common vacuum header that supplies chlorine or sulfur dioxide to the chlorinators and sulfonators, respectively. The chlorinators and sulfonators are located in a room separate from the gas storage room. The chlorinators and the sulfonators use vacuum injectors powered by the City water supply system to draw chlorine or sulfur dioxide gas through the chlorinators and sulfonators respectively. The chlorinators and sulfonators each consume approximately 20 to 50 gpm to create the vacuum necessary to achieve the full capacity of each chlorinator and sulfonator, respectively. The chlorine injection point and the side stream pumping system. The chlorine gas feed system is equipped with two 500 pound per day (lbs/d) Wallace and Tiernan vacuum chlorinators. One chlorinators is from the original installation in 1981 and other chlorinator (Wallace and Tiernan V 10K) was installed in 2004 as a replacement for the other original chlorinator. Chlorinators are arranged in parallel so that either chlorinator can supply the necessary chlorine dose for disinfection of secondary effluent. During normal operation one chlorinator is used and the other is isolated as a backup. The redundant chlorinator allows uninterrupted chlorine feed during planned maintenance and servicing. The backup chlorinator must be activated manually in the case of a system failure.

The chlorine solution conveyed through a pipe manifold system outfitted with rotameters to distribute the chlorine solution to the rapid mix tank at head end of the each of the two chlorine contact tanks. Each chlorinator (i.e., chlorine dose) is controlled by a Siemens Controller (Wallace and Tiernan SFC SC) that uses a feedback control loop based on continuous measurement of the oxidation reduction potential (ORP) of the chlorinated wastewater approximately 15 feet downstream of the chlorine rapid mix tank. The ORP set point is static but is operator adjustable based on an empirically determined correlation between ORP and total residual chlorine. The chlorine distribution manifold also provides for the metering of chlorine solution to the other areas of the plant including the outlet box of the aeration basins and to the sludge recycle wells at Clarifiers 1 and 2.

A third chlorinator was installed in in 1981 as part of the original plant design to chlorinate the gravity thicker sludge feed; the plant manager reports the gravity thickener feed chlorination system was not used so the chlorinator was converted to a sulfonator in 1998 when the City commenced practicing dechlorination of the plant effluent. The dechlorination system consists of two vacuum sulfonators. City water to drive the sulfur dioxide injectors. One of the sulfonators has a capacity of 450 lbs/day and the other has a capacity of 250 lbs/day. The 250 lbs/d sulfonator was installed in 2004. Sulfonators-injector systems are arranged in parallel so that either sulfonator-injector can supply the necessary sulfur dioxide dose for dechlorination of the disinfected secondary effluent. During normal operation one sulfonator-injector system is used and the other is isolated as a backup. The redundant sulfonator-injector system allows uninterrupted chlorine feed during planned maintenance and servicing. The backup chlorinator must be activated manually in case of a system failure. The sulfur dioxide solution conveyed through a pipe manifold system outfitted with rotameters to distribute the aqueous sulfur dioxide solution to the head boxes of the effluent flow metering structure (i.e., parshall flumes). Both sulfonators (i.e., sulfur dioxide dose) are controlled by a US Filter Model 960 Water Disinfection Controller that uses a feedback control loop based on continuous measurement of the oxidation reduction potential (ORP) of the disinfected effluent contained in the tail box of the effluent flow metering structure down (i.e., downstream of the Parshall flumes). The ORP set point is static but is operator adjustable based on an empirically determined correlation between ORP and total residual chlorine.

Each room of the gas storage room and the chlorinator/sulfonator room are constantly ventilated with chlorine gas detectors that can activate the automatic shut-off valves as well as initiate local and SCADA alarms. The currently the gas storage and chlorinator/sulfonator rooms are not equipped with fire sprinkler systems. Chlorine usage, and ORP at the head end of the chlorine contact chamber and ORP after dechlorination are continuously monitored by the City's SCADA system.

CURRENT AND HISTORICAL CHEMICAL CONSUMPTION AND PERFORMANCE

Table 1 summarizes WWTP influent flows, effluent flows, chlorine consumption and sulfur dioxide consumption for the period of 2013 to 2016.

Figure 1 shows influent flow and chlorine consumption on a monthly basis. The chlorine consumption data includes that used for other purposes (chlorination of RAS, etc.) As shown, peak month chlorine consumption was approximately 110 lb/day; however, typical consumption was in the range of 35 to 80 lb/day.

Figures 2 and 3 show fecal coliform concentrations as a function of daily flows and chlorine dose, respectively, for the period of January 2013 through April 2016. As shown, at higher flows, there is a much greater probability of higher fecal coliform levels. When daily flows exceed 10 mgd, the data indicate that there is a 38 percent probability (6 out of 16) that the fecal coliform level would exceed 100 cfu/100 mL.

In 2016, there were two excursions where fecal coliform levels exceeded 100 org./100 mL:

- June 23, 2016, 1900 org./100 mL fecal coliform, west chlorination chamber off-line, 2.301 mgd daily flow
- October 14, 2016, 1300 org./100 mL fecal coliform, high flows, 12.419 mgd daily flow

TABLE 1 - WWTP Flows, Chlorine Consumption and Sulfur Dioxide Consumption

		2013	2014	2015	2016	Average	Maximum
Influent							
Minimum Day Flow	mgd	1.75	1.68	1.61	1.83	1.72	1.83
Ave. Day Flow	mgd	3.25	3.62	3.45	3.99	3.58	3.99
Max. Day Flow	mgd	10.98	13.15	20.60	20.50	16.31	20.60
Effluent							
Minimum Day Flow	mgd	1.75	1.71	1.42	1.56	1.61	1.75
Ave. Day Flow	mgd	3.15	3.66	3.47	3.84	3.53	3.84
Max. Day Flow	mgd	10.54	13.35	20.60	20.50	16.25	20.60
Ave. Cl2 used	lb/d	68	57	67	55	62	68
Max. Cl2 used	lb/d	200	160	320	240	230	320
Ave. SO2 used	lb/d	53	42	41	48	46	53
Max. SO2 used	lb/d	180	170	210	240	200	240

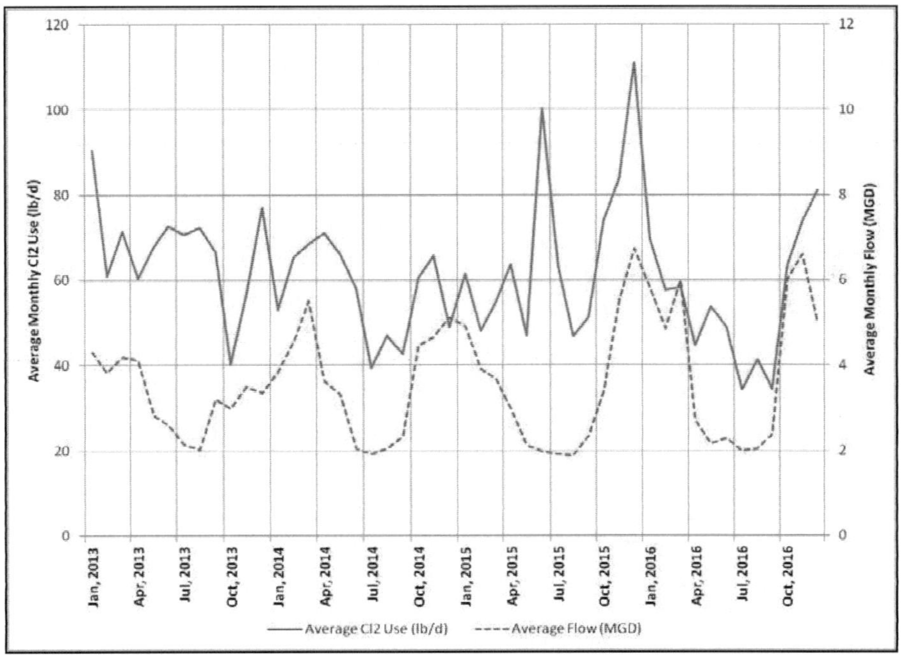

FIGURE 1 - Monthly Chlorine Consumption – Aberdeen WWTP (2013-2016)

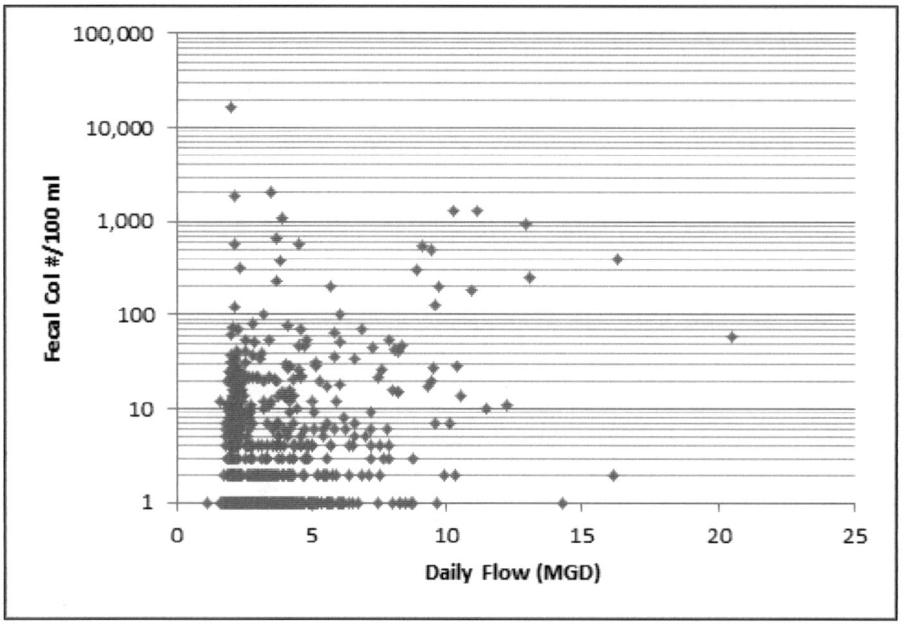

FIGURE 2 - Fecal Coliform Concentration vs. Daily Flow – Jan. 2013 to Apr. 2017

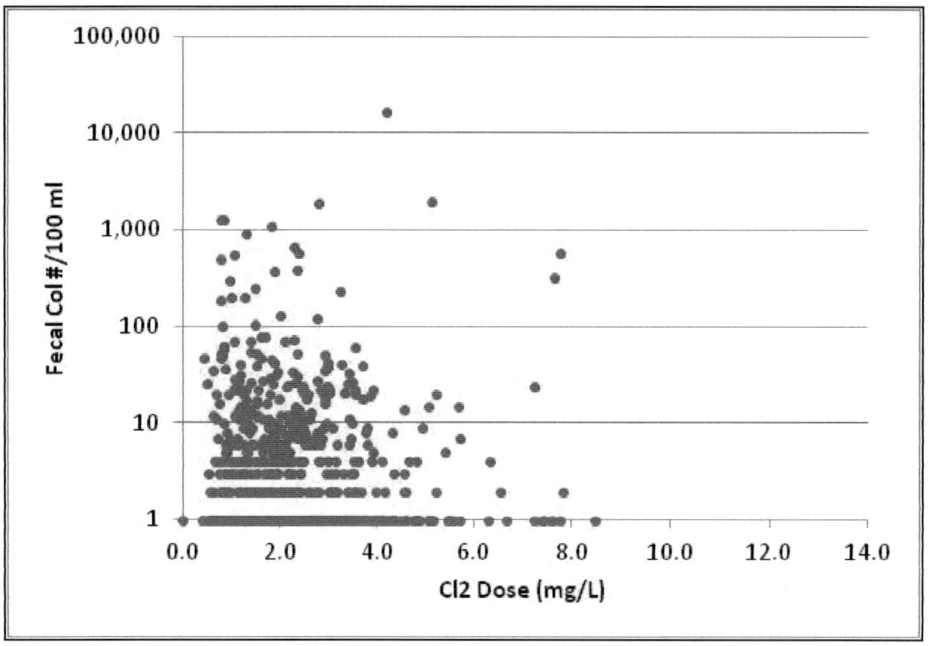

FIGURE 3 - Fecal Coliform Concentration vs. Cl_2 Dose – Jan. 2013 to Apr. 2017

DISINFECTION ALTERNATIVES

Alternative 1 – Gas Chlorination/Dechlorination System With In-Kind Equipment Replacement ("No Action" Alternative)

With this alternative, the City would retain the existing gas chlorination and gaseous sulfur dioxide dechlorination system. It is assumed that the City would replace the disinfection equipment with in-kind replacements as the equipment approaches the end of its useful life.

Because the WWTP does not nitrify consistently, retaining a chlorination system would potentially have issues with virus removal. Non-nitrified wastewater effluents contain significant levels of ammonia. Ammonia combines with chlorine during disinfection to form chloramines (monochloramine NH_2Cl, dichloramine and nitrogen trichloride $NHCl_2$ and NCl_3). No free chlorine is generally measurable until breakpoint is achieved whereby the chlorine dose is about 7.6 times the ammonia nitrogen concentration. Free chlorine residual measures the concentration of hypochlorous acid (HOCl), which is a product of combining water and chlorine. HOCl is a highly effective disinfectant. Chloramines are formed when chlorine is added to water that already contains ammonia (NH_3) or when both chlorine and ammonia are added to water. Chloramines are stable and effective disinfectants, but are less effective than HOCl in the short term. Chloramines require roughly 100 to 200 times more contact time than HOCl for the same residual level to achieve similar bacterial or virus inactivation, respectively. When nitrification is near complete then the chlorine dose required to achieve breakpoint is lower and a free chlorine residual can be achieved along with its stronger germicidal characteristics.

<u>Advantages and Disadvantages</u>

The "No Action" alternative would minimize capital and operating costs, but would not address existing significant issues of safety and reliability.

Alternative 2 – Gas Chlorination/Dechlorination System with Upgrades

This alternative maintains the existing gas chlorination/dechlorination system and adds scrubbers to mitigate risks to City staff and the general public. Several strategies are available to address OSHA's 1992 PSM regulation (29 CFR 1910.119) and the possible need for an RMP under 40 CFR 68.

For the Aberdeen wastewater treatment facility, either a scrubber for the entire chlorine/sulfur dioxide storage room or a containment vessel for each chlorine and sulfur dioxide container could be provided to achieve compliance with the IFC. The capacity of the scrubber required would need to adequate to contain and neutralize the full contents of the largest single storage vessel—in this case, 1 ton cylinder of chlorine or sulfur dioxide gas. Caustic soda would be the neutralizing agent, since it is capable of treating both chlorine and sulfur dioxide. A 3,000 cfm (air capacity) scrubber would be required to neutralize a 1-ton cylinder. The existing chlorination building HVAC system would have to be modified so that all ventilation to the atmosphere would be shut down in the event of a leak and all gases would be routed to the scrubber.

The components of a scrubber system include a caustic soda tank and feed system, a fan, controls and instrumentation, and various piping and ducting. The caustic soda is fed into the ejector via a recirculation pump and mixes with the incoming gas to be neutralized. Liquid droplets interact with and absorb the gas. Cross-flow packed-bed sections are used to maximize the contact between gas and the sprayed caustic solution. A clean gas outlet outfitted with a mist eliminator would be provided for treated gas discharge. A maintenance consideration is that the local fire marshal may require the City to replace the caustic soda annually.

The scrubber can be located outdoors, but an enclosure to provide protection from the elements is recommended by manufacturers. Heat tracing of tanks and piping would be required to prevent solidification by freezing of the caustic solution. The entire system should be connected into the City's standby electrical generator system in the event of a combined power outage and gas leak.

Advantages and Disadvantages

Alternative 2 would have relatively low capital and operating costs and address most of the existing safety issues and consequently would be safer than the "No Action" alternative. However, there would still be safety risks, environmental concerns, and reliability issues associated with the supply of gaseous chlorine and sulfur dioxide because they are very hazardous chemicals.

Alternative 3 – Bulk Hypochlorite

Due to the multiple and increasingly stringent regulatory requirements and the high cost of installing and maintaining scrubbers or containment vessels, many utilities have transitioned from gas chlorination to using bulk sodium hypochlorite for disinfection and from gaseous sulfur dioxide to SBS or CTS for dechlorination. Sodium hypochlorite does not have the same potential for catastrophic release as gas chlorination. Hypochlorite systems have significantly less stringent requirements for storage, leak detection and treatment/containment, OSHA, and RMP measures associated with gas chlorination.

Sodium hypochlorite (NaOCl) is the only liquid hypochlorite form in current use. There are several grades available. Sodium hypochlorite, often referred to as *liquid bleach*, is commercially available only in liquid form, typically in concentrations between 5 and 15 percent available chlorine.

Calcium hypochlorite ($Ca(OCl)_2$) is the predominant dry form. Sometimes referred to as *powder bleach*, calcium hypochlorite is a dry material typically consisting of 65 percent available chlorine. Often called *high-test hypochlorite* (HTH), 1 pound is equivalent to 0.65 pounds of elemental chlorine. Because of the higher cost and relatively low popularity (and performance issues) for tablet systems for larger plants, the focus of this section is on liquid sodium hypochlorite.

Bulk Liquid Sodium Hypochlorite

Sodium hypochlorite is a safer alternative to gaseous chlorine in regard to storage and transport of the chemical. Bulk sodium hypochlorite typically is delivered in solution with 12.5 percent chlorine content. However, design considerations should take into account the decay of sodium hypochlorite solution strength and natural off-gassing during decomposition that can bind piping and pumps. Additionally, decomposition of sodium hypochlorite can result in the formation of sodium chlorate.

A sodium hypochlorite disinfection system generally is composed of a storage tank, metering pumps, and associated ancillary equipment. Dechlorination can be achieved via the existing sulfur dioxide system or by employing the use of a non-gaseous quenching alternative such as sodium bisulfite, calcium thiosulfate, or sodium metabisulfite.

Many cities and utilities, including King County, Seattle Public Utilities, and the City of Portland, have replaced their chlorine gas systems with a liquid sodium hypochlorite system. Liquid sodium hypochlorite solution is available at a concentration by weight of 12 to 15 percent. A 12.5 percent solution has an equivalent chlorine content of about 1 pound of chlorine per gallon of solution. Use of liquid sodium hypochlorite results in a slight increase in the pH of the disinfected water.

The primary benefit of sodium hypochlorite is that it is a safer system than chlorine gas because liquids have a reduced risk of catastrophic release that could harm operators or nearby residents. This option is reliable when backup storage and pumping capabilities are included.

The main drawbacks to bulk sodium hypochlorite are higher cost (compared to chlorine gas) and its tendency to degrade, which is affected by product concentration, temperature, and exposure to sunlight. Degradation decreases the amount of effective hypochlorite disinfection in the liquid and increases the formation of byproducts such as chlorate ion (ClO_3^-). Additional monitoring of the stored product is recommended for such systems. Dilution of the delivered product will reduce the rate of degradation. Some facilities dilute 12.5 percent sodium hypochlorite on-site and store it at a concentration of 6 to 8 percent, if it will be stored for more than 30 days. Containment around the storage tanks is required in the event of a spill or leak. Another drawback is that bulk sodium hypochlorite is a corrosive liquid that requires double-contained piping, additional operator handling precautions, and frequent maintenance of the feed equipment.

Positive displacement diaphragm pumps are commonly used for metering sodium hypochlorite. Other pumps, such as venture siphon pumps, peristaltic pumps and gear pumps, have also been used successfully. Because sodium hypochlorite at these concentrations is a corrosive liquid even at a concentration of 1 percent, it is important that suitable non-metallic materials, such as PVC, PVDF, and PE for tubing and pipe, and Viton, Teflon, or Kalrez, should be specified for pump seals and O-rings. Special solvent cement is required to assure the long-term performance of solvent cement welded PVC pipe joints. Hypochlorite also tends to generate off-gases through the formation of oxygen during degradation. The off-gassing can cause vapor locking of the dosing pumps so proper venting is essential. Therefore, it is important that venting of these gases be provided immediately before the pumps and that the metering pumps and system be designed to handle the off-gassing.

Additional improvements such as a secondary containment area and climate-controlled storage are necessary due to the chemical's corrosive nature and high decomposition rate; sodium hypochlorite has a relatively short shelf life that is dependent on sunlight exposure, temperature, and initial concentration. Therefore, the chemical should be stored in a cool location in a corrosion-resistant tank.

The predominant cost associated with operation and maintenance of a sodium hypochlorite feed system will be the cost per gallon of bulk chemical consumed.

<u>Advantages and Disadvantages</u>

Using bulk sodium hypochlorite is equally as effective as using gaseous chlorine and is a commonly used and proven disinfection technology. It is especially advantageous with respect to availability, safety, and power consumption.

Bulk sodium hypochlorite when applied to wastewater adds alkalinity/increases pH, but Cl_2 gas application consumes alkalinity and tends to lower the pH.

Less equipment and lower power consumption are required compared to other alternatives. However, operational and maintenance costs are significantly higher compared to gas due to the higher chemical cost. The purchase price of sodium hypochlorite can be three to four times the cost of gaseous chlorine.

Sodium hypochlorite systems still require the use of chemicals for dechlorination

Alternative 4 – On-Site Generation of Sodium Hypochlorite

On-site generation of sodium hypochlorite reduces some of the hazards associated with using bulk 12.5 percent sodium hypochlorite. On-site generation systems use salt, water, and electricity to produce a sodium hypochlorite solution. The 0.8 percent sodium hypochlorite solution is most common for on-site generation and is less hazardous than 12.5 percent sodium hypochlorite, and it does not degrade in strength as rapidly as higher concentrations. However, systems generating as high as 16 percent sodium hypochlorite are available and increasingly popular. All systems require the addition of salt on a periodic basis and use additional power to produce the sodium hypochlorite solution. Even though an on-site generation system has a higher capital cost than other chlorine alternatives, it does have lower operating costs. Several manufacturers are now available.

On-site hypochlorite generation is gaining increasing acceptance and is being installed at facilities across the country and in the Northwest. Like sodium hypochlorite, on-site generation is considerably safer than chlorine gas for use at a WWTP. Advantages over liquid hypochlorite are that there is no concern about product degradation due to lower concentration and shorter holding times, and shippers and operators must generally only handle dry salt rather than liquid hypochlorite. Reliability of these systems has generally improved in the last several years. During normal operation, minimal operator attention is required.

On-site hypochlorite generation has gained increased acceptance and is being installed at many facilities across the country and in the Northwest. Like sodium hypochlorite, on-site generation is considerably safer than chlorine gas for use at a WWTP. Advantages of on-site generation over bulk delivery of liquid hypochlorite are:

- There is no concern about product degradation due to lower concentration and shorter holding times; and
- Shippers and operators generally only handle dry salt rather than liquid hypochlorite.

Reliability of these systems has generally improved in the last several years. During normal operation, minimal operator attention is required.

On-site hypochlorite generation produces a dilute (0.8 to 1 percent) hypochlorite solution that is fed as the primary disinfectant. The process uses water, salt, and electricity to produce the hypochlorite solution. Each pound equivalent of chlorine requires approximately 3.5 pounds of salt, 15 gallons of water, and 2.5 kilowatt-hours (kWh) of electricity.

An on-site generation system consists of a brine tank/salt saturator, a rectifier, PLC, generating units, and a product tank. Metering pumps for the brine and hypochlorite are also required. High-quality vacuum-dried salt can be delivered in 50-pound bags, 2,000-pound super sacks, or delivered in bulk and blown into salt storage tanks. The water used for preparing the brine is typically softened to prevent scaling in the system, requiring disposal of brine waste from the softener to the sanitary system. The hypochlorite solution produced is transferred to a storage tank from which it is metered to the system. The sodium hypochlorite storage tank typically provides a minimum of 1 day reserve of the feed product as well as a place to vent hydrogen gas, a byproduct of the hypochlorite generation process. If a longer shutdown of the generating unit is anticipated, then bulk sodium hypochlorite can be delivered to, diluted, and fed from the product tank.

Advantages and Disadvantages

Using sodium hypochlorite generated onsite is equally as effective as using gaseous chlorine or bulk hypochlorite and is an increasingly used and proven disinfection technology. It eliminates the hazards associated with handling and storage of chlorine gas and bulk 12.5 percent sodium hypochlorite.

On-site generated sodium hypochlorite also adds alkalinity/increases pH while Cl_2 gas application consumes alkalinity and tends to lower the pH.

The system would be more complex operationally than some of the other systems (including bulk liquid hypochlorite) and require significantly more power than conventional gaseous chlorination but less than ultra violet light disinfection. The system also requires considerably more maintenance than bulk sodium hypochlorite.

All sodium hypochlorite based disinfection systems require the use of chemicals for dechlorination.

Alternative 5 – Ultraviolet Disinfection

Ultraviolet (UV) light as a disinfection method has become very popular in recent years because of the safety issues and disinfection byproduct issues concerning chlorination. Additionally, costs of utilizing UV systems have decreased due to system design developments as well as reliability improvements. A UV disinfection system transfers electromagnetic energy from an ultraviolet lamp to denature organism's genetic material, thus halting the pathogen's ability to reproduce. This system is composed of mercury arc lamps, reactor(s), ballasts, and control systems.

Effectiveness of a UV system is dependent upon flow rates and wastewater characteristics; one of the key characteristics is ultraviolet transmittance (UVT). Water with low UV transmittance (UVT) absorbs more UV light and therefore requires larger dose of UV light to achieve a desired level of inactivation as compared to a water with a high UVT; hence, the capital and operating and maintenance costs are higher. The lamp spacing also may need to be reduced, which would result in higher head loss across the UV system. Sometimes the UVT can be improved by enhanced removal of constituents in the secondary process and/or tertiary filtration. (WERF, 2007). The UVT of the wastewater to be disinfected is one of the most important parameters in determining the efficacy of UV disinfection. Therefore, frequent monitoring of UVT is critical to successful UV system operation in plants that have variable effluent transmittance. If a UV disinfection system were designed to perform under a water transmittance of 65 percent and an actual transmittance of 55 percent was observed, the system may underperform by 33 percent or more. Industrial discharges (with recalcitrant dissolved UV-absorbing constituents) and wet-weather events (leading to increased effluent turbidity) often exhibit significant adverse effects on UVT.

Figure 4 summarizes UVT and turbidity data collected in May 2017 for Aberdeen WWTP effluent. There is insufficient data to conclusively size the UV system, with only 1 month of data over a relatively dry period. However, based on the data to date, it appears that a design UVT of 55 to 60 percent may be appropriate. Fifty-five percent is used for this analysis as a conservative estimate; however, this design UVT should be verified with additional testing, especially at high flow conditions.

The fecal coliform concentration in the effluent produced by the secondary treatment process was characterized to be 100,500 CFU/100 mL during the collection of UVT and turbidity data. This value is consistent with a low sludge age activated sludge treatment system effluent consistent with the operation of the Aberdeen WWTP. Therefore, a minimum of a 2.7 – log reduction would be required to achieve the discharge criterion of 200 CFU/100 mL.

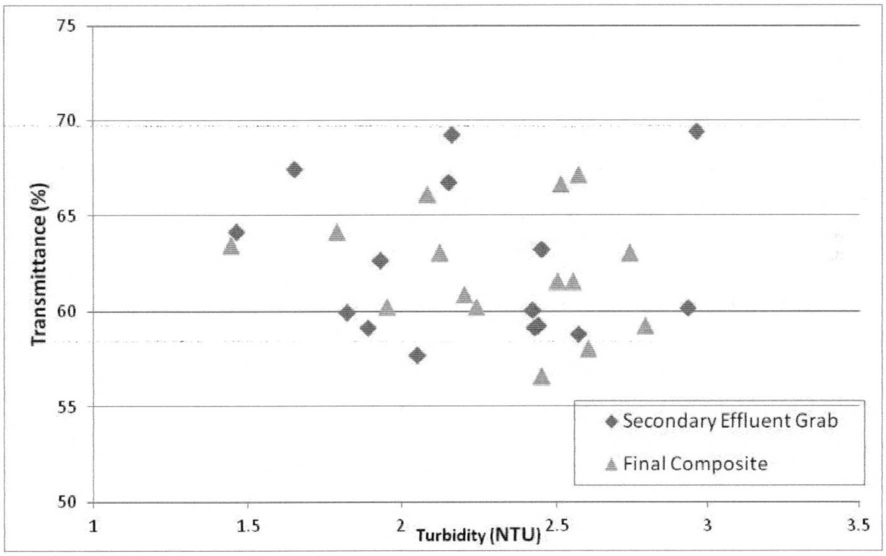

FIGURE 4 - Aberdeen WWTP – Effluent UVT and Turbidity

For implementation at this facility, UV disinfection equipment was evaluated in a single facility, with dual, parallel UV disinfection channels. It is not recommended that the new UV channels be constructed within the existing under-launder chlorine contact basins, as there would be significant access and maintenance issues with such a configuration. Each channel contains (at a minimum) one duty-bank of UV lamps (for average day flows), with a redundant bank (for peak flow operation and operation during routine maintenance on the duty-bank) of an equivalent number of lamps.

UV disinfection systems are a proven technology for disinfection of wastewater effluent. Available UV disinfection systems include low-pressure systems, medium-pressure systems, and newer, low-pressure, high-intensity systems.

In UV disinfection, specific wavelengths of electromagnetic radiation are used to inactivate microorganisms through the denaturing of their DNA (and/or RNA). Wavelengths ranging from 250 to 270 nanometers (nm) are readily absorbed by DNA/RNA, effectively inactivating certain pathogens found in municipal wastewater by rendering them unable to replicate. The dose of UV light is measured as the product of intensity and exposure time, as milliWatt-seconds per square centimeter (mW-s/cm^2), which is equal to one milli-Joule per square centimeter (mJ/cm^2).

Typical design doses for municipal wastewater are in the range of 20 to 45 mJ/cm^2, which should be verified during predesign through bench- and/or pilot-scale testing. In the design phase, it is recommended that collimated beam testing be performed to verify dose requirements; however, for the purposes of this evaluation, it was assumed the design UV dose is 30 mJ/cm^2 at a UV transmittance of 55 percent or greater (as demonstrated by bioassay testing on MS-2 coliphage, an emerging indicator organism standard); which would ensure 99.9 percent inactivation of fecal coliform in effluent.

Other parameters are significant as well: suspended solids can shield organisms from exposure to UV, and color and organics can absorb UV radiation, reducing its effectiveness. Other important parameters are those that can foul the UV lamps' protective quartz sleeves, such as hardness and iron. Both hardness and iron can form a coating on the quartz sleeves, impeding the transfer of UV into the bulk wastewater.

One of the main benefits of UV disinfection is its lack of chemicals. Other than for occasional cleaning of lamp sleeves (and the mercury present in the lamps), no chemicals are required. However, power costs, and in some cases maintenance costs, are increased. If the City were to install a UV system for effluent disinfection, then sufficient backup power should be designed into the system or an emergency bulk hypochlorite dosing system and dechlorination system would be required to address power outages.

A UV disinfection system consists of a power supply, lamps, reactor chambers, cleaning equipment, flow control, and controls and instrumentation. Reactor chambers (either open or enclosed channels, including in-pipe reactors) are set up with UV lamps in a horizontal or vertical configuration. Lamps are protected from the effluent by a quartz sleeve. Effluent flows through the reactor chambers and is exposed to UV light. Dose is measured through measurement of intensity at various monitoring points in the channel. In an open-channel system, effluent weir control is necessary to maintain the submergence of the lamps.

A major maintenance task for a UV system is cleaning the surface of the quartz sleeves. Sleeves require periodic cleaning due to the tendency of constituents in wastewater to foul the surface of the lamp. Cleaning requirements are determined based on the reduction in measured intensity. A monitoring system with sensors at the point of lowest dose, typically on the wall of the reactor chamber, measures the intensity of UV radiation. After the intensity drops below a set point, the system control will indicate the need for cleaning. Cleaning mechanisms, as explained in the following sections, vary with the type of UV system; most current systems now have automated cleaning systems to reduce operator maintenance requirements.

Low-Pressure, Low-Intensity UV System

Low-pressure, low-intensity (LP/LI)UV systems used to be the most common type of UV system. A LI/LP UV system consists of a power supply, ballasts, lamps, a reactor, cleaning equipment, and controls and instrumentation. The LP/LI lamps emit a near-monochromatic radiation of 253.7 nm. Approximately 40 to 60 lamps are required per million gallons per day of effluent flow, with an energy requirement of 3.8 to 4.2 kW/mgd. Lamp output is fixed; to conserve energy, the number of lamps in service can be paced to the flow and transmittance of the effluent stream. Most LP/LI UV systems do not provide for automatic cleaning of the lamps; therefore, provisions for manually cleaning the lamps must be designed into the system.

LP/LI lamps operate at temperatures around 40 to 50 degrees C and typically have a lifespan of 8,000 to 10,000 hours. Their temperature range is close enough to the temperature of the wastewater that there is no significant change in the temperature of the wastewater. Conversion of applied power into UV radiation is approximately 35 to 40 percent. Most LP/LI UV systems for wastewater consist of lamps suspended horizontally or vertically in an open channel. LP/LI systems have been especially cost-effective in retrofits of existing plants, where an existing chlorine contact basin can be converted into an open-channel reactor at minimal cost. The existing chlorine contact tanks would be difficult to impossible to retrofit with an open channel UV systems because top access to the tank is significantly obstructed by the existing clarifier launder trough and the tank is curved.

Due to the high number of lamps required, cleaning of a low-pressure system usually entails removal of the bank of lamps and dipping the bank into a chemical solution.

Medium-Pressure UV System

Medium-pressure UV systems produce a high-intensity, broad-spectrum light (200 to 300 nm), with bands of strongest emission at 265 and 254 nm. They consist of a power supply, transformers, capacitors, lamps, a reactor, and controls and instrumentation. Due to the high intensity of the lamps, only approximately four lamps are required per million gallons per day. Medium-pressure lamps are approximately two to three times less efficient than low-pressure lamps (due to the fact that much of their broad spectrum emission is not germicidal); they range from 15 to 25 percent conversion from applied power to UV radiation. The overall energy requirement is 7 to 10 kW/mgd. However, the reduced number of lamps facilitates cleaning of the system, and most systems have capacity for self-cleaning using wiper mechanisms (which, as noted below, is especially necessary because the lamps operate at higher temperatures, resulting in inorganic scaling).

Medium-pressure systems typically have variable power settings, allowing variation of the lamps' power use to 35 to 60 percent of the maximum during periods of low flow and/or high effluent transmittance. This permits more energy-efficient operation and can extend lamp life. Medium-pressure lamps typically have a lifespan of 4,000 to 8,000 hours, which is lower than the lifespan for low-pressure lamps, and operate at higher temperature ranges than low-pressure systems.

Medium-pressure systems are available in closed-chamber (in-pipe), closed-channel, and open-channel designs, with an automated wiper system for lamp cleaning. The most common systems currently being installed are closed-channel, piped-flow systems. Typically, the automatic cleaning cycle directs wipers to travel along the length of the lamp at timed intervals to remove deposits before they become encrusted onto the lamp surface. Chemical addition to prevent fouling is another means of maintaining the lamps; it can be used in combination with wipers.

Vendors for this technology include Calgon Carbon, ETS, Aquionics, and Trojan.

Low-Pressure, High-Intensity UV System

Within the past 10 years, low-pressure, high-intensity lamp (LP/HI) UV systems have become very popular. These systems emit the same monochromatic emission of traditional LP/LI but at greater intensity or output. LP/HI lamps are nearly as energy efficient as the LP/LI lamps, but require significantly fewer lamps—approximately two to four lamps per million gallons per day of effluent flow. The energy demand is about 15 percent greater than LP/LI systems, ranging from 4.2 to 4.8 kW/mgd. These LP/HI systems typically have self-cleaning wiper mechanisms to keep the lamps clean. Lamp life typically range from 8,000 to 12,000 hours, depending on the intensity of use.

LP/LI lamps can be controlled to generate a UV dose paced to the transmittance and flow of effluent. Open- or closed-channel configurations are available. Like LP/LI, the LP/HI systems operate at approximately 60 degrees Celsius. Like medium-pressure systems, LP/HI systems can disinfect low-transmittance waters.

Advantages and Disadvantages

UV as a disinfection method is a reliable, proven technology for wastewater applications. It involves very little chemical handling/storage (for lamp sleeve cleaning), provides a safe method of disinfection, and does not create additional byproducts that are introduced into the effluent. The UV bulbs contain mercury which is hazardous and therefore the bulbs need to be disposed of appropriately. Operation and maintenance costs are moderate, but there is no need to dechlorinate the effluent. UV has a relatively high energy consumption, and ongoing lamp and ballast replacement costs.

ALTERNATIVE 6 – PERACETIC ACID

Peracetic acid (CH_3COOOH) (PAA), also known as peroxyacetic acid, ethaneperoxoic acid, or acetyl hydroxide, is a strong oxidant. Peracetic acid is generated by creating a mixture of acetic acid and hydrogen peroxide. PAA can be generated on site or transported.

PAA is a strong oxidant and inactivates pathogenic microorganisms by disrupting the cell membrane. A PAA disinfection system typically is comprised of a storage tank, metering pumps, contact tank, and associated ancillary equipment. Contact time for the chemical is typically between 10 and 30 minutes, and the dosage is normally between 1 and 3 mg/L. Because PAA is a newer technology, several facilities across the Northwest are conducting bench and pilot testing to ensure that it would work with their wastewater. The addition of PAA biocide to the effluent may require the addition of a sulfur-based reducing agent (i.e., sodium bisulfite or sodium thiosulfate) to PAA before discharge to quench the residual.

Because of its bactericidal effectiveness, the use of peracetic acid as a disinfectant for wastewater effluents has been drawing increasing interest in recent years. PAA decomposes to harmless acetic acid, oxygen, and water; and thus does not yield harmful DBPs, but the organic residue of the acetic acid imparts BOD and provides substrate for bacterial growth downstream of the disinfection zone.

<u>Advantages and Disadvantages</u>

The advantages of PAA for wastewater disinfection are the potentially low capital costs (especially when converting from chlorination), strong bactericidal effectiveness, absence of persistent toxic or mutagenic residuals or byproducts, little dependence on pH, and short contact time.

Peracetic acid has similar advantages to sodium hypochlorite: minimal operation and maintenance and fewer safety concerns in regard to storage and transport of the chemical. Additionally, PAA's oxidation potential is higher than hypochlorite's and the oxidant demand of PAA does not produce highly toxic or carcinogenic chlorinated species.

Major disadvantages associated with PAA disinfection are the residual acetic acid and thus the potential for elevated effluent BOD and microbial regrowth (acetic acid is already present in the mixture and also is formed after peracetic acid decomposition), as well as reduced effectiveness for inactivation of viruses. Another disadvantage of the use of peracetic acid is its high cost, which is partly a result of limited production capacity worldwide.

PAA has been used mostly for discharges that have bacterial effluent limits, such as fecal or total coliform or E. coli, and less commonly for virus inactivation. The PAA may not be cost-competitive when high doses are required to virus removal requirements. Independent tests have indicated that the same disinfectant dose and contact time yields efficient inactivation of *Escherichia coli* or fecal coliform, but minimal inactivation of viruses (including the coliphage virus that EPA believes is a better indicator of the disease-causing potential) was observed.

PAA is not a common method of wastewater disinfection in the U.S., and is therefore, disadvantageous in regard to both chemical availability and bulk chemical costs.

Some formulations of PAA have reportedly exhibited explosive decomposition at elevated temperatures or if exposed to shock/friction. The chemical must be stored in closed containers in a cool, well-ventilated area.

Alternative 7 – Ozonation

Although ozone and chlorine were both developed as disinfection methods at roughly the same time in the late nineteenth and early twentieth centuries, chlorine became the dominant wastewater disinfection technology due to its substantially lower capital cost. Moreover, many early ozone disinfection systems were abandoned because of maintenance and operational problems (Leong et al., 2008). However, concerns about the production of chlorinated disinfection byproducts (DBPs) have caused WWTPs to look again at alternative treatment technologies. Improvements in ozone-production technologies, combined with the capacity of ozone to oxidize micropollutants and color- and odor-causing compounds, has brought ozone back into consideration as a disinfection method, especially in applications involving water reuse.

Leong et al. (2008) reported that as of 2007, there were seven major WWTPs in the United States using ozone with a median design flow of 10 mgd and a range of 3 to 34 mgd. As of 2016, ozone disinfection systems are in use or under construction at approximately 20 U.S. WRRFs, many of which cite color removal or the destruction of trace organics as deliberate supplementary goals (International Ozone Association, 2016).

Ozone is a very strong oxidant and an active biocide when dissolved in water. Disinfection using ozone is accomplished by oxidation and destruction of the organism's cell wall resulting in cell lysis. The effectiveness of disinfection is dependent upon such factors as characteristics of the target organism, ozone concentration, and contact time. The system itself is composed of a feed gas preparation facility, power supply, ozone generation facility, contact basin, and an exhaust gas destruction facility. The aforementioned power supply or energy usage is mainly attributed to air preparation, ozone generation, and ozone contacting. On-site ozone generation with liquid oxygen is significantly more efficient than with ambient air, and so the proposed ozone generation system requires on-site liquid oxygen storage and feed.

The ozone disinfection process at this facility would include a new ozone generation facility including liquid oxygen storage tank, ozone dissolution reactors, conveyance of concentrated ozone solution to the existing chlorine contact basins, and retrofit (covering) of the existing chlorine contact basins for efficient capture of any ozone off-gas and installation of a catalytic or thermal off-gas treatment system.

Advantages and Disadvantages

Ozone disinfection is utilized at medium to large facilities typically after secondary treatment and presents several advantages. Ozone is an effective disinfectant, surpassing chlorine and PAA in its ability to destroy viruses and bacteria. It also requires a shorter contact time. Use of ozone, relative to PAA or chlorine, reduces persistent residuals due to its rapid decomposition rate. It also contributes to the dissolved oxygen concentration of the effluent. Ozone is the only conventional wastewater disinfectant that also simultaneously oxidizes an array of microconstituents (unregulated trace organics including pharmaceuticals). However, this method of disinfection also exhibits a few significant disadvantages. Ozone generation requires significant additional training to operate as well as adequately understanding the safety requirements. This disinfection method is not economical for wastewater with high levels of suspended solids; capital costs and power consumption can be relatively high. In addition, ozone itself is highly corrosive and toxic, which necessitates corrosion-resistant storage and chemically compatible conveyance materials. Off-gas treatment system are expensive and are required.

Alternative 8 – Chlorine Dioxide

Chlorine dioxide rarely is used as a disinfectant for secondary or tertiary wastewater effluent in the United States, as a result of traditional concerns associated with its unstable nature and the handling of hazardous chemicals required for its generation. However, advancements and introduction of chlorine-dioxide-generating systems that produce the disinfectant on site, by combining sodium chlorite and hydrochloric acid, have helped to reduce these concerns.

Some testing indicates that chlorine dioxide is only slightly superior to chlorine as a bactericide, but is a much superior viricide (WEF, 2010). Chlorine dioxide also was shown to be more effective than chlorine against protozoa, including the cysts of *Giardia* and the oocysts of *Cryptosporidium* in wastewater streams. A feasibility study published by the EPA in 1980 reported that the operation cost of a chlorine dioxide disinfection process on secondary effluent from a municipal wastewater treatment plant would be two to five times the cost of conventional chlorination.

Chlorine dioxide is produced by activating sodium chlorite with an oxidizing agent or an acid source. Sodium chlorite is converted to chlorine dioxide through a chlorine dioxide generator and applied as a dilute solution. Chlorine dioxide solutions should be applied to the processing system at a point and in a manner which permits adequate mixing and uniform distribution. The feed point should be well below the water level to prevent volatilization of the chlorine dioxide. Precautions should be taken to avoid coincident feeding of chlorine dioxide with lime or powdered activated carbon.

A recent Water Environment Research Foundation (WERF) (Alexandria, Virginia) survey found that no major municipal WWTPs were using chlorine dioxide as a primary disinfectant (Leong et al., 2008).

Advantages and Disadvantages

Disinfection of wastewater by chlorine dioxide offers additional advantages over chlorination, as follows:

- It is a highly effective disinfectant, exhibiting relatively constant biocide power in a pH range between 6 and 9.
- It dissolves readily in water, and may enhance coagulation.
- It reduces formation of organohalogenated compounds, including THMs.

The disadvantages of using chlorine dioxide include the following:

- It can form potentially toxic DBPs, such as chlorite and chlorate.
- The production cost is high, and needs to be produced onsite because it is an explosive gas.
- Chlorine dioxide decomposes in sunlight.

Alternative 9 – Hybrid UV/Chlorination System

In recent years, the sequential or simultaneous use of two or more disinfectants in wastewater treatment has been gaining attention. Using multiple disinfectants has the following potential benefits: increased effectiveness, synergistic effects, reduced formation of DBPs, potentially reduced costs, and operational flexibility (Leong, 2008). For Aberdeen, with the very high peak flows and limited hydraulic profile, there are challenges in implementing sequential effluent disinfection; however, a simultaneous hybrid system would have some advantages. A combined UV-hypochlorite system could be provided that would provide UV treatment at typical flows, and a hypochlorite system (either bulk or on-site generation) for higher flows. For the purposes of this screening evaluation, it is assumed that the UV system would be sized for 5 mgd; flows exceeding 5 mgd would be diverted to the chlorination-dechlorination system.

Advantages and Disadvantages

This hybrid system would reduce use of chemicals on site, increasing safety and the formation of disinfection byproducts that are introduced into the effluent. However, the implementation of a hybrid system would add significant operational complexity, requiring that two systems be operated and maintained. Another disadvantage of the hybrid system would be storing the bulk sodium hypochlorite; the strength of 12.5 percent sodium hypochlorite degrades with time stored

DISINFECTION ALTERNATIVES SCREENING

The technologies were rated with respect to multiple weighted criteria within four broad groups: (1) community/social; (2) technical; (3) natural environment; and (4) costs. The highest scores (10) were provided for the technologies that best met criteria. Following the scoring in each of the four major groups, the scores were analyzed by weighting the groups four different ways as a sensitivity analysis. Table 2 summarizes the criteria.

The following alternatives were selected for a more detailed evaluation:

- Alternative 2 – Chlorine Gas with Upgrades;
- Alternative 3 – Hypochlorite (Bulk);
- Alternative 4 – Hypochlorite (On-Site Generation);
- Alternative 5 – Ultraviolet Disinfection; and
- Alternative 9 – Hybrid Chlorine/UV Disinfection.

TABLE 5 - Disinfection Technology Scoring Criteria and Comments

Criteria	Goal	Explanation
Community/Social		
1 Air Emissions	Minimize the potential for air emissions.	In general, toxic gas alternatives were rated the lowest in this category, and the non-chemical alternative (UV) the highest.
2 Aesthetics	Minimize visual impact from new facility to neighboring areas.	Technologies that require most new higher-profile facilities were generally rated lowest.
3 Noise	Minimize the potential for noise during construction and operation.	Technologies that had the highest potential for noise were rated the lowest.
4 Traffic	Minimize truck traffic during construction and operation.	Technologies that required frequent deliveries rated the lowest.
5 Health and Safety	Minimize potential risk or liability to community health and safety from exposure to explosions, fire, chemicals, gaseous emissions, etc.	In general, toxic gas alternatives were rated the lowest in this category, and the non-chemical alternative (UV) the highest.
6 Footprint	Minimize facility footprint.	Technologies that required the biggest footprint were rated the lowest.
7 Public Perception	Maximize potential for the project to receive public support and acceptance.	In general, toxic gas alternatives were rated the lowest in this category, and the non-chemical alternative (UV) the highest.
Technical		
1 NPDES Permit Compliance	Minimize the potential for violations of the City's NPDES permit.	High scores were provided to chlorination systems and UV systems since they have the longest track record of success.
2 Ease of Implementation	Maximize ease of implementation from technical and practical basis.	Higher scores were provided for technologies that provided the least disruption to existing operations (e.g., chlorine gas alternatives).
3 Operational Complexity	Minimize the risk to existing operational procedures and monitoring requirements.	All UV alternatives were provided high scores. Chlorine gas ranked the lowest due to need for dechlorination system and gas scrubbing systems, sodium hypochlorite alternatives ranked lower than UV due to need for dechlorination, age degradation, and off-gas vapor lock issues in dosing systems.
4 Reliability	Minimize risk that process may fail or need frequent or complex maintenance.	Lower scores were provided with more complex controls or requiring more frequent maintenance.

Criteria	Goal	Explanation
5 Energy Use	Minimize net energy use.	Energy costs are minimal for all alternatives, except on-site hypochlorite generation, ozone, and UV systems. Low-pressure UV systems require approximately 4 kW/mgd, and medium-pressure systems 2 to 3 times that. On-site generation systems require approximately half the energy of low-pressure UV systems.
6 Flexibility for Expansion	Maximize flexibility of new infrastructure to fit into potential future treatment plans (including regionalization).	Scoring reflects likelihood of fitting into future treatment plans, which could be manifested through: (1) construction of modular systems (e.g., tanks or process equipment) expandable for potential future regional flows; (2) selection of equipment that could be utilized in a new regional plant or reconfigured existing plant; and/or (3) provision of relatively low capital cost improvements initially until future plans are more concrete. Scoring for this criterion will be more fully evaluated as top options are considered in the Engineering Report.
7 Adaptability to Changing Influent Quality	Maximize ability to meet disinfection requirements over a range of influent quality.	Chlorine alternatives scored higher than UV, as variable transmittance (UVT) can impact UV adaptability. (As more effluent UVT data is collected, the range of UVT will be better understood and a more appropriate design UVT established.) Ozone and peracetic acid have issues with variable oxidant demand.
8 Regulatory Approvals	Minimize the need and time to obtain permits and approvals, and minimize risk that future regulations will make alternative less viable.	Gas chlorination and sulfur dioxide dechlorination are covered by a host of regulatory requirements. As mentioned previously, the IFC requires facilities with over 150 pounds of chlorine or sulfur dioxide to have treatment capacity (either containment vessels or a scrubber). There are a number of ventilation and safety requirements as well. RMPs must also be prepared for facilities storing or using over 2,500 pounds of chlorine gas and 5,000 pounds of sulfur dioxide. Finally, the ongoing OSHA PSM requirements, which have even lower thresholds, must be met.
Natural Environment		
1 Air Quality Protection	Minimize the potential for air contamination.	In general, toxic gas alternatives were rated the lowest in this category, and the non-chemical alternative (UV) the highest.
2 Health and Safety	Minimize potential risk or liability to staff or public from exposure to fire, explosions, or hazardous materials (including reportable quantity spills).	In general, alternatives with high toxicity and explosion/fire risks (e.g., chlorine, chlorine dioxide) were rated the lowest in this category, and the non-chemical alternative (UV) the highest. Chlorine and ozone alternatives also rated low due to formation of toxic and carcinogenic DBPs.

	Criteria	Goal	Explanation
3	Shellfish Protection	Minimize potential for negative impacts to shellfish industry.	Technologies with higher and more reliable removals (at typical doses of disinfectant) for enteroviruses and other pathogens associated with shellfish contamination rated the highest. (Chlorination alternatives would rate more highly if free chlorine residual were maintained.)
4	Spill/Release Potential	Minimize potential for spills along the travelled routes.	Technologies that required frequent deliveries of toxic chemicals rated the lowest.
Costs			
1	Operation and Maintenance Cost Control	Minimize variability and uncertainty of costs.	Use of chlorine dioxide, ozone, peracetic acid, along with the "No Action" alternative involve considerable uncertainty due to the impact of existing and possible future compliance and safety issues.
2	Capital Cost	Minimize capital cost.	The chlorine gas alternatives rated the highest (followed by bulk hypochlorite), as they involve the fewest capital improvements. All of the other improvements are expected to have significantly higher capital costs.
3	Operation and Maintenance Cost	Minimize operation and maintenance cost (e.g., energy, transportation cost, contingency, etc.).	The chlorine gas alternatives rated the highest (followed by bulk hypochlorite). Operation and maintenance costs for UV are expected to be somewhat higher,but will depend on the type of system selected and the UVT values during operation.
4	Life Cycle Cost	Minimize life cycle cost.	The chlorine gas alternatives rated the highest (followed by bulk hypochlorite).

DETAILED ALTERNATIVES ANALYSIS

The 5 alternatives were subject to a more thorough evaluation, including detailed life cycle cost evaluations and non-monetary criteria rankings. The results of the cost evaluation are provided in Table 3. As shown, bulk hypochlorite has the lowest capital and life cycle costs. An upgraded gas chlorination system and on-site generation system are the next lowest cost alternatives respectively.

Chlorine Gas with Upgrades

Since the gas chlorination/dechlorination system is approaching the end of its useful life, full replacement of the existing system is recommended, with the addition of an enhanced safety system. The equipment to be replaced includes the chlorinator, sulfonator, scales, regulators, gas monitor, chlorine analyzers, ORP probe, ORP controller, and flash mixer. All of the replacement items were selected to be as similar as possible to (or modern upgrades to) the existing equipment. A chlorine/sulfur dioxide scrubber was sized to contain all of the gas in a full 1-ton chlorine or sulfur dioxide tank. The scrubber selected uses a dry media that is non-hazardous and is generally replaced every 10 years. Since the existing chlorination system was overhauled in 2009, the replacement units are not expected to be needed until 2019 (+10 years), and the net present value of the project reflects this delay. The scrubber, process water pumps and filter, and flash mixers, will be installed in year 1.

Because chlorine gas is hazardous, the chlorination equipment would be housed in an upgraded, isolated room with a continuous ventilation system, automatic cylinder shut-off valves, automatic sprinkler system, and alarm system. The chlorine scales and ambient chlorine monitor would be connected to the facility SCADA system. For the purposes of this analysis, it has been assumed that the facility would be designed as an H-3 occupancy.

Bulk Hypochlorite

For Aberdeen, a 12.5 percent bulk sodium hypochlorite system would include two 2,500-gallon polyethylene storage tanks with secondary containment and four dosing pumps mounted on a metering skid. The metering pump skid includes a calibration cylinder, off-gas relief valve, pulsation dampener, pressure relief valve, and backpressure valve. This system would provide for sufficient storage to allow up to monthly deliveries of hypochlorite for average 20-year demands and deliveries every 2 weeks for peak 20-year demands. The pumps would be controlled by variable frequency drives controlled based on ORP measurements.

On-Site Generation of Sodium Hypochlorite

The on-site hypochlorite generation alternative (on-site) can be divided into five major components: (1) the hypochlorite generation system, which consists of a pump, water softeners, an electrolyzer, associated electronic conditioning equipment and controls, and hydrogen gas dispersion fans; (2) a brine storage tank and salt-loading crane; (3) hypochlorite storage tanks; (4) the hypochlorite dosing system, which includes chlorine analyzers, dosing pumps, ORP probes, and ORP controllers; and (5) the dechlorination system, which includes a CTS storage tank, CTS dosing pumps, and ORP probe, and an ORP controller.

Due to the size of the system it has been assumed that salt would be delivered in bulk and blown into the brine tank. Salt deliveries would need to occur every 1 to 2 months. The metering pump skid includes a calibration cylinder, off-gas relief valve, pulsation dampener, pressure relief valve, and backpressure valve. In the event of a loss of power, the plant can continue to dose hypochlorite from the reserves in its storage tanks. If the volume of hypochlorite in the storage tanks is within 2 day's estimated demand, a pallet (tote) of bulk hypochlorite can be delivered within 24 hours, diluted into the storage tanks and dosed as usual. One 275-gallon tote of 12.5 percent hypochlorite will create 4,400 gallons when diluted to 0.8 percent. These emergency operational procedures permit the on-site alternative to operate with standby power only for the hypochlorite and dechlorination dosing systems, which require less than 5 kW of power.

Although attractive in concept, many on-site generation systems have had operational problems. As an on-site generation system has more equipment than the other chlorine alternatives, it potentially requires more maintenance. According to equipment manufacturers, the electrolytic cells typically have a 7 to 14-year service life and cost approximately $70,000 to replace. In addition, regular cleaning of the electrolytic cells is required; however, the larger systems as would be provided for Aberdeen can have automated provisions for cell cleaning. Other operations and maintenance tasks include regeneration of water softeners and rebuilding of brine pumps.

UV Disinfection

Dozens of ultraviolet disinfection systems have been successfully installed and operated in Washington State over the past 20 years. However, there are some challenges for implementing UV disinfection at Aberdeen:

1. High peak flows (high peaking factor), requiring high UV equipment costs;
2. Plant hydraulics requiring the provision of effluent pumping at high flows with UV disinfection;
3. Low transmittance at times, due to the high peak flows and the acceptance of UV-absorbing leachate;
4. Energy consumption and the need to provide backup power for the UV system;

5. Lack of an existing structure appropriate for use in installing an open channel system, or above-ground piping in a location appropriate for an in-pipe system;

6. Difficulties in expanding a new UV system to accommodate potential regionalization.

These issues are reflected in the high costs presented in Table 3.

Hybrid UV / Hypochlorite System

For Aberdeen, with the very high peak flows and limited hydraulic profile, a combined UV-hypochlorite system could be provided that would provide UV treatment at typical flows, and a hypochlorite system (assumed to be a bulk sodium hypochlorite system with calcium hypochlorite for dechlorination) for higher flows. For this evaluation, it has been assumed that the UV system would be sized for 5 mgd; for flows exceeding 5 mgd, the entire flow would be treated by the chlorination-dechlorination system.

The flow split for bulk hypochlorite and UV was determined by evaluating the average length of the summer dry season, defined as the longest period of the year in which there are no hourly flows greater than 3,500 gpm (5 mgd). Between 2013 and 2017, the dry season accounted for 40 percent of all days and the wet season accounted for 60 percent of all days. It was assumed that the flows during the dry season represented the lowest flows for the year. Based on the flow distribution curve for the same time period, the 40th percentile daily average flow represents 23 percent of all flows. Therefore, the UV system, which would operate only during the dry period, would handle 23 percent of the design annual average flow of 4.19 mgd, which is an amount equivalent to 0.964 mgd. The bulk hypochlorite system will handle 77 percent of the design annual average flow, which is an amount equivalent to 3.23 mgd.

This hybrid system would reduce use of chemicals on site, increasing safety and the formation of disinfection byproducts that are introduced into the effluent. In addition, since flows through the UV system would be limited to 5 mgd, a new effluent pump station would not be required (as was needed for the full flow low-pressure, high intensity open channel option.) However, the implementation of a hybrid system would add significant operational complexity, requiring that two systems be operated and maintained. Another disadvantage of the hybrid system would be storing the chlorination chemical.

SUMMARY

Estimated costs are provided in Table 3. Ultraviolet disinfection, which usually is very competitive in these assessments, had high capital costs and consequently high life cycle costs, due to high hydraulic peaking factor, the need for effluent pumping due to increased head loses, periodic low transmittance of the effluent to UV light, the need to provide backup power for the UV system, and lack of an existing structures appropriate for use in installing UV systems.

TABLE 3 - Estimated Capital, Annual O&M And 20-Year Life Cycle Costs

Alternative	Estimated Capital Costs(1)	Estimated Annual Operation & Maintenance Costs	Estimated Net Present Value of O&M and Capital Replacement Costs	Estimated Life Cycle Cost(1)
2 – Chlorine Gas with Upgrades	$2,189,900	$164,500	$2,708,500	$4,898,400
3 – Bulk Hypochlorite	$1,070,300	$214,400	$3,262,900	$4,333,200
4 – Hypochlorite, On-Site Generation	$2,061,700	$183,200	$2,906,100	$4,967,800
5A – Ultraviolet Disinfection (Low Pressure, High Intensity, Open Channel)	$13,853,000	$112,000	$1,666,000	$15,519,000
5B – Ultraviolet Disinfection (Medium Pressure, Closed Vessel)	$18,258,500	$154,210	$2,294,000	$20,552,500
9 – Hybrid Chlorine/UV Disinfection	$6,098,200	$162,918	$2,502,500	$8,600,700

In addition to the capital and life cycle cost assessment, the technologies were rated with respect to multiple weighted criteria within four broad groups: (1) community/social, (2) technical, (3) natural environment, and (4) costs. The technologies with the highest ratings depended on the weighting applied to each group. Given the magnitude of the capital and life cycle costs, the significant differences in life cycle costs, and potential economic impacts of this project for the City, it is appropriate to weight costs significantly in the overall assessment. With this weighting (Scenario 2 in Figure 5), bulk hypochlorite is the highest rated alternative, as it combines improved safety and reliability with the lowest capital and life cycle costs.

	Importance Score (1 to 10)	Weighted Importance %	A. Chlorine Gas (with upgrades)		B. Hypochlorite (bulk)		C. Hypochlorite (on-site generation)		D. Ultraviolet Light		E. Hybrid Chlorine/UV	
Analysis 1			Score	Weighted Score	Score	Weighted Score	Score	Weighted Score	Score	Weighted Score	Score	Weighted Score
Community Social	10	25.0%	3.91	0.978723	5.55	1.388298	5.49	1.37234	6.53	1.632979	5.06	1.265957
Technical	10	25.0%	6.31	1.578358	6.07	1.518657	5.27	1.317164	5.99	1.496269	5.39	1.347015
Natural Environment	10	25.0%	2.60	0.65	4.00	1	4.60	1.15	8.20	2.05	5.40	1.35
Costs	10	25.0%	6.68	1.671053	7.37	1.842105	6.68	1.671053	4.42	1.105263	5.11	1.276316
Final Score	40	100%		4.88		5.75		5.51		6.28		5.24

	Importance Score (1 to 10)	Weighted Importance %	A. Chlorine Gas (with upgrades)		B. Hypochlorite (bulk)		C. Hypochlorite (on-site generation)		D. Ultraviolet Light		E. Hybrid Chlorine/UV	
Analysis 2			Score	Weighted Score	Score	Weighted Score	Score	Weighted Score	Score	Weighted Score	Score	Weighted Score
Community Social	2	10.5%	3.91	0.412094	5.55	0.584546	5.49	0.577828	6.53	0.68757	5.06	0.533035
Technical	5	26.3%	6.31	1.66143	6.07	1.598586	5.27	1.386489	5.99	1.57502	5.39	1.41791
Natural Environment	2	10.5%	2.60	0.273684	4.00	0.421053	4.60	0.484211	8.20	0.863158	5.40	0.568421
Costs	10	52.6%	6.68	3.518006	7.37	3.878116	6.68	3.518006	4.42	2.32687	5.11	2.686981
Final Score	19	100%		5.87		6.48		5.97		5.45		5.21

	Importance Score (1 to 10)	Weighted Importance %	A. Chlorine Gas (with upgrades)		B. Hypochlorite (bulk)		C. Hypochlorite (on-site generation)		D. Ultraviolet Light		E. Hybrid Chlorine/UV	
Analysis 3			Score	Weighted Score	Score	Weighted Score	Score	Weighted Score	Score	Weighted Score	Score	Weighted Score
Community Social	10	33.3%	3.91	1.304965	5.55	1.851064	5.49	1.829787	6.53	2.177305	5.06	1.687943
Technical	5	16.7%	6.31	1.052239	6.07	1.012438	5.27	0.878109	5.99	0.997512	5.39	0.89801
Natural Environment	10	33.3%	2.60	0.866667	4.00	1.333333	4.60	1.533333	8.20	2.733333	5.40	1.8
Costs	5	16.7%	6.68	1.114035	7.37	1.22807	6.68	1.114035	4.42	0.736842	5.11	0.850877
Final Score	30	100%		4.34		5.42		5.36		6.64		5.24

	Importance Score (1 to 10)	Weighted Importance %	A. Chlorine Gas (with upgrades)		B. Hypochlorite (bulk)		C. Hypochlorite (on-site generation)		D. Ultraviolet Light		E. Hybrid Chlorine/UV	
Analysis 4			Score	Weighted Score	Score	Weighted Score	Score	Weighted Score	Score	Weighted Score	Score	Weighted Score
Community Social	5	16.7%	3.91	0.652482	5.55	0.925532	5.49	0.914894	6.53	1.088652	5.06	0.843972
Technical	10	33.3%	6.31	2.104478	6.07	2.024876	5.27	1.756219	5.99	1.995025	5.39	1.79602
Natural Environment	10	33.3%	2.60	0.866667	4.00	1.333333	4.60	1.533333	8.20	2.733333	5.40	1.8
Costs	5	16.7%	6.68	1.114035	7.37	1.22807	6.68	1.114035	4.42	0.736842	5.11	0.850877
Final Score	30.000	100%		4.74		5.51		5.318		6.55		5.29

FIGURE 5 - Disinfection Technology Scoring and Overall Ratings

REFERENCES

Leong, L. Y. C.; J. Kuo; C. Tang; 2008. *Disinfection of Wastewater Effluent—Comparison of Alternative Technologies*, Water Environment Research Foundation: Alexandria, Virginia.

Metcalf & Eddy, Tchobanoglous, Stensel, Tsuchihashi and Burton, 2014. *Wastewater Engineering: Treatment and Resource Recovery*, 5th edition.

Montazeri, Goettert, et al, 2015. *Pathogenic Enteric Viruses and Microbial Indicators during Secondary Treatment of Municipal Wastewater*, Applied and Environmental Microbiology, Vol. 81, No. 18, September.

Pouillot, Van Doren, et al., 2015. *Meta-Analysis of the Reduction of Norovirus and Male-Specific Coliphage Concentrations in Wastewater Treatment Plants*, Applied and Environmental Microbiology, Vol. 81, No. 14, July.

Swift, J. L.; J. P. Wilson; D. Welch; M. Johnson; P. Conley; B. Bowman; 2000. *An Assessment of O&M Costs for Ultraviolet Disinfection Systems*, WEFTEC.

Water Environment Federation, 2010. *Manual of Practice No. 8, Design of Wastewater Treatment Plants*.

Using Dynamic Dilution Methods to Derive More Accurate Effluent Total Residual Chlorine Limits for WWRFs

Robert Sharp[1*], Xiao Lin[2], Sarah Galst[2], Kathy Ammari[1], and Laura Grieco[3]

[1]Manhattan College, Bronx NY
[2]Hazen and Sawyer, New York
[3]New York City Department of Environmental Protection
*Corresponding Author: Robert.sharp@manhattan.edu

ABSTRACT

As many Water Resource and Recovery Facilities (WRRFs) across the United States face increasingly stringent effluent criteria, including Total Nitrogen, Total Residual Chlorine and Indicator Organisms, there is a need for a more rigorous and realistic approach for determining plant specific effluent total residual chlorine limits. This paper describes the development and application of a dynamic dilution and decay approach for re-calculating effluent TRC limits based on the actual processes that take place in the chlorine contact tank, the discharge pipe and the receiving water. This new approach was used to recalculate the effluent TRC limits for seven WWRFs, resulting in some plants having significantly higher effluent TRC limits.

Key Words: Total Residual Chlorine (TRC), disinfection, dynamic dilution, chlorine decay, effluent criteria

INTRODUCTION

The majority of WRRFs in the United States use chlorination to achieve required disinfection of wastewater prior to discharge. Increasingly, WRRFs are being required to meet more stringent effluent Total Residual Chlorine (TRC) permit limits. New draft permit criteria with revised TRC discharge limits were issued for all 14 NYC WRRFs. The new limits were significantly lower than the previous limits and would require some plants to make expensive capital improvements. Typically, TRC limits are determined using simple chlorine decay kinetics and receiving water dilution factors. This method ignores several processes related to the fate and transport of TRC in discharged chlorinated wastewater effluent. Such limits are believed to be overly conservative and not representative of actual receiving water conditions. A new TRC re-evaluation method was developed to more accurately mimic the actual fate and transport processes associated with the discharge of chlorinated wastewater into a receiving water. This novel method was used to re-calculate the TRC limits at seven NYC WRRFs.

OBJECTIVE OF STUDY

The purpose of this study was to accurately re-calculate new TRC limits for seven New York City DEP WRRFs. The water quality criteria for TRC or Chlorine Produced Oxidant (CPO) is 13 ppb (EPA 1984). NYC DEP received revised discharge permits that reduced the effluent TRC limit at all 14 plants by 0.40 mg/l. This proposed reduction would require expensive upgrades at multiple plants, including either dechlorination facilities or UV disinfection facilities. This new method was used to re-evaluate the TRC limits for seven of NYC's WRRFs.

DYNAMIC DILUTION AND DECAY APPROACH

The fate of chlorine between the time it is dosed to the wastewater and the time it reaches the edge of the outfall mixing zone can be described in four successive processes:

1. Instantaneous Decay/Demand in the Chlorine Contact Tank: Initial chlorine demand caused by the presence of dissolved and particulate organics, inorganic ammonia, metals and other chlorine demanding compounds that present in the treated wastewater matrix.

2. Early High Rate Decay: After instantaneous demand is realized, high rate chlorine decay occurs for a period of time which is a function of effluent water quality, original chlorine dose and environmental variables (temperature, sun light, etc.). In general, most of the high rate decay takes place in the chlorine contact tank (CCT), which should maintain a minimum 30-minute HRT at peak discharge flows. However, in some waters the high rate of decay can last well past 30 minutes if permitted in the CCT or the outfall piping.

3. Pipe Decay/Demand: Discharge pipe decay and demand is caused by natural decay of chlorine in the effluent as it moves through the discharge pipe (Ks) and chlorine demand reactions with the discharge pipe wall (Kp). This decay can be significant if the residence time in the discharge pipe is significant. Ideally, discharge pipe decay is determined via sampling at the end of the CCT and near the end of the discharge pipe. Where access to the end of the pipe is not possible, the literature provides several models for pipe decay and the corresponding pipe wall decay and diffusion rate constants (PVSC, Rossman et. al., Huang et. al., 1997). The over-all first order decay rate (K_d) of chlorine in the discharge pipe can be described as the added rates for wall effects and natural chlorine decay: $K_d = Ks + Kp$.

4. Dilution and Receiving Water Decay and Demand: The TRC concentration will decrease in the receiving water via progressive dilution with receiving water as the WW flow is incorporated with the receiving water in the designated outfall mixing zone. During this dilution, the TRC continues to decay and encounters additional demands from constituents present in the receiving water. Once the chlorinated wastewater enters the receiving stream there is the potential for the formation of other Chemically Produced Oxidants (CPO). Finally, at the edges of the mixing zone, the TRC concentration should not exceed the water quality-based limit of 13 ppb (assuming acute criteria).

The role that each of these four processes play in the decay/degradation of chlorine is plant and outfall specific and dependent upon: wastewater quality, wastewater flows, chlorine contact tank design, length and diameter of discharge pipe, receiving water quality, the magnitude and dynamics of outfall dilution, size and extent of outfall mixing zone, and environmental factors such as temperature. For these reasons it is important to evaluate dilution-based chlorine degradation on a plant-by-plant basis to ensure the resulting TRC limit is representative of the most current outfall conditions.

Each process described above plays a role in the fate and final TRC concentration at the edge of the prescribed mixing zone. A novel lab-based dilution and decay method was developed to consider these four processes, as well as the complex interactions between chlorine and the wastewater and receiving water constituents. The new method was developed based on dilution dynamics obtained from previous dye tests and CORMIX™ mixing zone modeling, where the impact of TRC uptake and decay during receiving water dilution, as well as in-pipe decay and wall uptake, are all accounted for in the testing. To evaluate this new method, a series of fed-batch tests were run where unchlorinated secondary effluent was chlorinated with a range of chlorine doses (see Figure 1). After the wastewater was disinfected, it was diluted with receiving water following the specific dynamic dilution pattern determined from CORMIX modeling. Dynamic dilution allows for the measurement of simultaneous dilution and decay, and considers the dilution hydraulics, and the complicated chlorine chemistry and reactions with wastewater and receiving water constituents. Figure 1 shows how the fed-batch tests were carried out, along with the corresponding TRC decay processes being evaluated. During each test, time series samples were collected and analyzed for TRC/CPO and pH. Unchlorinated wastewater and receiving water samples were characterized for TSS, nitrate, nitrite, ammonia, pH, and COD.

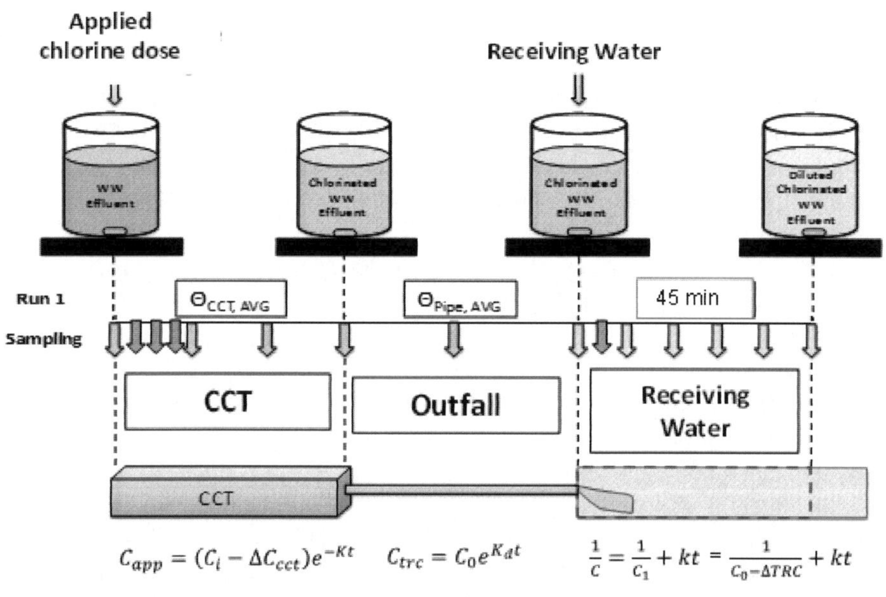

Figure 1 – Dilution and Decay TRC Degradation Batch Tests

Calculation of the effluent TRC limit using results from the dynamic dilution batch tests involved a two-step process (**Figure 2**). The first step was to back-calculate the allowable TRC concentration leaving the outfall pipe and entering the receiving water based on a second order decay model developed from the batch testing (Equation 1). The second step involved estimating the chlorine decay from the end of the discharge pipe to the end of the chlorine contact, which is the point of compliance for TRC (Equation 2). For this step, an average first-order decay rate for chlorine decay in the wastewater effluent and those rates found in the literature was applied using the hydraulic residence time in the outfall piping.

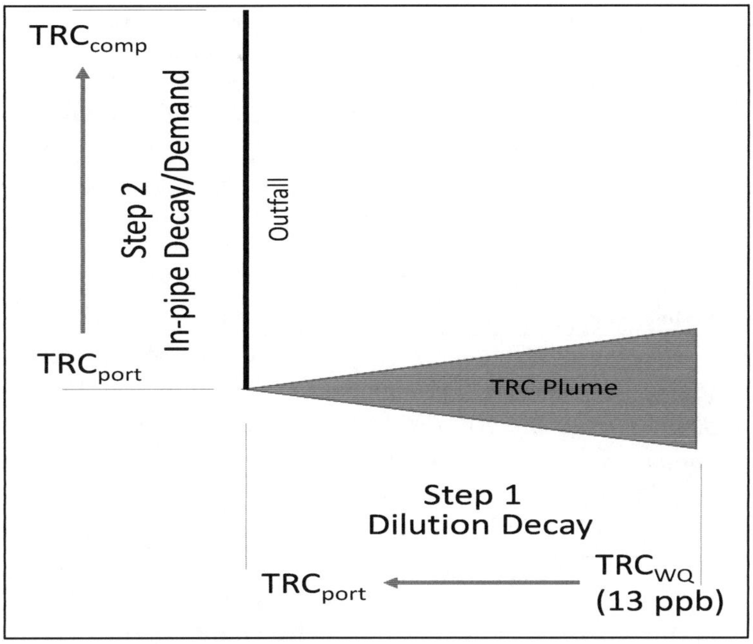

Figure 2 – Back calculation of TRC Effluent Criteria. Step 1: Back calculation from Acute Water Quality Standard of 13 ppb to the end of the discharge pipe based on **dynamic dilution and decay model;** Step 2: Calculation of TRC decay in discharge pipe.

Equation 1 shows the second-order relationship between the acute water quality TRC criteria of 13 ppb and the corresponding TRC concentration at the end of the discharge pipe (C_o). C_o represents the TRC concentration that would be allowed in the wastewater effluent at the end of the discharge pipe and just prior to entering the receiving water. C_1 represented the concentration after initial dilution and decay (C_o - ΔTRC). All other constants and variables in Equation 1 are defined below.

$$\frac{1}{C} = \frac{1}{C_1} + kt = \frac{1}{C_0 - \Delta TRC} + kt$$ ....**Equation 1**

Where: C = 13 ppb (Acute WQ TRC Criteria); t = 1800 seconds; C_1 = TRC after instantaneous uptake = (Co - ΔTRC); Co = initial TRC (TRC at end of discharge pipe); k = Second order dilution and decay rate; ΔTRC = change in TRC across the CCT prior to discharge

Equation 2 represents the decay of TRC in the discharge pipe and allows for the calculation of the TRC limit at the point of compliance, C_{TRC} (end of the CCT). The newly calculated C_{TRC} takes into account the dynamic dilution and decay in receiving water.

$$C_o / C_{TRC} = \exp(-K_d \bullet t) \rightarrow C_{TRC} = C_o \exp(K_d \bullet t)$$ .... **Equation 2**

Where: C_{TRC} = TRC limit at point of compliance (end of CCT); Co = TRC limit at end of discharge pipe via Eqn 1; Kd = (Ks + Kp) = Average over-all pipe first-order decay constant = 13.85/day; t = travel time in the discharge pipe

MATERIALS AND METHODS

Sample Characterization: Throughout the study, samples of both plant effluent (pre-chlorination) and receiving water were collected from each plant. The effluent and receiving water samples were characterized for pH, background TRC, TSS, conductivity, COD, nitrate, ammonia, and nitrite. Every set of samples collected was characterized prior to being tested to ensure the samples were representative of the source and to determine if any excess TRC uptake would be expected (i.e. excessive TSS, Nitrite).

Preliminary Chlorine Degradation Rate Studies: Preliminary Batch Decay Studies were carried out on the plant secondary wastewater effluent for each plant to allow for the estimation of residual TRC that would be leaving the Chlorine Contact Tanks (CCTs) and ensure it was a reasonable value (i.e. within the range of expected effluent TRC based on current and previous TRC effluent criteria). For plant secondary wastewater effluent decay batch testing, sodium hypochlorite was added to the secondary effluent sample and decay was measured for a period equal to the average CCT residence time (see Figure 1). A minimum of 36 preliminary chlorine decay batch tests were carried out for each WRRF in the study. Results from these batch studies were fitted using a best-fit mathematical model in which corresponding decay rates were obtained. These results were used to estimate the concentration of TRC that would be entering the receiving water (C_o), which is the initial TRC in the wastewater before receiving water dilution and decay begins. In general, the wastewater chlorine decay test results were used to establish the starting TRC concentration used in the dynamic dilution and decay studies was realistic and within the range of previous effluent TRC levels.

Determination of TRC decay in Discharge Piping: The decay/degradation of chlorine in the outfall discharge pipe can be significant if residence time in the outfall is significant. Equation 2 describes the first order in-pipe decay which was modeled as a first-order decay process with the over-all first order-decay constant (K_d) being the sum of the TRC decay rate in the effluent stream (K_s) and demand from the pipe wall (K_p). The pipe wall demand can contribute significantly to the overall in-pipe decay. Huang et. al. found that the pipe wall demand can account for up to 90% of the in-pipe chlorine decay. There were three K_d values found in the literature (Table 1) and an average over-all pipe first-order decay constant of 13.85/day was used in the development of TRC limits.

Table 1 – Values for Over-all In-Pipe Decay Constants

Plant	/Min	/Day
PVSC	0.0059	8
Cook & Fairey	0.00935	13
Huang et. al.	0.0136	20
Average	0.00962	13.85

Dynamic Dilution and Decay in Receiving Water: Figure 1 shows how the dynamic dilution and decay batch tests were carried out. A minimum of 36 receiving water/disinfected wastewater dynamic dilution and decay batch tests were carried out for each WRRF in the study. The results from these studies and those from the preliminary chlorine decay batch tests were used in conjunction with Equations 1 and 2 to determine a new effluent TRC limit for each of the plants in the study.

RESULTS AND DISCUSSIONS

Figure 3 shows the results obtained from the dynamic dilution and decay batch testing for one plant effluent/receiving water. The data obtained from the early part of the batch tests provided the first-order decay rate and initial uptake of chlorine for different doses of chlorine added to the wastewater effluent. The second set of data shown in Figure 3 shows the second order decay rate constant for combined dilution and decay in the receiving water. A total of 36 fed-batch tests were carried out for each plant. Curve fitting procedures were used to find the best fit kinetics for both the chlorine decay rates in the contact tanks and for the dynamic dilution and decay rates. The data and rates generated from these tests were used to back calculate the new TRC limit using Equations 1 and 2.

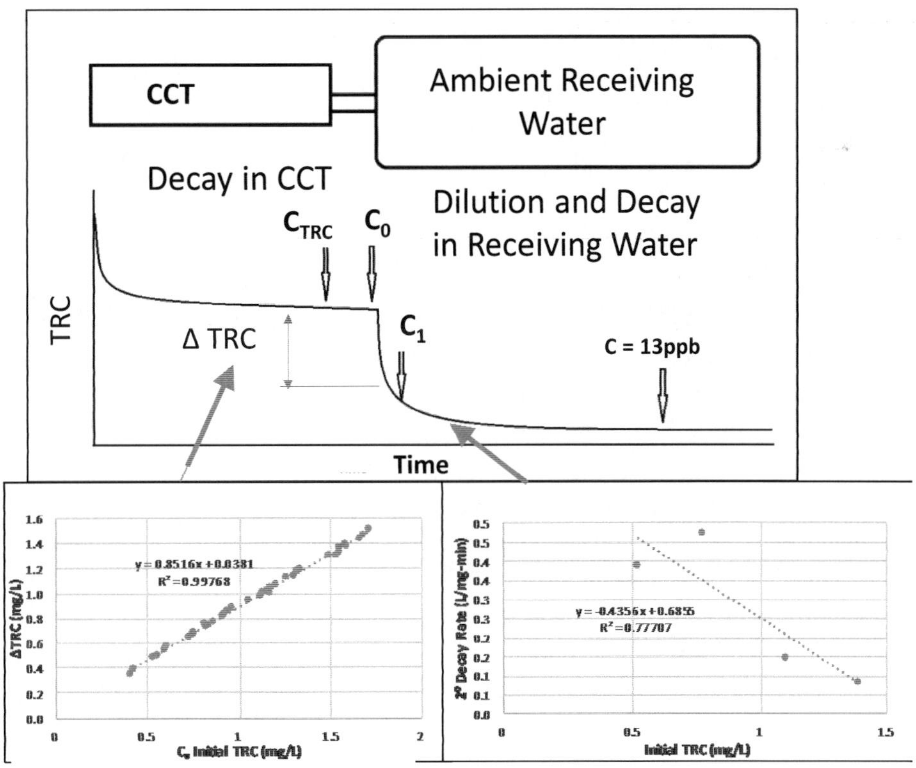

Initial TRC Uptake and Decay

2nd Order Decay/Diltution Rate Const

Figure 3 – Results from Dynamic Dilution and Decay Batch Testing for a Single Plant

Table 2 shows how the dynamic dilution and decay method impacts the TRC limit. Using this new method resulted in an average increase in TRC of 60% (36% to 93%) compared to the proposed TRC limits. The newly calculated limits will allow some plants to optimize their disinfection operations in lieu of constructing new disinfection facilities. In some cases, the results were used to help establish a new effluent TRC limit that allowed the plant to focus its efforts on disinfection optimization and avoid costly upgrades to the facility. This approach is currently being used at other municipalities to determine if a more realistic and achievable TRC limit may be appropriate.

Table 2 – Impact of Dynamic Dilution Decay Approach on TRC Discharge Limit

	Bowery Bay	**Hunts Point**	**Tallman Island**	**Jamaica**	**Port Richmond**	**Wards Island**	**26th Ward**
2013 TRC Permit Limit(mg/L)	0.59	0.57	0.47	0.53	0.53	0.7	0.45
Proposed TRC Permit Limit(mg/L)	0.19	0.17	0.07	0.13	0.13	0.3	0.05
Newly Calculated TRC Permit Limit (mg/L)	0.401	0.265	0.189	1.802	0.494	0.534	0.131
%Increase of TRC using Dynamic Dilution/Decay	**53**	**36**	**63**	**93**	**74**	**44**	**62**

CONCLUSION

A new approach for calculating TRC effluent discharge limits was developed and tested on seven WRRFs. The results indicate that this new method can be used to more accurately depict the actual fate of TRC from point of application to edge of the receiving water mixing zone. The new dynamic dilution and decay approach takes into account both the complex reactions and interactions between the applied chlorine, the constituents in the wastewater and receiving water, as well as the impact of progressive dilution that is representative of the receiving water. This method produced more accurate and realistic TRC limits that were specific to the plant's effluent quality and its receiving water dilution characteristics, while still meeting water quality criteria and protecting the receiving water ecosystem. For some plants, these results were used to help establish new TRC limits and allowed the WRRF to focus on disinfection process optimization and avoid costly facility upgrades.

References

1. Cook, J.B., Fairey, A.W. 1997 "Decay of Total Residual Chlorine and Reduction of Fecal Coliform in the Plum Island Outfall Pipe Prior to Discharge to the Charleston Harbor".
2. Huang, H., R.E. Fergen. J.P. Cook, and G.S. Fox 1997 "Effluent Total Residual Chlorine Decay in Outfall Pipe.". *Jour of Environ. Engineer*. pp. 813-816.
3. LMSE 1983 "Chlorine Plume Modeling Study" Prepared by LMSE for Utility Water Act Group.
4. GLEC. 1998 "Chlorine Produced Oxidant Decay and Disinfection Studies for the PVSC's Final Effluent". Prepared by Great Lakes Environmental Center for the PVSC
5. Rossman, L.A., Clark, R.M. and Grayman, W.M. (1994) "Modeling chlorine Residuals in drinking Water distribution systems. *Jour. Of Environ. Engineering*. 120(4), 803-820.